AF327473

# Automotive Engine Alternatives

**Edited by**
**Robert L. Evans**

*The University of British Columbia*
*Vancouver, British Columbia, Canada*

PLENUM PRESS • NEW YORK AND LONDON

Library of Congress Cataloging in Publication Data

International Symposium on Alternative and Advanced Automotive Engines (1986: Vancouver, B.C.)
Automotive engine alternatives.

"Proceedings of the International Symposium on Alternative and Advanced Automotive Engines, held August 11-12, 1986, in Vancouver, B.C., Canada"—T.p. verso.
"Sponsored by EXPO 86 and the University of British Columbia"—Pref.
Includes bibliographies and index.
Contents: How shall we power tomorrow's automobiles?/Charles A. Amann—The stratified charge engine concept/Duane Abata—The dual-fuel engine/Ghazi A. Karim—[etc.] 1. Automobiles—Motors—Technological innovations—Congresses. I. Evans, Robert L. II. Expo 86 (Vancouver, B.C.) III. University of British Columbia. IV. Title.
TL210.I47   1986                          629.25                          87-6949
ISBN 0-306-42549-1


Proceedings of the International Symposium on Alternative and Advanced Automotive Engines, held August 11-12, 1986, in Vancouver, B.C., Canada

© 1987 Plenum Press, New York
A Division of Plenum Publishing Corporation
233 Spring Street, New York, N.Y. 10013

PREFACE

This book contains the proceedings of the International Symposium on Alternative and Advanced Automotive Engines, held in Vancouver, B.C., on August 11 and 12, 1986. The symposium was sponsored by EXPO 86 and The University of British Columbia, and was part of the specialized periods program of EXPO 86, the 1986 world's fair held in Vancouver. Some 80 attendees were drawn from 11 countries, representing the academic, automotive and large engine communities.

The purpose of the symposium was to provide a critical review of the major alternatives to the internal combustion engine. The scope of the symposium was limited to consideration of combustion engines, so that electric power, for example, was not considered. This was not a reflection on the possible contribution which electric propulsion may make in the future, but rather an attempt to focus the proceedings more sharply than if all possible propulsion systems had been considered. In this way all of the contributors were able to participate in the sometimes lively discussion sessions following the presentation of each paper.

The internal combustion engine, as applied to automotive propulsion, is probably undergoing a more rapid rate of evolution than at any time since the early part of this century. This rapid change is due to several factors, not the least of which are regulations designed to improve fuel economy and reduce exhaust emissions. Both of these factors have led engine designers to be more innovative than in the past and have led to an increased market share for the diesel engine, which until recently was primarily limited to the heavy vehicle market, with one or two exceptions.

Advances in digital electronics in recent years, and in particular the micro-processor, have provided the engine designer with a whole new arsenal of techniques for providing engine control. Electronic control systems have been rapidly introduced, particularly on spark-ignition engines, with the result that remarkable improvements in both fuel economy and exhaust emissions have been achieved. This rapid development in conventional engine technology has meant that the developers of truly alternative engines, such as the Stirling engine, are faced with a moving target as they try to break into a market which has been dominated by the reciprocating internal combustion engine for some six decades.

It was in this exciting climate of rapid change in engine design and development that the symposium was convened. If one conclusion were to be drawn from the symposium, it would have to be that the conventional internal combustion engine is likely to be the major automotive prime mover to the end of the century. It may appear as either a spark-ignition or diesel engine, or perhaps part way in between as a stratified-charge engine, but it is unlikely to be replaced by a radically new design. Through evolution, rather than revolution, engine designers will be trying to meet the twin challenges of improved fuel economy and reduced exhaust emissions.

R.L. Evans
Vancouver, B.C.

CONTENTS

# HOW SHALL WE POWER TOMORROW'S AUTOMOBILE?

Charles A. Amann

Engine Research Department
General Motors Research Laboratories
Warren, Michigan

## ABSTRACT

The thrust toward further gains in the fuel-utilization
efficiency of the passenger-car engine, without sacrificing
its many other desirable attributes, is continuing.  The
search for an attractive alternative powerplant has always
been included in such efforts, but so far none has emerged.
Prominent on the list of contenders today are the Stirling
engine, the gas turbine, and the advanced diesel, including
uncooled versions incorporating structural ceramics.  Large-
scale production of none of these is projected for passenger
cars in the foreseeable future.  Meanwhile, improvements
continue to be made to indicated thermal efficiency,
mechanical efficiency, and volumetric efficiency of the
spark-ignition engine.  Supercharging and variable engine
geometry are additional options, and the advent of
electronic controls has proven beneficial.  The spark-
ignition engine promises the ability to operate on the
leading alternative fuels.  Given the evolving scenario,
that engine is expected to remain dominant in passenger cars
to the end of this century.

## INTRODUCTION

In just 15 years we will be into a new century.  What kind of engine
will power our passenger cars for the rest of this one?  That is a very
intriguing question.

The gasoline-fueled homogeneous-charge spark-ignition engine has
dominated the field for the past 2/3 of a century, but that has not always
been so.  In 1900 that powerplant was in third place, behind the steam
engine and the battery-electric system.  And there is no guarantee that
today's spark-ignition engine will retain its preeminent position in the
future.  Certainly there is no shortage of competing alternatives.

The possibilities are indicated on the heat-engine tree of Fig. 1.
Both continuous-combustion and intermittent-combustion engines are
candidates.  In both categories, combustion may occur either internal or
external to the engine.  Of the resulting four possibilities, however, an

"

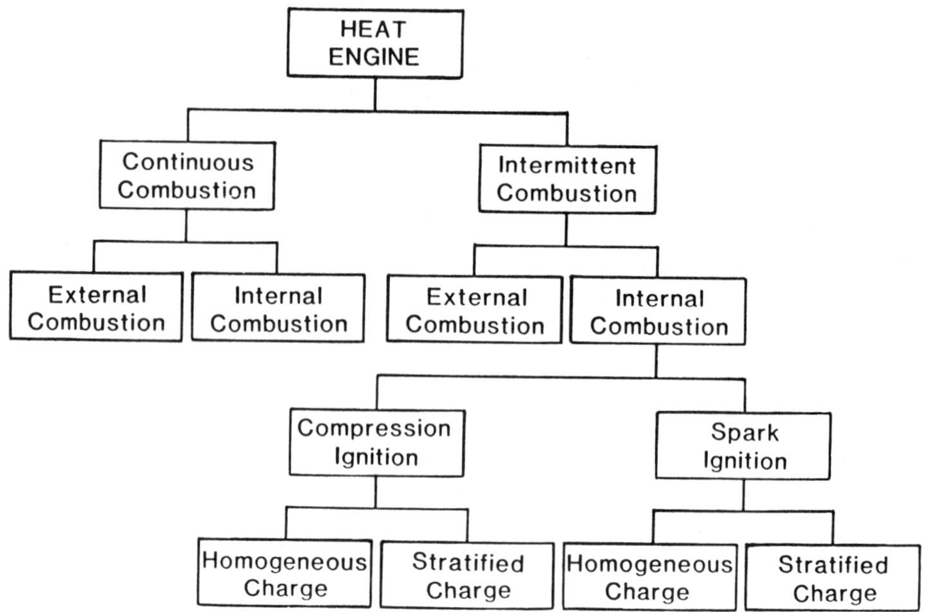


Fig. 1.   Heat-engine tree.

attractive intermittently burning external-combustion engine has yet to be identified.

In the continuous-external-combustion class, the Rankine-cycle steam engine has the longest history behind it.  In recent years the substitution of various organic fluids for water in the steam engine has provided an interesting variant.  The Stirling-cycle engine, which has existed since the early 1800s, is a continuous-external-combustion engine that uses a gaseous working fluid, avoiding the phase change that characterizes the Rankine cycle.

The leading continuous-internal-combustion candidate is based on the Brayton cycle.  The reciprocating Brayton-cycle engines was in use during the last half of the 1800s.  It has even been proposed for passenger-car propulsion in modern times [1], but the leading option in this engine class is the non-reciprocating version, the gas turbine.

The intermittent-internal-combustion engine requires frequent periodic ignition of gulps of cylinder charge.  That ignition may be accomplished either spontaneously, as the result of compression of the charge, or by some forced means, usually by an electrical spark.

Going a step further in Fig. 1, ignition in the intermittent-internal-combustion engine may be effected in either a homogeneous or a stratified charge of fuel and air.  The compression-ignited stratified-charge engine is recognized as the diesel.  Homogeneous-charge compression-ignition has been employed successfully in reciprocating engines [2,3], but limitations placed on operating speed and load by reaction kinetics make it an unlikely candidate for the passenger-car application [4].  In contrast, when ignition is forced by a spark, both homogeneous and stratified charges constitute viable possibilities for the automobile.

------

[1]Numbers in brackets designate references found at the end of this paper.

From this review of Fig. 1, the leading challengers to the spark-ignition engine are the steam and Stirling engines, the gas turbine and the diesel.  The operating principles of these alternatives have been reviewed previously [5].  For any of them to succeed in the passenger car, they must be judged alongside the spark-ignition engine with respect to a long list of attributes that includes good fuel economy, low emissions, compatibility with available materials and fuels, low cost, compact size, low mass, brisk performance, easy starting, low noise, and good durability with minimal maintenance.

The first two attributes on this list are necessary, but not sufficient, qualities for success in the U.S. passenger car.  In these times of concern over the future outlook for the availability of quality fuel, the successful alternative powerplant must at least match today's spark-ignition engine in fuel economy.  The ability of the powerplant to operate satisfactorily on alternative fuels is another issue that deserves attention.  In the near term, however, alternative fuels are unlikely to be available in sufficient quantities to accommodate more than a small fraction of the vehicle fleet.  Until an alternative fuel is readily available and widely distributed, the economics of producing an alternative powerplant solely on the basis of its ability to operate on that alternative fuel must be subject to scrutiny.

A convenient method for assessing the fuel economy of an engine as applied to the passenger car is to measure the fuel economy of the car on the U.S. EPA combined urban and highway transient driving schedule. However, car fuel economy depends not only on the efficiency of the engine, but also on the efficiency of the drivetrain, the rolling resistance of the tires, the aerodynamic drag of the car, and especially the mass of the car. In the decade starting in 1975, the fuel economy of the average U.S. car increased sufficiently to reduce its fuel consumption by 41%, but a good share of this gain was attributable to a 24% reduction in test weight [6].

It can be demonstrated that a useful way to normalize vehicle mass out of vehicle fuel economy is to deal with a fuel economy index that is equal to the product of volumetric fuel consumption on the EPA combined driving schedule and car test weight [7].  This index gives a more reasonable assessment of the efficiency of the engine and its drivetrain than does vehicle fuel economy.

Although a step in the right direction, fuel economy index is not perfect, however.  It can be inflated by underpowering the car.  Installing an alternative engine that lacks adequate power to provide the brisk performance previously listed among the desirable attributes, as has sometimes been done, can provide an unreasonably favorable evaluation of that engine based on its fuel economy index.  For instance, the band in Fig. 2 shows how fuel economy index can be influenced by the time it takes to accelerate from a standing start to 97 km/h (60 mi/h).  This band encompasses data points from a sample of 29 1985-86 production cars with automatic transmissions from twelve different manufacturers, ranging in test weight class from 2750 to 4250 lb and in engine displacement from 1.6 to 5.0 L.  (The slope of this band is typical of homogeneous-charge spark-ignition engines and may not be applicable to other powerplant types in correcting their fuel economy indices for deficiencies in acceleration performance due to inadequate power.)  Despite this shortcoming, fuel economy index remains a much better basis on which to evaluate the fuel economies of alternative powerplants than the fuel economy of the car in which it is installed.

The second essential attribute of an alternative automotive powerplant is compliance with emission standards.  Over the past two decades,

remarkable progress has been made in controlling the tailpipe emissions
from U.S. passenger cars. A 1981 survey of in-use vehicles indicated that
it then took ten cars to emit as much unburned hydrocarbon (HC) and carbon
monoxide (CO) as came from a single automobile in pre-control days, and the
emission of oxides of nitrogen (NOx) was only a quarter of what it had been
before emission control [7]. Retrogression in exhaust emissions in an
alternative powerplant is not only unacceptable, it may be illegal.

While effort proceeds on alternative powerplants, improvements
continue to be made to the spark-ignition engine. It presents a moving
target. Two purposes of this paper are to provide an update on the status
of the leading alternative powerplants, and to indicate areas in which the
spark-ignition engine is being further improved.

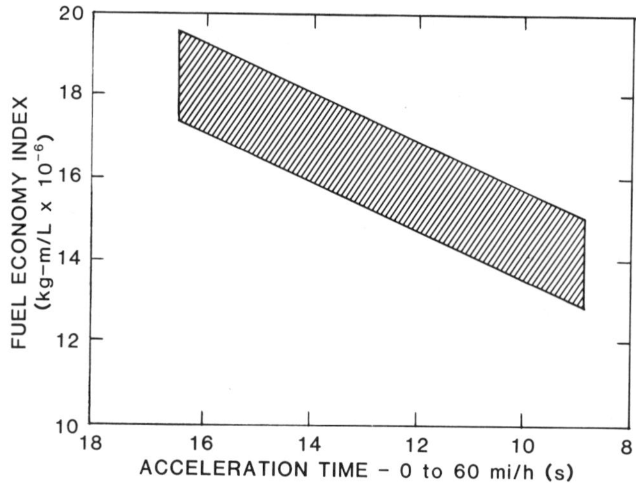


Fig. 2. Effect of performance on fuel economy index.

## LEADING ALTERNATIVE POWERPLANTS

An indication of how the fuel economy indices of the salient
alternatives relate to the index for the U.S. fleet average is provided in
Fig. 3. Only the diesel engine betters the spark-ignition engine, by a
margin of about 40%. The diesel enjoys an inherent advantage of about 12%
over the gasoline engine because of the higher energy content of a gallon
of diesel fuel relative to gasoline. Another portion of its margin of
superiority is attributable to the generally weaker performance of diesel
passenger cars. Even after allowance for these factors, however, the
average diesel engine retains an advantage in brake thermal efficiency over
the average gasoline engine. Further comments on each of the options
illustrated in Fig. 3 are offered in turn below.

### The Steam Engine

During the decade beginning in the late 1960s, no less than eight
different steam-engine powered cars were sponsored and demonstrated in the
U.S. [8]. These included two by General Motors, one by the predecessor to
the U.S. Department of Energy (DOE), two by the U.S. Department of
Transportation, two by the State of California, and one by a private
individual. The fuel economy of each fell short of contemporary production
cars with spark-ignition engines. The single representative point shown

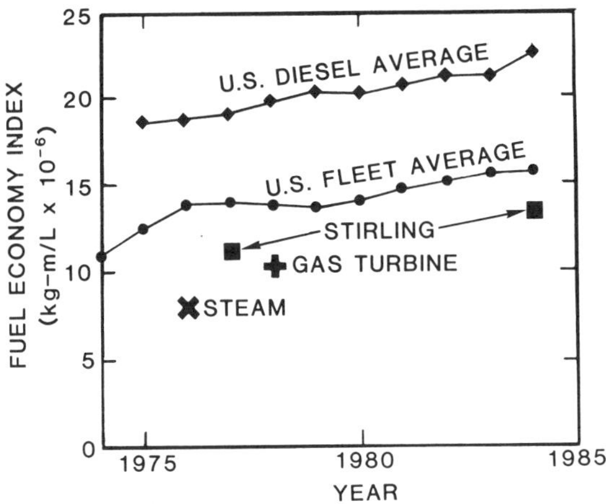


Fig. 3.  Fuel economy index for various engine types.

for the steam engine in Fig. 3 is for the steam-powered 1975 Dodge Monaco developed by Scientific Energy Systems as a part of the highly visible government Alternative Automotive Propulsion Systems Program.  Of the six steam cars in this sample of eight for which emissions data were made available, three met the emissions goals of 0.41/3.4/0.4 g/mi HC/CO/NOx at low mileage, but in two of these the margin of safety was unrealistically small.  In all probability, the fuel economy and emissions of any of these steam cars could have been improved with additional effort, but the fuel-economy shortfall was sufficient to discourage continuation.  Today the steam car is no longer considered a serious contender.

## The Stirling Engine

The modern Stirling engine was brought to the U.S. by General Motors from the N. V. Philips Company of the Netherlands at the beginning of the 1960s.  A decade of collaborative progress ensued, during which the principal interest in the engine was for non-automotive applications. During the 1970s the Ford Motor Company, under contract to DOE, built and installed a 134-kW swash-plate Stirling engine in a Torino.  The 1977 Stirling point in Fig. 3 indicates the fuel economy achieved with this initial installation.  Both HC and NOx emissions exceeded targets (which were the same as those listed above for the steam engine) by about 40%, but the CO target was met at low mileage.  Performance fell short of the objective.

DOE continued its Stirling-engine program with Mechanical Technology Incorporated (MTI), which has acquired engine technology from United Stirling AB of Sweden.  MTI has progressed through a series of engines and engine improvements.  In 1984, as a part of DOE's Industry Test and Evaluation Program, tests were conducted by General Motors on MTI's experimental installation of a 53-kW Stirling engine in an American Motors Spirit [9].  As received, the Spirit belonged in the 3250-lb test weight class (TWC).  It was determined that for the power available from the engine, it belonged more appropriately in a 2500-lb TWC Chevette if acceptable performance was to be achieved.  However, the engine was too large to fit into a Chevette.  Consequently, it was tested in the Spirit on a chassis dynamometer as if it were installed instead in a 2500-lb TWC car.  Because the Stirling engine was heavier than the spark-ignition engine it replaced, this practice favored the Stirling engine by not charging its greater mass against it.  (Alternatively, this practice could

be interpreted as corresponding to a test on a future Stirling engine
having a mass lowered to the level of the equal-power spark-ignition
engine.  It is not yet clear that such a mass reduction is possible.)

The 1984 Stirling point in Fig. 3 represents the result of this
evaluation.  Considerable progress has been demonstrated since 1977, but
the U.S. fleet average did not stand still, and the Stirling engine still
showed a shortfall in fuel economy.  In other tests it was found that at
the 2500-lb test weight, the Stirling Spirit offered acceleration
performance competitive with the Chevette at speeds up to 48 km/h, but was
somewhat deficient in high-speed performance.  Cooling is always a concern
in a closed-cycle engine like the Stirling because all of the cycle
inefficiency must be accommodated by the cooling system, whereas in the
internal combustion engine, a significant part of the cycle inefficiency is
carried directly into the atmosphere with the exhaust gas.  With a radiator
triple the volume of the one in the Chevette, the Stirling Spirit performed
satisfactorily in wind-tunnel tests at temperatures up to 36°C (95°F).  The
MTI Stirling engine was quiet, running with the fan inoperative a good
share of the time.  Low-mileage emissions met the current U.S. standards of
0.41/3/4/1.0 g/mi HC/CO/NOx.

Of special interest in this evaluation was why the Stirling engine,
which showed superior fuel economy in steady-state tests on an engine
dynamometer, did not do better in transient vehicle fuel-economy tests.
The fuel economy of the Stirling Spirit actually matched that of the
Chevette on the highway schedule, but it fell 21% short on the urban
schedule.  Two major contributing factors to this discrepancy in urban
driving were found to be a 75% higher idle fuel rate compared to the
Chevette, and the extra fuel consumed in warming up the Stirling engine.
On the cold start, an extra 80 s of engine operation was required before
the heater tubes reached the specified temperature of 700°C.  During the
10-min hot soak specified as part of the urban schedule, the engine was
allowed to continue running on the 6- to 7-min capacity of residual heat
stored in the engine in order to protect certain static seals from
overheating.  Then on the hot restart, an extra 30 s of engine operation
was required to restore the specified heater-tube temperature.

These findings were brought to the attention of MTI and are being
corrected to the extent possible in the design of a new engine.  This new
powerplant is projected by MTI to be smaller and lighter than the one
tested at General Motors, and to exceed the U.S. fleet average fuel economy
while meeting emission constraints.  The new engine is not yet ready for
evaluation, however.

Any extra warmup time required for the closed-cycle Stirling engine
will be an unwelcomed annoyance to the consumer.  The slow leakage of
hydrogen from the system is another aggravation that needs to be overcome.
Excessive manufacturing cost remains a barrier to acceptance.

## The Gas Turbine

The regenerative gas turbine makes a desirable automotive powerplant
because it is free of the noise and vibration associated with intermittent
combustion and reciprocating pistons, it can be made small and light, and
it normally has an excellent torque-curve shape for road-vehicle use.  In a
laboratory setting, the ability to meet 0.41/3.4/0.4 g/mi HC/CO/NOx levels
has been demonstrated at low mileage [10].

Chrysler, Ford and General Motors have all had gas turbine programs,
with origins going back to the late 1940s.  Likewise, many automotive
manufacturers in Europe and Japan have been active.  The U.S. government

has long sponsored work in this field, currently under DOE.  The single
representative gas turbine point in Fig. 3, from a  turbine-powered car
built by Chrysler for DOE, illustrates a common shortcoming of past
passenger-car gas turbines -- non-competitive fuel economy.  The thermal
efficiency of the gas turbine engine depends heavily on turbine inlet
temperature, regenerator effectiveness, and component efficiencies, each
considered in turn below.

It is now generally conceded that if the automotive gas turbine is to
become competitive in fuel economy, it will have to incorporate structural
ceramics in its hot parts, including the highly stressed turbine rotor.
This is expected to permit raising turbine inlet temperature from today's
level of around 1050°C with high-temperature metal alloys to as much as
1350°C.  That is the direction in which most research on this engine is now
headed.

The turbine rotor places very difficult demands on ceramics.  Despite
over fifteen years of continuous effort in this area, unqualified success
has proven elusive.  Lack of ductility in ceramics is a major obstacle that
contributes to excessive scatter in physical properties.  In addition, the
costs of fabrication, processing and inspection remain excessive.  These
difficulties are expected to yield to additional research, but to what
extent remains a matter of conjecture.

The regenerator reduces the fuel required to reach a specified turbine
inlet temperature through preheating the burner inlet air by transferring
heat into it from the turbine discharge gas.  At the same time it cools the
engine exhaust gas to a tolerable level and helps to contain noise
emanating from the turbine that would otherwise enter the environment with
the exhaust.  For a given heat-transfer surface geometry, regenerator
volume increases rapidly with increasing effectiveness.  The periodic-flow
rotating regenerator has been developed to provide high effectiveness in a
reasonably compact space and is available in ceramic to accept future
higher operating temperatures.

Higher compressor and turbine efficiencies are always sought and are
important to the attainment of acceptable fuel economy.  Unfortunately the
efficiencies of these components are sensitive to size.  As ceramics pave
the way to higher turbine inlet temperatures, the engine airflow
requirement for a specified power rating decreases.  This puts an extra
burden on the turbomachinery engineer, who already faces the problem of
improving upon established efficiency levels when engine rated power is
decreased to match the requirements of smaller, lighter cars.

Nearly all automotive gas turbines demonstrated to date have been of
the free-shaft type, in which a compressor and a turbine mounted on a
common shaft serve as a gasifier to feed a supply of compressed hot gas to
a downstream power turbine.  This allows the compressor, which provides the
air for the cycle and therefore establishes the available power potential
of the engine, to run at a speed independent of the power-producing
turbine, which is connected to the vehicle driving wheels through gearing
and is therefore responsive to vehicle speed.  As a result of this
arrangement, the engine is able to deliver maximum torque when the output
shaft is stalled, a fete possible with the reciprocating internal-
combustion engine only if it is coupled to a torque converter.

When a car with such a free-shaft gas turbine engine is accelerated
normally from a standing start, however, that maximum torque is not
immediately available because the idling gasifier is typically running at
only half its rated speed and delivering only a fraction of rated airflow.
Before the engine can deliver maximum torque, then, the gasifier has to be

accelerated to its rated speed so that it can provide the airflow associated with full power. Unfortunately that takes time, on the order of a second. If that time is excessive, the driver senses an objectionable delay in throttle response.

Among other factors, the response of the gasifier is proportional to its polar moment of inertia. The switch from a metal to a ceramic turbine should be helpful in minimizing response time of the engine because the density of turbine ceramic is only about a third the density of high-temperature alloy. A force in the opposite direction is the trend from axial-flow to radial-inflow gasifier turbines in the interest of retaining high turbine efficiency as turbomachinery is scaled down in size, for the inertia of the radial turbine rotor is typically substantially greater than that of a low-inertia axial-flow rotor of the same material. Turbine designers will have to pay close attention to engine acceleration if an acceptable automotive gas turbine is to become a reality.

## The Diesel Engine

The diesel has been running cars for 50 years. Given current fuel prices, its superior fuel economy (Fig. 3) is not sufficient to outweigh its many negative factors relative to today's gasoline engine. At least that is the way the U.S. consumer has voted in dealer showrooms, where diesel car penetration peaked at about 6% in 1981. Since then it has steadily declined. Last year, it slid to about 1% of sales. Half of the dozen companies that once built diesel cars for the U.S. market have dropped out.

There are a variety of reasons behind the disaffection of the U.S. consumer for the diesel car. Among them are the decline in fuel prices, which had once been projected to be two to three times their current level, imposition of an extra tax on diesel fuel by the U.S. government, which makes diesel fuel less attractive relative to gasoline, the higher first cost of the diesel car, its generally weaker performance, and the higher noise level and the exhaust odor associated with today's diesel engine. Despite such shortcomings, the diesel remains the most fuel-efficient engine known that is suitable for passenger-car propulsion. Faced with the long-range prospects for petroleum, it is difficult to ignore this engine until a more efficient alternative is identified.

Conventionally cooled diesels. All passenger-car diesels on the market today are of the indirect-injection (IDI) type. Research is now primarily aimed at the direct-injection (DI) type because it promises 10 to 15% higher efficiency than the IDI version.

The first problem with the DI engine is emissions. It has proven difficult to bring HC and NOx into compliance with U.S. standards, although this hurdle may be overcome in small cars with small engines. This follows from the following relationship:

$$\text{Emissions (g/mi)} = \text{const.} \times \frac{\text{(EI)}}{\text{(FE)}} \tag{1}$$

where (EI) = emission index (g pollutant/kg fuel)
      (FE) = fuel economy (km/L or mi/gal)

Since U.S. standards are expressed in g/mi, it is seen that a vehicle that achieves superior fuel economy can tolerate a higher emission per unit of fuel consumed.

One of the ways of controlling NOx is to employ exhaust gas
recirculation (EGR), which has proven very effective.  Unfortunately,
introducing EGR results in an increase in particulate production.  Very
difficult particulate standards lie ahead for the U.S.  Research has shown
that this tradeoff between NOx, as controlled by EGR, and particulates is
inherent to the diesel combustion process [11].  EGR lowers flame
temperature, which is a well known way to decrease the production of NOx,
but the lower flame temperature inhibits the oxidation of the soot formed
during diffusion combustion, which becomes the major constituent of
particulate matter in the exhaust gas.

Promulgated U.S. particulate standards pose a serious threat to the
future of the diesel.  Decreasing the aromatic content of the fuel and
placing tighter control on top-end fuel volatility could help [12], but
there is understandable reluctance to accept the increased fuel cost
associated with more strict fuel specifications.  Removing the sulfur from
the fuel would also be beneficial [13], and this might happen for other
environmental reasons.  It is widely believed that exhaust particulate
trapping will become a necessity, however.

Unfortunately these particulate filters plug with trapped particulates
in fairly short driving distances, imposing unacceptable back pressure on
the engine.  They can be regenerated by burning off the trapped particulate
matter, but one cannot count on the exhaust gas to be hot enough often
enough to keep the trap acceptably clean.  This leads to such approaches as
igniting the trapped matter periodically, in which case a burning front
progresses slowly through the trap until it has been regenerated, and
introducing additives into the fuel on board the vehicle to lower the
ignition temperature of the particulates.  Introducing a fuel additive
raises fresh environmental questions.  At this stage the durability of the
trap is an unresolved issue.  Clearly, addition of a particulate trap, its
regeneration system and controls will add an unwelcome cost to the diesel
engine.

Low-heat-rejection diesels.  A comparatively recent innovation in the
diesel field is the low-heat-rejection (LHR) version.  It has often been
termed the "adiabatic" diesel [14], but complete adiabaticity is not
possible.  The idea behind the LHR diesel is to avoid the normal heat loss
to the liquid coolant by eliminating the coolant.  The resulting increase
in engine operating temperature dictates that certain critical engine parts
incorporate ceramics, either monolithic or in the form of coatings over
metal substrates.

One might expect from the first law of thermodynamics that eliminating
the heat rejection to the traditional liquid cooling system would be
rewarded with a substantial increase in energy available on the crankshaft,
but the second law of thermodynamics proclaims otherwise.  Most of the
energy thus conserved appears in the exhaust gas in the form of increased
temperature.  Some of this energy can be recovered by adding a bottoming
engine based on the Rankine, Brayton, or Stirling cycle, but an economic
assessment of several bottoming engines for a LHR engine in a heavy-duty
truck has suggested that with present technology, the payback period is
excessive [15].

In a study of a naturally aspirated LHR diesel for the passenger car
[16], it was found that the loss in volumetric efficiency resulting from
the hot uncooled cylinder walls made some form of supercharging desirable
to restore engine power output.  For the heavy-duty application, the
turbocompound arrangement has usually been favored.  This involves the
addition of a turbocharger in series with a downstream power turbine geared

to the crankshaft.  It has been found that for the light-duty passenger-car application, the inclusion of the power turbine is of questionable merit because of the generally low exhaust-energy level at light load and idle.

The high gas temperatures of the LHR diesel tend to increase the NOx emission, but there have been cases where that problem has been alleviated by retarding injection timing.  Published test data on LHR diesels for passenger cars are practically non-existent.  Steady-state dynamometer measurements of HC and particulate emissions compared to the conventional diesel are mixed.  A satisfactory means of lubricating the piston/cylinder wall interface has yet to be demonstrated.  More progress is also needed in the ceramics field in the areas of reliability and cost.  In addition, there is a conflict in ceramic requirements.  The thermal conductivity should be low for good thermal insulation but high for low thermal stress. The thermal-expansion coefficient should be high (for ceramics) at joining interfaces with metal parts but low for low thermal stress.  The future for the LHR diesel in cars and light-duty trucks is uncertain, but the engine is presently in such a primitive state of development that the concept merits further research.

## THE HOMOGENEOUS-CHARGE SPARK-IGNITION ENGINE

Given the status of the leading alternatives discussed, the spark-ignited reciprocating internal-combustion engine is expected to continue its domination of the passenger-car field for the rest of this century. But it certainly will not remain stagnant.  The thrust will be toward further advances in fuel economy, within legislated emission constraints, and for integration into a powertrain that is more pleasing to the consumer.

A parameter measuring the fuel economy of this engine is its brake thermal efficiency, which is given by the following expression:

$$\eta_b = \eta_i \left[ 1 - \frac{\bar{P}_f + \bar{P}_p}{\bar{P}_i} \right] \tag{2}$$

where $\eta_b$ = brake thermal efficiency

$\eta_i$ = indicated thermal efficiency

$\bar{P}_p$ = pumping mean effective pressure

$\bar{P}_f$ = friction mean effective pressure

$\bar{P}_i$ = indicated mean effective pressure

It is seen that brake thermal efficiency can be improved by increasing the indicated thermal efficiency developed by the gas within the cylinder, by lowering the work expended in pumping the gas through the cylinder during the intake and exhaust strokes, by decreasing the work lost to friction, and by increasing the indicated mean effective pressure developed during the compression and expansion strokes.  This expression provides a framework for discussing where today's passenger-car engine is headed.

<u>Indicated Thermal Efficiency</u>

First, consider indicated thermal efficiency.  Theoretically, it rises
with increasing compression ratio, as shown for the ideal fuel-air Otto
cycle in Fig. 4.  Experimentally, that trend has been confirmed out to a
compression ratio well beyond the 9 or so used today [17].  Historically,
the road to higher compression ratio has always been blocked by combustion
knock.  The commitment in the U.S. to unleaded gasoline of 91 research
octane number has reinforced this barrier.  However, with a given fuel,
small advances in allowable compression ratio are possible by designing the
combustion chamber to burn the charge faster.  The idea is to sweep the
flame across the chamber before the prereactions that trigger knock have
had time for completion.  Thus, faster-burning chambers have allowed a slow
progression to the right along the experimental curve of Fig. 4.

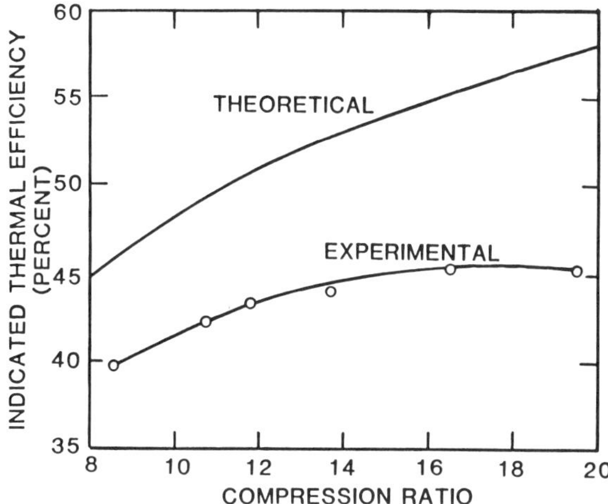


Fig. 4.  Effect of compression ratio on thermal efficiency.


That curve falls short of theory because of both heat loss and time
effects -- particularly the time required for combustion.  That means that
a faster-burning chamber can raise the level of this entire experimental
curve.

Theoretically, the indicated thermal efficiency can also be raised by
burning a more dilute mixture.  For example, the top curve of Fig. 5 shows
how the indicated thermal efficiency of the ideal fuel-air Otto cycle rises
as the mixture is diluted with excess air by increasing the air-fuel ratio.
The sharp break occurs at the stoichiometric ratio, to the left of which
unreleased chemical energy is rejected with the exhaust gas in the form of
carbon monoxide and hydrogen.  The experimental curves below follow the
trend of theory for rich mixtures.  With increasing dilution, however,
flame speed falls.  This increases the time loss enough to offset the
theoretical gains from charge dilution.  As illustrated, a faster burn rate
increases both the optimum dilution and its associated efficiency.

In the U.S. today, because NOx reducing catalysts are used that lose
their effectiveness in the presence of significant oxygen, engines are
close-loop controlled electronically to run at the stoichiometric ratio.
Dilution is then achieved by adding EGR to the air-fuel mixture.  When
diluting with EGR, the efficiency trends qualitatively resemble those in
Fig. 5 for diluting with excess air.  Interest exists in eliminating the

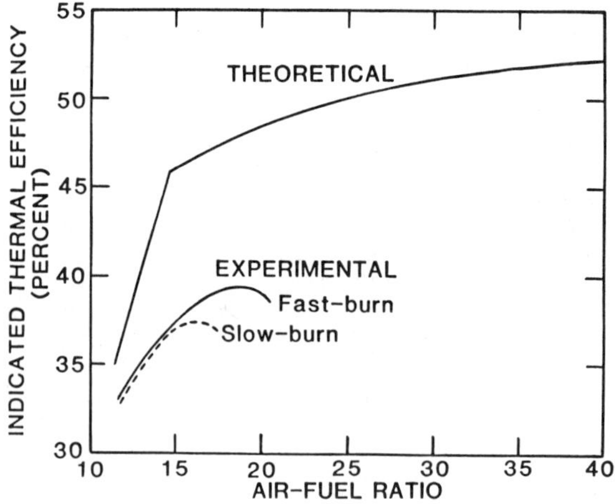



Fig. 5.   Effect of air-fuel ratio on thermal efficiency.


EGR system and doing the dilution with air in a lean-burn engine, but so far, poor driveability has discouraged that approach.

This situation is represented in Fig. 6 by tradeoff curves between brake thermal efficiency and NOx emission at constant speed and load.  The upper solid curve is for the lean-burn engine, which accomplishes dilution through an increase in air-fuel ratio.  The lower solid curve is for an engine controlled to the stoichiometric air-fuel ratio, dilution being accomplished with EGR.  The two trajectories share a common baseline point at the stoichiometric ratio without EGR.

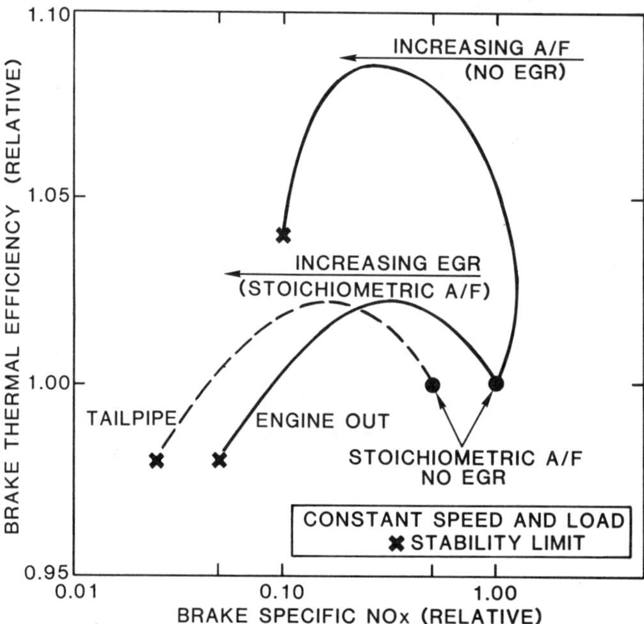



Fig. 6.   Efficiency-NOx tradeoff for lean-burn and stoichiometric
          engines.

As dilution is increased from the baseline point in the lean-burn engine, first NOx and efficiency are both seen to increase, then NOx decreases as efficiency continues to rise to a peak, then efficiency falls as NOx continues to decrease. The trajectory terminates at a point where combustion instability causes intolerable driveability.

As EGR is increased in the stoichiometric engine, NOx decreases monotonically as efficiency rises to a peak, then falls. The dashed curve illustrates the leftward shift effected by addition of a reducing catalyst for NOx control. Such a shift is not feasible for the lean-burn engine because of the unacceptance of significant exhaust oxygen by the reducing converter.

It is evident from Fig. 6 that the lean-burn engine offers better fuel economy, but the stoichiometric engine with reducing converter promises lower driveability-constrained NOx. Although the fuel-economy advantage shown for the lean-burn engine is about 6%, this margin depends on the engine design and its operating point and is generally held to range from 2 to 7%. The question is whether fast-burn technology can extend the driveability-constrained NOx emission of the lean-burn engine to a sufficiently low level without the help of an NOx catalyst. The evolution of a durable lean sensor for closed-loop mixture control could help in this regard. Of course, compliance with HC and CO standards is essential with either option.

<u>Fast burn</u>. Three reasons for increasing the burn rate in the combustion chamber have now been identified, all leading to higher indicated thermal efficiency. These are tolerance of a higher compression ratio, a closer approach to the theoretical limiting efficiency, and tolerance of greater charge dilution. The next logical question is how to achieve faster burn rates. There are two approaches -- one geometric and the other fluid mechanical.

Geometrically, the idea is to configure the chamber to burn the charge as soon after the spark is struck as possible. For example, in the simple representation of a pancake chamber in Fig. 7, it is clear that the peripheral spark plug yields the longest burn duration. The central spark plug provides the shortest duration, but when a pair of valves is placed in the head, a central spark plug is impractical. Two spark plugs arranged as shown prolong the burn a trifle, but the early part of the burn is hastened, which is thermodynamically beneficial.

The fluid-mechanical approach to fast burn involves turbulence enhancement. By correlating measurements of in-cylinder turbulence in a motoring engine with burning rates deduced from cylinder-pressure measurements and a computer simulation of engine combustion [18], it was established that the burning velocity of the developed flame followed a relationship approximated by

$$S_b = K_1 S_1 + K_2 u' \tag{3}$$

where $S_b$ = turbulent burning velocity

$\quad S_1$ = laminar flame speed

$\quad u'$ = rms turbulence intensity

$K_1, K_2$ = constants

Because laminar flame speed falls as the mixture is diluted, it becomes increasingly important to increase in-cylinder turbulence as that path to

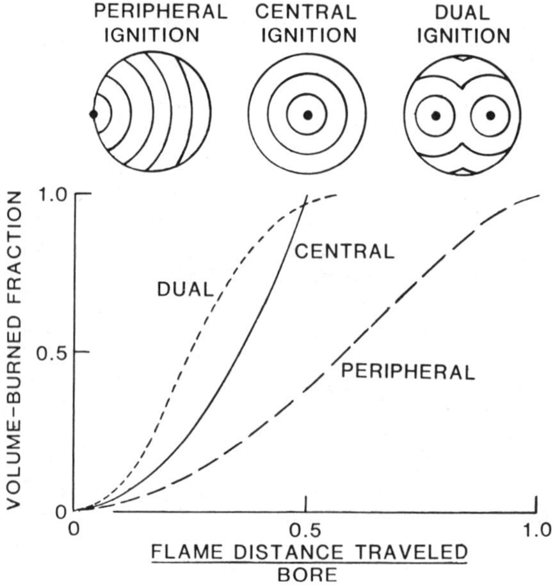


Fig. 7.   Effect of spark-plug location on burn duration.

greater efficiency is followed.

One way this has been accomplished effectively is by introducing swirl into the intake air, e.g., with a swirl port such as illustrated in Fig. 8. Another method of providing swirl involves shrouding the valve with a projection from the underside of the head, as illustrated in Fig. 9.   The large-scale bulk swirling motion established during intake subsequently decays into finer-scale turbulent motion.

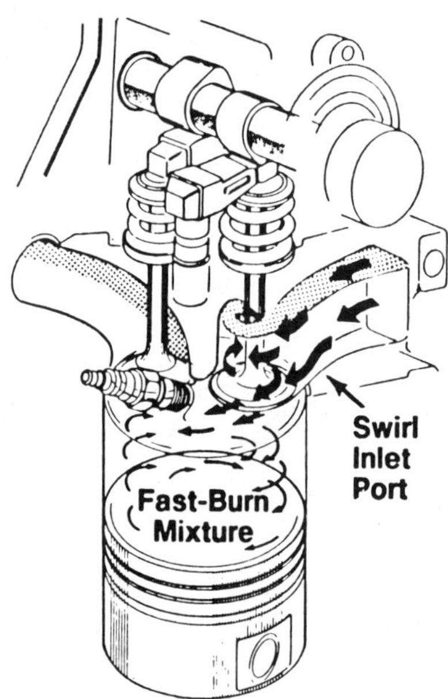


Fig. 8.   Intake swirl port.

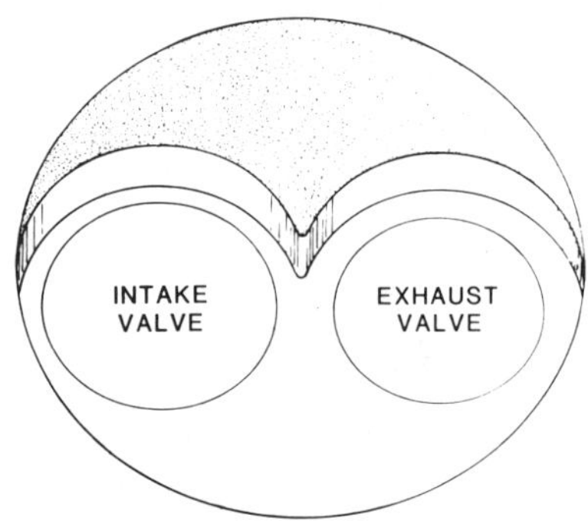


Fig. 9.   Intake-valve shrouding
in cylinder head.

Moderate swirl has proven very beneficial in a number of engines, but too much swirl can be counterproductive because it increases cylinder heat loss and impairs engine breathing. The simplified two-dimensional intake-valve representation of Fig. 10 illustrates the latter point. The mass flow entering the cylinder is proportional to the product of the radial velocity component ($U_r$) and the valve curtain area ($\pi dh$), but the pressure drop across the valve varies as the square of the actual velocity ($U$). As angle $\beta$ is increased to increase the swirl component ($U_t$), for the same pressure drop (same $U$), the radial component ($U_r$) is decreased and cylinder airflow reduced.

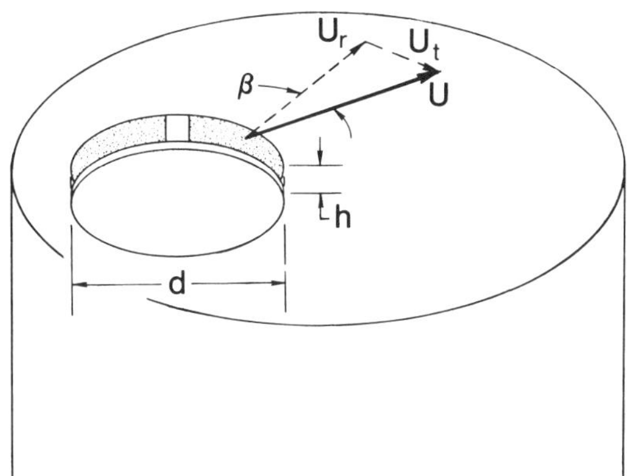

Fig. 10.  Swirling flow through a two-dimensional intake valve.


Other cylinder motions can contribute turbulence. For example, tumbling is illustrated in Fig. 11 and squish in Fig. 12. Swirl and tumbling flows are established during the intake stroke, and the resulting fluid motions have the opportunity to decay during the compression stroke. In contrast, squish builds in strength during the compression stroke. Adopting a simple two-dimensional model of squish, for a given cylinder geometry the velocity of the gas escaping the squish area as the piston ascends is found to vary as the ratio of the instantaneous piston velocity to the corresponding height of the squish area.

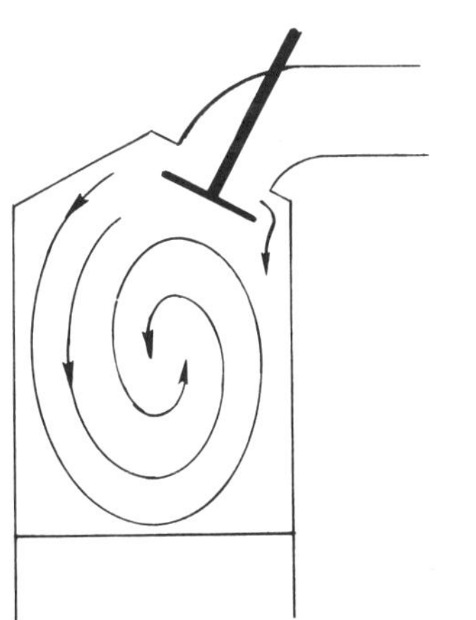

Fig. 11.  Tumbling motion.

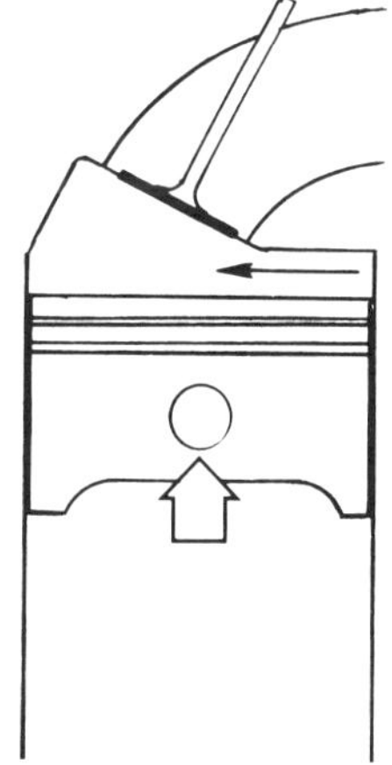

Fig. 12.  Squish motion.

The nature of the squish-velocity variation with crank angle is shown
in Fig. 13 for two different top-dead-center squish clearances that differ
by a factor of three. Peak squish velocity is very sensitive to that final
squish-area clearance, as well as to the fraction of the cylinder cross-
sectional area devoted to squish. Peak squish velocity calculated for a
motoring engine as in Fig. 13 may be prevented from developing by the
expansion of the burning gases ignited by a spark that is timed well before
top dead center.

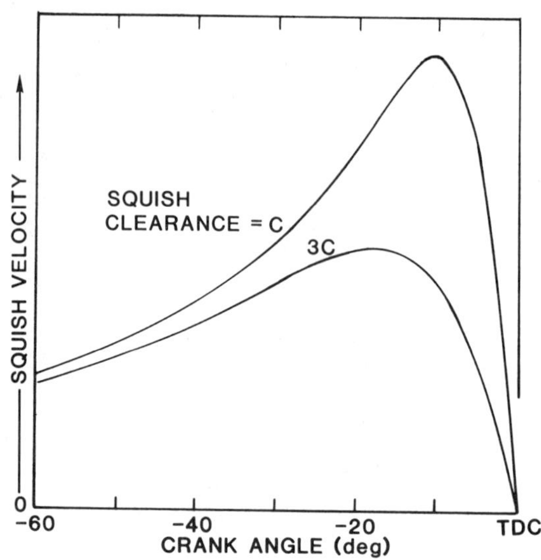


Fig. 13. Trend of squish velocity with crank angle.

Flame initiation and combustion variability. Although maintenance of
a high burning velocity in the developed flame is important, the flame
initiation period preceding flame development is equally significant. This
involves the growth of the flame kernel that is formed as a result of the
spark discharge. In order to grow, the kernel must release heat through
the engulfment of fresh mixture on its periphery faster than it loses heat
to the spark plug electrodes, the surrounding gas, and the combustion
chamber walls. Failing that, the kernel collapses and the engine misfires.

Cycle-to-cycle variability in combustion can be a troublesome problem
in burning dilute mixtures. Such variability can often be traced to
inconsistencies in the growth of consecutive flame kernels. A moderately
swirling flow can be helpful in this regard by exposing the kernel to a
directionally consistent flow pattern.

The nature of combustion variability is illustrated by the plots of
apparent mass-burned fraction in Fig. 14 for four different cycles, all
with the same spark timing. In the upper left, the solid trace is for a
faster-than-average cycle. Near full load this is the cycle likely to
knock because the spark timing, which is set for the average cycle, is
overadvanced for this one. An excessively long initiation period
characterizes the solid trace in the lower left. For this late-burning
cycle, the spark timing is too retarded. The solid trace in the upper
right indicates only a partial burn because of its low maximum amplitude.
The cycle in the lower right illustrates a misfire. The flame kernel grows
to some small maximum size but then extinguishes. It does not take many
partial-burns and misfires in a collection of consecutive cycles to cause

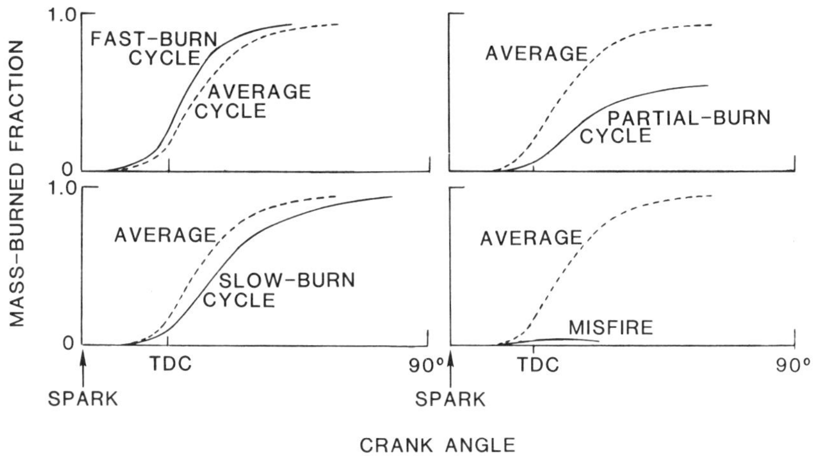


Fig. 14.  Types of heat-release schedules.

trouble with the emission of unburned HC.

If the average burn duration is prolonged, e.g., by high charge dilution or by using a slow-burn chamber design, cycle-to-cycle variability is often worsened by variability in the duration of the flame-initiation period.  A long burn duration calls for more spark advance, meaning that the spark must be struck earlier on the compression stroke.  At such early crank angles the cylinder flow pattern has had less time to become established, and the lower cylinder-gas temperature dictates a slower laminar flame speed.  Neither of these characteristics is favorable to consistent growth from the flame kernel into a developed flame front.

<u>Pumping Loss</u>

The second way to improve brake thermal efficiency in Eq. 2 is to reduce the pumping loss during the intake and exhaust strokes.  Pumping work is represented by the shaded area on the cylinder pressure-volume diagram of Fig. 15.  It depends fundamentally on three factors:  the magnitude of the vacuum in the intake manifold during the intake stroke, which affects the floor of the pumping diagram, the back pressure in the

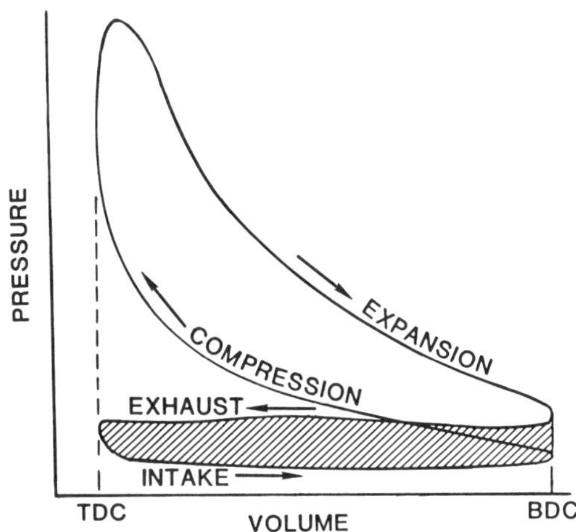


Fig. 15.  Pumping-work area on cylinder pressure-volume diagram.

exhaust system during the exhaust stroke, which affects the ceiling of the
pumping diagram, and the degree of flow restriction imposed by the intake
and exhaust ports and valves themselves.

Charge dilution decreases the pumping loss at a given speed and load,
whether with EGR or with excess air.  In either case the throttle must be
opened further at a given speed and load to induct the required fuel.  That
raises the floor of the pumping diagram toward atmospheric pressure, thus
decreasing the shaded area, and hence the pumping loss.

Exhaust-system design is important to the ceiling of the pumping
diagram, especially at high speed and load.  In this regard, catalytic
converters with lower flow restriction are seeing more widespread use.

## Mechanical Friction

The third route to higher brake thermal efficiency in Eq. 2 is through
lower friction.  A breakdown of friction in a representative engine is
shown in Fig. 16 as a function of car speed.  The distribution among the
various sources differs from one engine design to another, but the
principal contributor is normally the piston-and-ring assembly.  This
component can be decreased by reducing piston-ring tension.  Another
approach is to decrease the number of rings from three to two.  In both
cases the tradeoff between reduced friction on the one hand, and increased
oil consumption and blowby on the other, must be thoroughly examined.

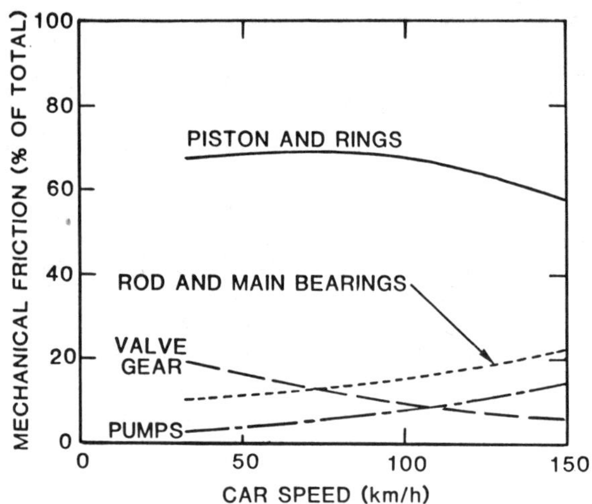


Fig. 16.  Breakdown of friction loss in a representative engine.

Another important factor is the fit between the piston and the
cylinder bore.  There is more to this than simply manufacturing round
cylinder holes in the block and truly cylindrical pistons to fit within
them.  At operating temperature the piston crown is hotter than the rest of
the piston, so the piston assumes a new shape when running.  Distortion of
the cylinder bore from bolting the head to the block during assembly, and
from temperature differences around the bore and/or mechanical constraints
during engine operation, can change the shape of an initially round bore.
Specially contouring the piston skirt for tribological reasons and
decreasing its area can be helpful.

A lower-mass piston and connecting rod can contribute by reducing piston side thrust and altering bearing loads. Under some circumstances it may prove possible to decrease the diameter of the crankpin journal and gain on bearing friction. Attention to crankshaft counterweighting is important to this approach.

The new generation of low-friction oil has provided a fuel-economy gain on the order of 2%, depending on driving conditions. Further gains in this area are being sought.

The valvetrain has a number of friction-producing rubbing contacts. Lighter valves allow weaker valve springs for reduced cam loads. Ceramics are being investigated because of a perceived potential for reducing friction at the cam-and-lifter interface. However, the biggest gain in this area comes from switching to roller followers. This gain is especially significant in the low-speed range.

The parasitic loads of the oil and coolant pumps can be decreased by designing more efficient units. It is also important to ensure that they are not overdesigned in terms of capacity. Driving the cooling pump electrically so it can be deactivated when not needed has even been considered as a way to improve engine efficiency. On many current cars the cooling fan is already operated in that fashion.

Indicated Mean Effective Pressure

The fourth parameter cited in Eq. 2 as affecting brake thermal efficiency is indicated mean effective pressure (IMEP). Fixing cylinder geometry, at optimum spark timing and a specified air-fuel ratio, IMEP tends to vary in direct proportion to the air mass inducted per cycle. Indicated power is proportional to the product of air per cycle and engine speed. From the standpoint of air mass per cycle, speed tends to be limited by the pressure drop across the intake-valve and port restriction. Increasing the effective area of that restriction allows a higher engine speed for the same pressure drop, hence more indicated power.

At the top of Fig. 17 is a pancake chamber with vertical valve stems. Below, the valve axes have been inclined to include a 30-deg angle. This provides a 5% increase in intake-valve area. In Fig. 18 the plan view of a two-valve pancake chamber is shown at the top. Below, the number of valves has been doubled. Intake-valve area is increased by 13%. Inclined valves and multiple valves are appearing in increasing numbers in modern engine designs.

Additional valve area and higher operating speeds might be good for indicated power, but operating at higher speeds can adversely affect brake thermal efficiency because of increased friction. Friction mean effective pressure (FMEP) is plotted against speed for a representative engine in Fig. 19. It is all right to run at higher speeds on those rare occasions when maximum power is demanded, but most of the time the engine should be operated at lower speeds, to the left in Fig. 19, to keep FMEP low.

The ratio of engine speed to car speed has always been an important parameter in establishing vehicle fuel economy. In accordance with the above point, a move toward additional gear ratios in the transmission has been taking place in the U.S. to ensure low engine speeds during cruise operation, yet retain the ability to draw on the full high-speed power capability of the engine when it is needed. Increasing the engine speed for maximum power in the future is going to continue the pressure on transmission designers for broader ratio coverage.

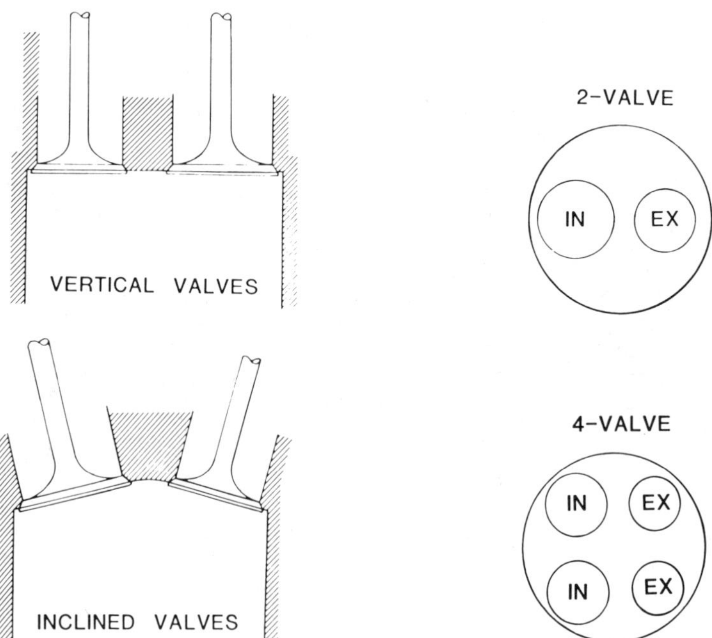


Fig. 17.   Vertical and inclined valve axes.

Fig. 18.   Two- and 4-valve heads.

High power at high engine speed is fine for the race-car driver, but in a passenger car it is desirable to preserve strong full-load torque at low engine speeds.  This has focused greater attention on manifold tuning, especially in the intake system.  As illustrated in Fig. 20, the cylinders, the intake plenum, the air cleaner, and the environment represent containers of air that are joined together by pipes of optional length.  By choosing these container volumes and pipe dimensions properly, near-maximum full-throttle torque can be extended over a broad range of engine speeds. Tuned intake systems have also been conceived in which a valve is incorporated to change the effective length or diameter of some part of the system as a function of engine operating condition [19].

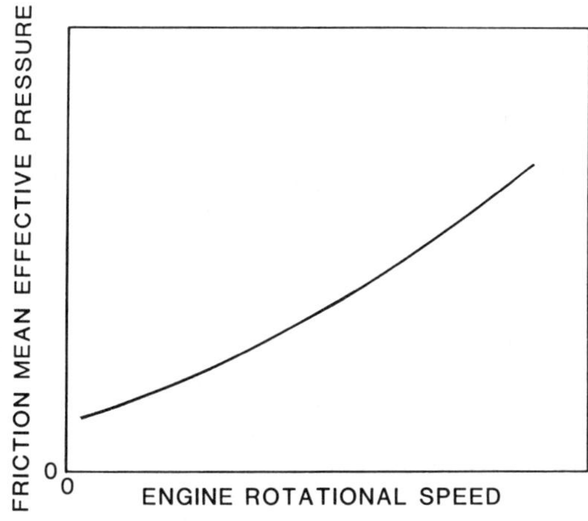


Fig. 19.   Effect of speed on friction mean effective pressure.

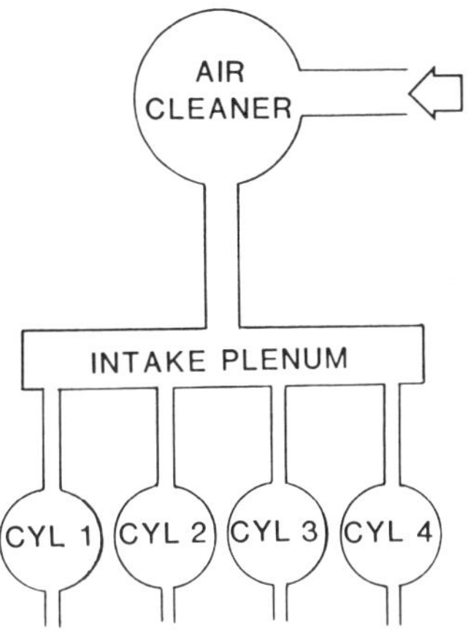

Fig. 20.  Engine intake system.

Most of the time, though, the passenger-car engine runs heavily
throttled rather than at full throttle.  That puts pressure on the designer
of a free-breathing high-speed engine.  The contribution of in-cylinder
turbulence to burn rate has already been discussed.  In a basic engine,
that turbulence originates primarily from the high-velocity flow through
the intake-valve restriction.  If that restriction is enlarged to extend
the rated speed of the engine for higher maximum power, then at the low-
speed cruise condition the intake area is so generous that the turbulence-
producing velocity through the intake restriction is inadequate to maintain
consistent combustion, and near the idle condition the residual-gas
fraction is high.  The result is cycle-to-cycle variability in IMEP, which
leads to uneven torque impulses on the crankshaft that can be reflected as
poor vehicle driveability.

The benefit of intake swirl in this instance has already been
discussed, as has the conflicting adverse effect of excessive swirl on
engine breathing.  Part-time swirl offers a way out of this dilemma.

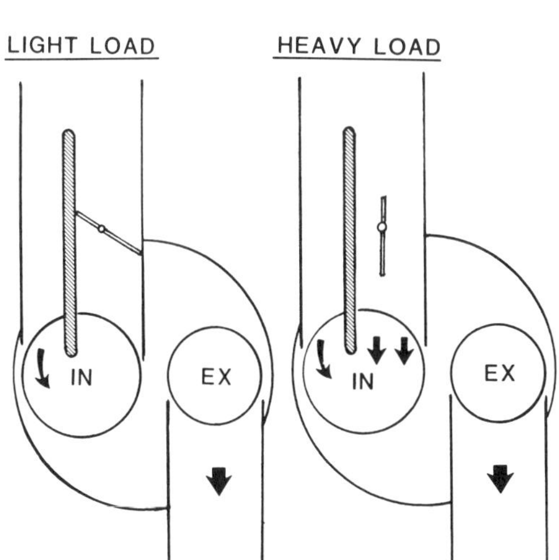


Fig. 21.  Swirl with
divided intake port.

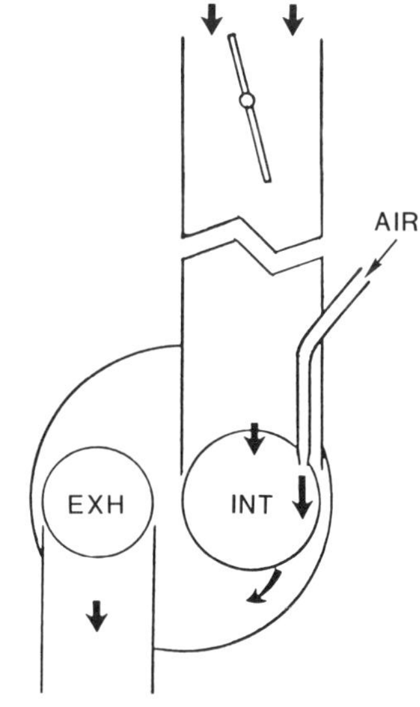


Fig. 22.  Swirl with
air induction tube.

With the arrangement of Fig. 21, a portion of the intake port is closed off at light load to provide swirl. At high loads the valve is opened for free breathing.

In Fig. 22 a tube leading directly from the air cleaner, upstream of the throttle, is mounted off center in the intake port. When the throttle is closed, the vacuum in the intake port draws maximum airflow through the tube, inducing swirl through momentum exchange. As the throttle is opened, the vacuum into which the tube is discharging decreases, the velocity of the air leaving the tube falls, and swirl is decreased.

In an engine with two intake valves per cylinder, one can be closed off at part load to encourage swirl, as illustrated in Fig. 23. When full power is demanded, both ports are opened for free breathing.

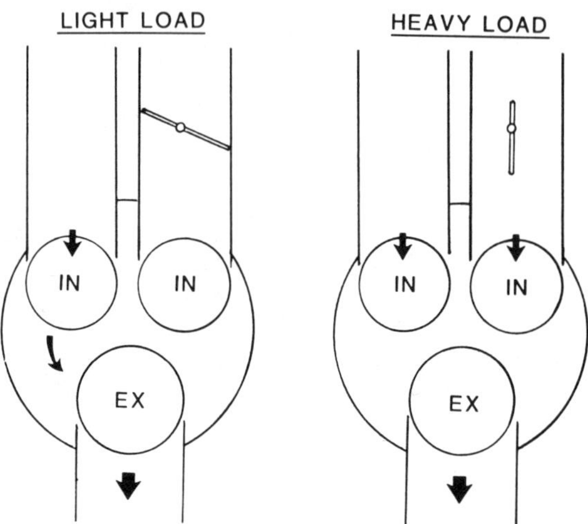


Fig. 23. Swirl production with two intake valves.

## Power Boosting

Closely related to the subject of IMEP and the cylinder airflow so fundamental to its determination is power boosting, which involves forcing more air into the cylinders by compressing it ahead of the intake manifold. Supercharging a spark-ignition engine in this manner encourages engine knock. The advent of the knock sensor, which detects incipient knock and retards the spark to avoid it, has made boosting a viable option. The intercooler now appearing on some turbocharged cars also helps to alleviate knock by decreasing the temperature of the charge fed to the cylinders.

The exhaust-turbine-driven supercharger has been used on passenger-car engines for some time. It has often been suggested that turbocharging a small engine combines small-engine economy with large-engine performance, but that promise has frequently not materialized. First, a strong-performance speed ratio for the drivetrain is normally chosen to cover up the delayed response of the turbocharger speed to sudden throttle opening. That causes the engine to run faster at a given car speed, which adversely affects the brake thermal efficiency of the engine. Second, the compression ratio is often reduced somewhat to increase the increment of power boost available from the turbocharger before combustion knock is encountered. Third, the thermal capacity of the extra mass in the exhaust system ahead of the catalytic converter delays catalyst lightoff. This may

necessitate a compromise in engine calibration in order to meet emission
standards. For these reasons, the turbocharger is generally sold as a
performance option, not an economy enhancer.

Skillful matching of engine, turbocharger and drivetrain has done much
in recent years to improve the responsiveness of the turbocharged car to
sudden increases in the demand for engine power. For further improvement,
ceramic turbine rotors are on the horizon. The lower density of a ceramic
rotor, compared to contemporary rotors of high-temperature alloy, is
beneficial in this regard.

Variable-geometry turbos are also making their appearance. The one
illustrated in Fig. 24 changes effective turbine flow area through movement
of a partition in the inlet scroll. The resulting ability to act like a
small turbocharger at low speeds and a large one at high speeds
substantially improves the full-throttle torque curve, which is normally
very poor at low speeds in a turbocharged engine.

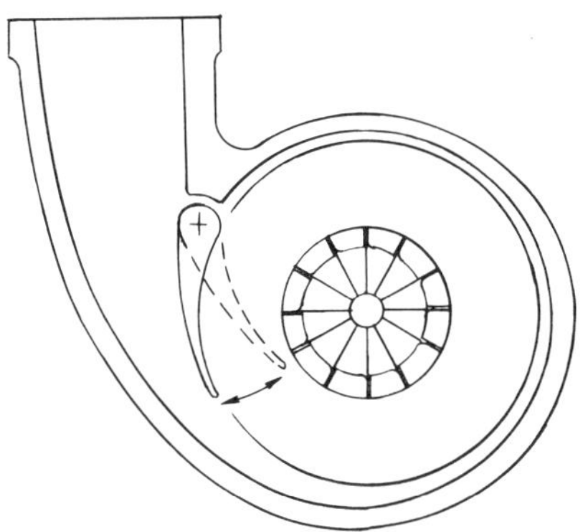

Fig. 24. Variable-geometry turbocharger turbine.

The mechanically driven supercharger overcomes the response lag of the
turbocharger and has no direct effect on catalytic-converter temperature.
However, now the power to drive it comes directly from the crankshaft
instead of coming almost free from the exhaust gas. As a result, boosting
a small engine for economy by this means does not work unless the
supercharger is transparent to the engine during most driving. This
involves either bypassing the supercharger or declutching it, then bringing
it into action only when power boost is needed.

Turbocharging and supercharging must always face a harsh economic
reality. Once a commitment is made to bore out, say, four cylinders in a
block, it costs little more to make them slightly bigger, or to lengthen
the stroke a bit, for more power. The number of valves and spark plugs
stays the same. In contrast, getting the extra power by supercharging
means more hardware and more money. For this reason, power boosting seems
destined for a limited market segment.

Electronic Controls

The halving of the U.S. NOx standard at the beginning of this decade
ushered in the era of electronic control. It was then that the NOx
catalyst came into widespread use. The conversion efficiency of the NOx
catalyst falls sharply in the presence of oxygen, thus ruling out the lean-

burn approach.  While the NOx catalyst accepts rich mixtures, that approach
is discouraged by the increase in HC and CO emissions and by deteriorating
fuel economy.  Running between these two extremes, very close to the
stoichiometric ratio, and diluting the cylinder charge with EGR was the
path selected.  The traditional carburetor could not hold the air-fuel
ratio sufficiently close to the stoichiometric value under all driving
conditions, however.  This led to the now standard approach of closed-loop
control of air-fuel ratio with an on-board computer receiving its signal
from an oxygen sensor in the exhaust.

Once the computer is on board, it can be used for many other control
functions.  For example, it is applied to idle-speed governing.  That
manages a stable idle speed independent of ambient pressure and
temperature, or changing accessory loads, or increased friction in an aging
throttle linkage.  The computer has made unnecessary the choke valve, with
its possibility of sticking and its imprecise and sometimes slow response
to temperature.  With electronic control of the choking function, the fuel
system can readily respond to differing engine warmup rates under varying
conditions of operation.  On some engines the computer, working in
conjunction with a knock sensor, detects incipient knock and retards spark
timing just enough to avoid it.  The computer has a memory capable of
learning from previous operation and adapting engine control accordingly.
It can facilitate improved coordination between the engine and
transmission.  It has self-diagnostic capabilities, storing information on
subsystem malfunctions for subsequent interrogation by the engine service
person.

The stoichiometric oxygen sensor in common use today responds to
exhaust-gas stoichiometry like a switch, indicating that oxygen either is
or is not present.  Lean sensors are under development and even in limited
use that respond instead to the actual amount of oxygen in the exhaust.
This work holds the promise of closed-loop control of air-fuel ratio as a
function of engine speed and load for lean-burn engines, improving their
chance of meeting emission standards and driveability targets.

The undesirability of variability in IMEP from one cycle to the next
has been mentioned.  Variability can also occur from one cylinder to the
next.  This may result from such factors as manufacturing tolerances,
engine deposits, deterioration, and maldistribution of mass flow or mixture
ratio.

Maldistribution of air-fuel ratio among individual cylinders happens
despite today's closed-loop control of air-fuel ratio to the stoichiometric
level because that control monitors only engine average air-fuel ratio, and
also because EGR may not be evenly distributed.  As illustrated in Fig. 25,
a cylinder that runs more dilute than the engine average needs additional
spark advance but does not receive it because traditionally, spark timing
is set for the average cylinder.  Thus the output from the weak cylinder is
doubly disadvantaged -- once because it is more dilute than average and
again because its spark timing is retarded from where it should be.  The
resulting uneven torque impulses threaten driveability.

Maldistribution can also cause a problem with knock.  Near full
throttle, maldistribution may cause one cylinder to knock before the rest.
In many implementations today's knock sensor detects this and retards spark
timing in all cylinders -- knocking and non-knocking alike.  This
unnecessarily penalizes engine power.

A cure for such characteristics is in sight.  Signals from each
cylinder on each cycle can be fed to the on-board computer, which, given
today's advanced distributorless ignition system, can adjust spark timing

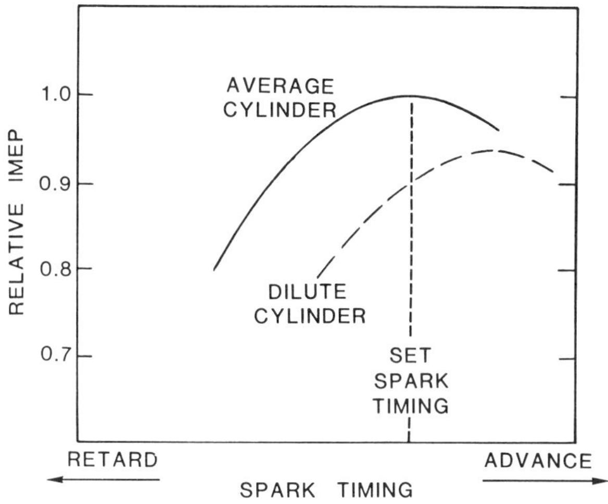


Fig. 25.  Interaction between mixture distribution and spark timing.


to the appropriate value in each cylinder individually.

Two approaches for extracting the needed signals from the cylinders have been advanced.  One is based on the measurement of cylinder pressure.  The other detects flame arrival at the far end of the combustion chamber by means of an ion probe installed in each cylinder.

These techniques also show promise of measuring mixture dilution.  With individual-port fuel injection, this would facilitate closed-loop control of mixture strength in each cylinder.

## Variable Engine Geometry

Variable engine geometry has been investigated, and in some cases applied in production, to improve engine characteristics.  The three most common goals have been to vary displacement, to vary compression ratio, and to alter valve timing as a function of speed and/or load.

<u>Variable displacement</u>.  The object of varying piston displacement is to avoid the need for intake throttling at part load by decreasing the volume of mixture inducted.  A variety of mechanisms has been conceived for varying piston stroke.  If a broad load range is to be covered, some unfavorable bore-stroke ratios result at one or both extremes of the range.  The variable-stroke engine has been simulated experimentally [20].  The inherently poor combustion-chamber geometry led to high HC emissions.  A substantial gain in fuel economy was indicated for the U.S. combined cycle, but only if the variable-stroke mechanism involved added no friction.  With additional mechanism friction, the projected fuel-economy gain could be rapidly eroded.

An alternative way to vary displacement is to render some cylinders inoperative by deactivating their valves.  One way of accomplishing this, illustrated in Fig. 26, is to move the rocker-arm fulcrum over the valve tip.  Cylinder deactivation has been applied in the U.S. in a production V-8 engine to gain a modest improvement in vehicle fuel economy.  Application to smaller engines with fewer cylinders makes it more difficult to meet driver pleasability objectives because of the resultant low cylinder-firing frequency.

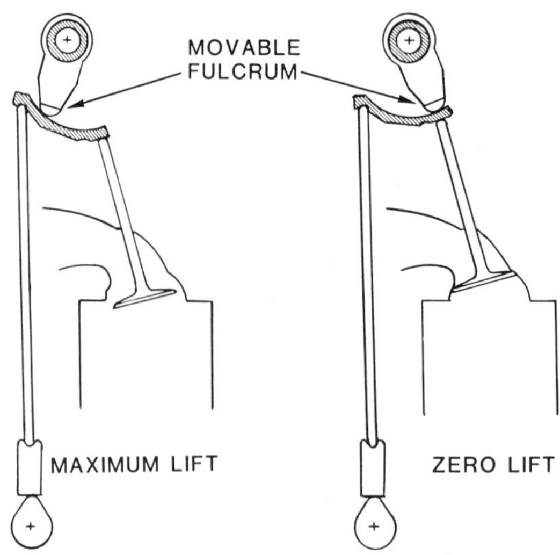


Fig. 26. Pivoted rocker arm for control of intake-valve lift.


<u>Variable compression ratio</u>. In a variable-compression-ratio engine, the compression ratio is set at a high level for superior efficiency at light load and reduced as load is increased to avoid knock. Methods for accomplishing variable compression ratio [21] include (1) telescoping the connecting rod, (2) moving the cylinder head relative to the crank/connecting rod/piston assembly, (3) moving the piston crown relative to the piston-pin axis, and (4) changing the clearance volume by means of a movable plug in the head. The first of these is awkward to do in the face of normal stresses and space restrictions. The second is embodied in the single-cylinder CFR octane-test engine but is difficult to transfer to a multicylinder engine. The third has been applied to a diesel engine with a two-piece piston by separating the piston crown from the piston base with a controllable-volume cavity that is filled with circulating oil [21]. Varying clearance volume with a movable plug was demonstrated long ago in a spark-ignition engine [22].

Recently a car that met European emission standards without a catalyst was tested on the U.S. urban schedule [23]. When powered by an engine having a movable plug to vary compression ratio from 9.5 to 15, it showed 15% better fuel economy than the baseline engine with fixed compression ratio. However, the HC emission was higher. Meeting the more difficult U.S. standard for HC could be a problem at high compression ratios because first of all, the engine-out emission is higher, and secondly, the lower exhaust-gas temperature accompanying high compression ratio raises concern about keeping the catalytic converter hot enough to maintain high conversion efficiency.

<u>Variable valvetrain geometry</u>. The principal objectives of variable valvetrain geometry are to improve the shape of the full-load torque curve, to enhance combustion at part load, and/or to improve light-load fuel economy. Each is considered below in turn.

Two full-load curves of torque versus speed are shown in Fig. 27 for the same engine with two different valve timings. The high-speed timing is characterized by late intake-valve closing. Power, proportional to the product of torque on the ordinate and speed on the abscissa, peaks at 6000 r/min. The low-speed torque curve is for earlier intake-valve closing. It

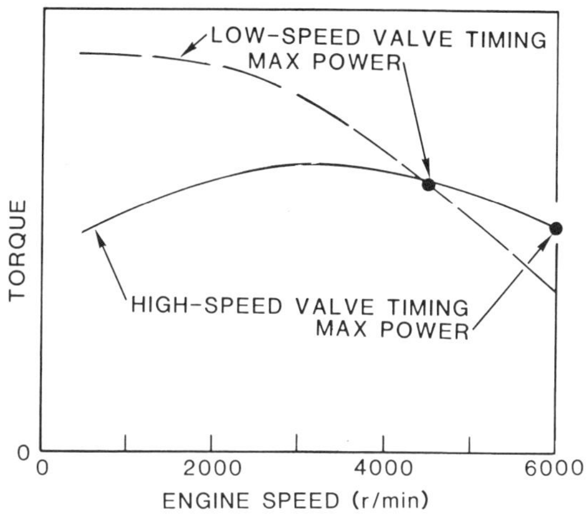


Fig. 27.  Effect of valve timing on torque.

provides substantially more torque at low speeds, but its power peaks at only 4500 r/min and is 12% less than the power achieved with the high-speed timing.

By re-indexing the intake cam on the camshaft as a function of speed, it becomes possible to provide a torque curve approaching the envelope of the two shown in Fig. 27.  Greater flexibility in valve timing is provided by an elongated three-dimensional cam that slides axially along the camshaft [24].  The cam follower is then exposed to whichever slice through the cam the axial-position controller has selected for it.

Independent control of all four valve events is even more desirable. Electric and hydraulic valve actuators, in place of the traditional camshaft, are possibilities.  However, proponents of such mechanisms encounter three hurdles to overcome -- an approach to zero valve-seating velocity for noise control, acceptable power consumption, and a favorable cost-benefit tradeoff.

Part-load combustion variability may often be linked to valve-timing choices made in considering full-load torque.  It is usual in a high-speed engine for the exhaust-closing and intake-opening events to overlap considerably near top dead center between the exhaust and intake strokes. When an engine with this cam timing is run at low speeds and light loads, a high residual fraction is retained in the cylinder at intake-valve closing. This dilution can contribute to excessive combustion variability that is correctable with low valve overlap.  However, with the low overlap, high-speed performance suffers.  To resolve this conflict, a passive variable-valve-overlap lifter has been demonstrated that changes overlap as a function of engine speed [25].

Another aid to part-load combustion is provided by intake-valve throttling [26].  By rotating the fulcrum of Fig. 26 toward the valve stem as load is reduced, any desired fraction of full lift can be provided. This feature makes the throttle unnecessary because part-load throttling can now be accomplished solely by decreasing valve lift.  At very light loads, with the entire throttling pressure drop occurring across the valve, the mixture attains sonic velocity in entering the cylinder.  This energetic entering flow promotes a fast burn rate.  Residual-gas fraction is also reduced, further enhancing burn rate.

To improve the light-load fuel economy, pumping loss can be reduced by
varying the intake-valve closing angle.  Both late intake-valve closing
(LIVC) and early intake-valve closing (EIVC) have been investigated.  With
LIVC, the throttle is held open while load is decreased by progressively
delaying intake-valve closure until later during the compression stroke
[27].  This allows the piston to expel part of the inducted mixture back
into the intake manifold until compression actually begins.  With EIVC, the
throttle is held open while load is decreased by closing the intake valve
progressively earlier during the intake stroke [28].  The trapped mass is
then expanded during the latter part of the intake stroke and recompressed
on the succeeding compression stroke.  In the comparison conducted, EIVC
was judged superior to LIVC for improving fuel economy, with the gain
amounting to less than 10% and being sensitive to any increase in friction
incurred from the valve mechanism.


## THE STRATIFIED-CHARGE SPARK-IGNITION ENGINE

The advantage to fuel economy of the lean-burn engine using a
homogeneous charge was discussed in the preceding section.  One limitation
to this approach was identified as the generally slower combustion
encountered in a dilute mixture.  If the mixture is stratified, richer than
the cylinder average near the ignition source and leaner than average or
preferably free of fuel in the rest of the chamber, then it becomes
possible to use an overall leaner mixture than can be managed with the fuel
and air mixed homogeneously, and in principle still achieve the
thermodynamic benefits and reduced pumping loss associated with the dilute
mixture.

There are a large number of ways to accomplish such stratification and
combust the resulting mixture.  The three illustrated in Fig. 28 serve as
examples.  From left to right, the fuel is introduced through carburetion,
through port fuel injection, and through direct cylinder injection.  Each
is discussed in turn below.

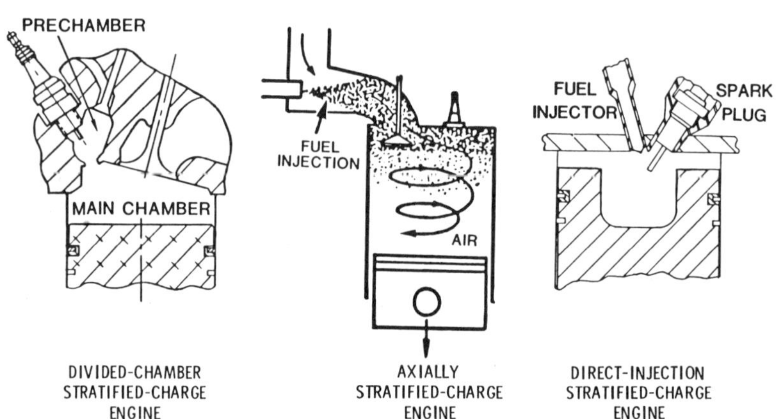


Fig. 28.  Three approaches to charge stratification.


## Charge Stratification with Carburetion

In the divided-chamber engine on the left in Fig. 28, a lean mixture
is carbureted into the main chamber, and a rich mixture is carbureted into
the prechamber through a separate intake valve.  Ignition of the rich
prechamber mixture propels a flaming torch of gas into the main chamber,
providing a powerful ignition source for the lean mixture.

Some typical emission results for such an engine run at fixed load and speed are presented in Fig. 29. They typify a recurring problem in many lean-burn engines, whether homogeneous or stratified. It is seen from the figure that the engine was able to run at quite lean air-fuel ratios, and that the NOx emission was quite low at those ratios. This is indeed essential because of the inability to use a reducing catalyst for NOx control at such lean mixtures. Unfortunately, the HC emission concurrently rose to unacceptable levels. Catalytic treatment of this emission with an oxidizing catalyst is made difficult at such lean mixtures by the associated low exhaust-gas temperature. An evaluation of this concept led to the conclusion that when U.S. emission standards had to be met, this engine offered no advantage over the conventional homogeneous-charge engine [29].

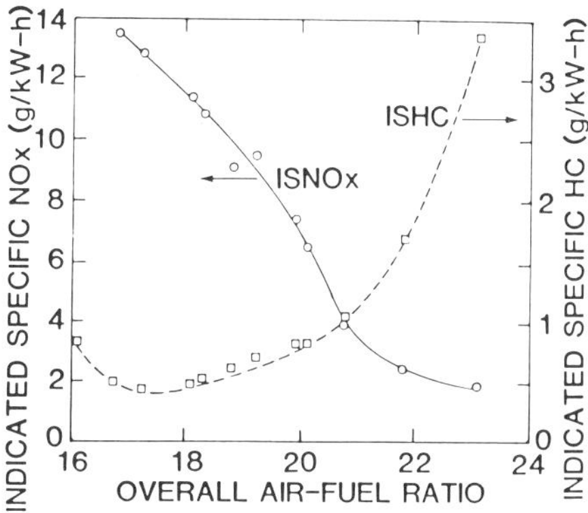


Fig. 29. Emissions versus overall air-fuel ratio for divided-chamber stratified-charge engine.

## Stratified Charge with Port Injection

In the axially stratified-charge version in the middle of Fig. 28, fuel injected into the intake port is timed to maintain a rich mixture at the top of the chamber, near the spark plug, and leave air layered against the piston. This is an appealing approach because it requires minimal modification to existing production engines. In an experimental installation of an engine embodying this concept, current U.S. emission standards were met at low mileage using an open-loop-controlled near-stoichiometric mixture with EGR rather than air as a diluent, and an oxidizing converter for control of HC and CO [30]. A 6% fuel economy gain was recorded relative to the conventional engine from which the stratified-charge version was adapted. However, the margin between the measured emissions and the standards was not yet wide enough to justify volume production. An engine of this class incorporating variable geometry in its intake system and a lean sensor for closed-loop control of air-fuel ratio has achieved market status in Japan [31], where completely different emission regulations are in place.

## Stratified Charge with Direct Injection

In the direct-injection stratified-charge engine on the right in Fig. 28, the fuel is sprayed directly into a piston bowl to form a central cloud of combustible mixture surrounded by air. Two different versions of this DISC (Direct-Injection Stratified-Charge) engine have been around for many years.

In the early-injection (E-DISC) version [32], injection begins comparatively early during the compression stroke and terminates before ignition. This allows some mixing time before combustion.

In the late-injection (L-DISC) version [33], injection starts much later during compression. The spark is fired during the injection period, meaning that fuel is sprayed into an existing flame as in the diesel engine.

Both of these engines encounter difficulty meeting U.S. emission standards for HC and/or NOx. Both require an oxidizing catalyst. Both are subjects of ongoing research because of their attractive fuel-economy potential, which exceeds that of the homogeneous-charge lean-burn engine. The L-DISC engine has a multifuel capability, but at high loads it is smoke-limited like the diesel.


## ALTERNATIVE FUELS

Petroleum is a non-renewable resource. Eventually, reasonably recoverable oil will be depleted. Experts estimate that date at sometime in the next century. As we approach it, the quality of crude oil will deteriorate and gasoline prices must rise. Attention is already being devoted to the operation of automotive vehicles on alternative fuels.

The acceptance of an alternative fuel depends on the answers to several questions. How available is it to the consumer, both in quantity and price? In what engine(s) will it perform satisfactorily? How does it compare to today's fuels in terms of on-board storage, engine durability, emissions, safety, etc.? It is the question of matching the engine with the alternative fuel that is appropriate to this review.

The external-combustion Stirling engine is the most omniverous of the engines meriting consideration. It has already demonstrated its ability to run on such atypical fuels as wood chips and rice husks. The gas turbine has shown its ability to run on powdered coal and 80-proof tequila. Such demonstrations are impressive. For automotive use, however, the proper question is how the various engines perform on gaseous fuels and alcohol, for those are the alternative fuels most likely to pass the test of availability. If the established intermittent-combustion spark-ignition and diesel engines can manage respectably with those fuels, it is unreasonable to expect that an alternative powerplant is going to displace them solely on the basis of its fuel tolerance.

Hundreds of thousands of cars around the world are already operated on gaseous fuels -- both natural gas, which is mostly methane, and LPG, which is mostly propane. Natural gas is particularly plentiful in some parts of the world. Large-scale automotive use of natural gas tends to be localized in the geographical regions where it is found. This minimizes the cost-escalating fuel transportation problem. Because it is not dispensed from every publicly accessible service station, natural gas is presently more likely to be used in captive fleets operating within a radius from a central servicing site.

Natural gas and LPG both enjoy high octane ratings, thus suiting them well to the homogeneous-charge spark-ignition engine. They exhibit low exhaust emissions. However, the volume occupied by the gas displaces air, lowering volumetric efficiency and consequently decreasing power. In addition, the on-board storage of gaseous fuel involves bulky, heavy tanks compared to gasoline, or alternatively, reduced range on a tank of fuel. The safety of gaseous fuel is also a consideration.


30

Given the high octane rating of natural gas, one might expect it to be a poor fuel for a compression-ignition engine. By itself, it is, but diesel-powered buses have been run successfully in passenger service by inducting natural gas mixed with air on the intake stroke and igniting it with pilot injection of compression-ignited diesel fuel. Increased HC emission can be a concern with this approach.

Alcohol also enjoys a high octane rating. Both methanol and ethanol are presently blended into some U.S. gasolines in small amounts. Their concentrations in unleaded gasoline are limited by regulation of the U.S. Environmental Protection Agency. Regulations also require the inclusion of additives to counter the phase separation encouraged by the presence of less than 1% of water in methanol/gasoline blends. The vapor pressure of the fuel increases as small amounts of alcohol are added to gasoline, threatening a serious problem with evaporative-emission regulations in the U.S.

Ethanol is normally made from agricultural products and will never be available in sufficient quantities in the U.S. to serve as more than a blending agent. In some other countries, however, neat ethanol makes more sense. For example, ethanol from sugar cane has been widely used in Brazil.

Methanol is made principally from natural gas, although some day it might come in significant quantities from coal. That prospect has helped to stir interest in the long-range use of methanol as a common automotive fuel.

With an octane rating around a hundred, methanol is able to tolerate a higher knock-limited compression ratio than gasoline for greater thermal efficiency. This gain is not enough to compensate for the low energy content of methanol, though, which is only about half that of gasoline. Roughly speaking, the choice is between twice the fuel-tank size and half the range. The cold-starting characteristics of methanol are dismal. Not surprisingly, driveability is inferior during warmup. Methanol burns well in lean mixtures and produces low NOx emissions, but typically it yields high emissions of unburned fuel and troublesome aldehydes. The methanol-air mixture in a partially filled tank is flammable at normal temperatures, and methanol burns with a nearly invisible flame. These two characteristics have been considered safety hazards. Methanol attacks many of the materials in today's fuel system, and increased rates of cylinder wear have been encountered.

The energy available from the current annual production of methanol is but a tiny fraction of that provided by gasoline. In fact, last year less than 1% of the energy consumed by U.S. cars and light-duty trucks came from methanol and ethanol combined. To bridge the gap between the present, when the quantity of methanol available is miniscule, and the distant future, when it might become a significant automotive fuel, experimental cars are under development that will sense the composition of the fuel in the tank and adjust the engine controls accordingly. Such a "Flexible-Fuel Vehicle" promises to cost more than one developed for a specified blend of methanol and gasoline. At the same time its flexibility prevents it from making optimal use of methanol because, barring variable geometry, its compression ratio must be low enough to accommodate gasoline.

As with natural gas, the high octane rating of methanol suggests its unsuitability as a diesel fuel. The heavy-duty two-stroke diesel has proven capable of operating on methanol, however [34]. By bypassing its scavenging blower as appropriate, the two-stroke engine has the unique capability of retaining a high fraction of hot residual products within the

cylinder.  Such a high charge temperature has been shown to reduce the
ignition delay of methanol significantly, turning it into a fuel with
acceptable autoignition characteristics over most of the operating range.
For light loads, a glow plug can be used as an ignition assist.

Both the E-DISC and L-DISC engines operate well on methanol.  There is
reason to expect that the difficult cold-start problem with methanol in the
homogeneous-charge engine can be overcome in the DISC engine.  However,
compliance with emission standards has yet to be demonstrated.


THE FUTURE OUTLOOK

It is concluded from this review that given its balance of positive
attributes, the spark-ignition engine will remain dominant in the
passenger-car field for the remainder of this century.  The lean-burn
version, perhaps with charge stratification, is a distinct future
possibility.  The fuel economy of the spark-ignition engine will continue
to be improved in an evolutionary manner, for it has already reached a
sophisticated level of development.  Many of the improvement items
discussed offer only a small fuel-economy gain on the order of, say, 2%.
That is a significant increment, but disappointingly small because it is
within the repeatability band for many test facilities.  What is often
overlooked, however, is that by incorporating five such improvements into
an engine, its fuel economy can be improved a very respectable 10%.

The rotary-engine variant of the homogeneous-charge spark-ignition
powerplant has earned a niche in the specialty-car market.  It offers small
size and low mass, low vibration level and a smooth power delivery.  When
tuned to U.S. emission standards, however, superior fuel economy has not
been counted among its advantages [35].  Developments to improve its fuel
economy, including charge stratification, are under way, just as they are
for the traditional reciprocating spark-ignition engine.

Forthcoming particulate standards, coupled with severe NOx standards,
threaten the future of the passenger-car diesel.  Research on the low-heat-
rejection diesel incorporating ceramics will continue, at least until its
promises and problems are better defined.

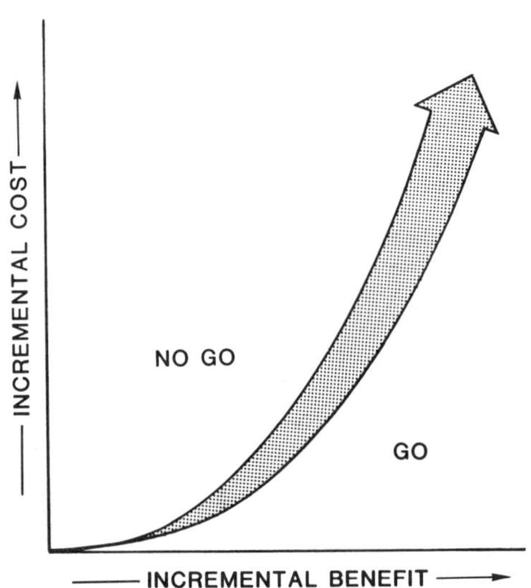



Fig. 30.  The cost-benefit tradeoff.

Meanwhile, research is likely to continue in the U.S. on two high-risk options, the Stirling engine and the gas turbine, primarily under government sponsorship. An independent study completed 11 year ago concluded that one or both of these could be ready for production at least by 1985 [36], but despite a decade of effort, neither seems perceptibly closer to that status today than it was then. For the passenger-car application, the Stirling engine is hampered by its size. The gas turbine awaits success with ceramics. Both engines are still attempting to catch up with the production engine in fuel economy. Both appear costly. An alternative engine that carries a premium price but merely matches the production engine with respect to the attributes desired by the consumer cannot be expected to succeed.

In spite of the progress made on the efficiency of the spark-ignition engine over the years, a large number of bits and pieces of technology remain to be exploited to improve its fuel economy further. As in the past, the acceptability of each technology item will be judged on the basis of a cost-benefit tradeoff conceptually illustrated in Fig. 30. For example, an innovation that increases the cost of the engine by a factor of two but offers only a 2% gain in fuel economy is difficult to justify. Despite the discipline of the cost-benefit tradeoff, their are plenty of ideas for improvements that will find their way into the product. When the year 2000 arrives, the latest-model engine will relegate today's best effort to the museum.



REFERENCES

1. G. B. Warren and J. W. Bjerklie, "Proposed Reciprocating Internal Combustion Engine with Constant Pressure Combustion." SAE Paper 690045 (1969).

2. S. Onishi, S. H. Jo, K. Shoda, P. D. Jo and S. Kato, "Active Thermo-Atmospheric Combustion (ATAC) -- A New Combustion Process for Internal Combustion Engines." SAE Transactions 88, 1851-1860 (1979).

3. M. Noguchi, Y. Tanaka, T. Tanaka and Y. Takeuchi, "A Study on Gasoline Engine Combustion by Observation of Intermediate Reactive Products During Combustion." SAE Transactions 88, 2816-2828 (1979).

4. P. M. Najt and D. E. Foster, "Compression-Ignited Homogeneous Charge Combustion." SAE Transactions 92, 1.964-1.979 (1983).

5. C. A. Amann, "Why Not a New Engine?" SAE Transactions 89, 4561-4593 (1980).

6. R. M. Heavenrich, J. D. Murrell and J. P. Chang, "Light Duty Automotive Trends Through 1986." SAE Paper 860366 (1986).

7. C. A. Amann, "The Powertrain, Fuel Economy and the Environment." International Journal of Vehicle Design 7, 1-34 (1986).

8. S. Luchter and R. A. Renner, "An Assessment of the Technology of Rankine Engines for Automobiles." U.S. Energy Research and Development Report ERDA-77-54 (1977).

9. W. H. Haverdink, F. E. Heffner and C. A. Amann, "Assessment of an Experimental Stirling-Engine-Powered Automobile." Proceedings of the 22nd Automotive Technology Development Contractors' Meeting, 151-166 (1985).

10. J. S. Collman, C. A. Amann, C. C. Matthews, R. J. Stettler and F. J. Verkamp, "The GT-225 -- An Engine for Passenger-Car Gas Turbine Research." SAE Transactions 84, 690-712 (1975).

11. S. L. Plee, T. Ahmad and J. P. Myers, "Flame Temperature Correlation for the Effects of Exhaust Gas Recirculation on Diesel Particulate and NOx Emissions." SAE Transactions 90, 3738-3770 (1981).

12. H. A. Burley and T. L. Rosebrock, "Automotive Diesel Engines -- Fuel Composition Versus Particulates." SAE Transactions 88, 3112-3123 (1979).

13. J. C. Wall and S. K. Hoekman, "Fuel Composition Effects on Heavy-Duty Particulate Emissions." SAE Transactions 93, 5.1030-5.1071 (1984).

14. R. Kamo and W. Bryzik, "Adiabatic Turbocompound Engine Prediction." SAE Transactions 87, 213-223 (1978).

15. M. M. Bailey, "Comparative Evaluation of Three Alternative Power Cycles for Waste Heat Recovery from the Exhaust of Adiabatic Diesel Engines." NASA TM-86953 (1985).

16. D. C. Siegla and C. A. Amann, "Exploratory Study of the Low-Heat-Rejection Diesel for Passenger-Car Application." SAE Transactions 93, 259-283 (1984).

17. D. F. Caris and E. E. Nelson, "A New Look at High Compression Engines." SAE Transactions 67, 112-124 (1959).

18. J. N. Mattavi, E. G. Groff, J. H. Lienesch, F. A. Matekunas and R. N. Noyes, "Engine Improvements Through Combustion Modeling." Combustion Modeling in Reciprocating Engines, 537-579 (Eds. J. N. Mattavi and C. A. Amann), Plenum Press, New York (1980).

19. I. Matsumoto and A. Ohato, "Variable Induction Systems to Improve Volumetric Efficiency at Low and/or Medium Engine Speeds." SAE Paper 860100 (1986).

20. D. C. Siegla and R. M. Siewert, "The Variable Stroke Engine -- Problems and Promises." SAE Transactions 87, 2726-2736 (1978).

21. W. A. Wallace and F. B. Lux, "A Variable Compression Ratio Engine Development." SAE Transactions 72, 680-694 (1964).

22. W. B. Paul and I. B. Humphreys, "Humphreys Constant Compression Engine." SAE Transactions 6, 259-286 (1952).

23. H. Oetting and P. Walzer, "Alternative Approaches to Lean Burn Concepts." (in German) Proceedings of the Symposium on Lean Burn Engines, 255-272, VDI Berichte 578 (1985).

24. A. Titolo, "The Fiat System for Load Control by Variable Valve Timing." FISITA Conference on the Automotive Spark-Ignition Engine in a New European Context, Aachen (1985).

25. R. J. Herrin and D. J. Pozniak, "A Lost-Motion, Variable-Valve-Timing System for Automotive Piston Engines." SAE Transactions 93, 2.717-2.729 (1984).

26. D. L. Stivender, "Intake Valve Throttling - A Sonic Throttling Intake Valve." SAE Transactions 77, 1293-1303 (1968).

27. J. H. Tuttle, "Controlling Engine Load by Means of Late Intake-Valve Closing." SAE Transactions 89, 2429-2441 (1980).

28. J. H. Tuttle, "Controlling Engine Load by Means of Early Intake-Valve Closing." SAE Transactions 91, 1648-1662 (1982).

29. D. L. Dimick, S. L. Genslak, R. E. Greib and M. J. Malik, "Emissions and Economy Potential of Prechamber Stratified Charge Engines." SAE Transactions 88, 1653-1667 (1979).

30. M.-F. Chang, M. P. Nolan, J. H. Rillings and A. A. Quader, "The Axially Stratified-Charge Engine: Control, Calibration, and Vehicle Implementation." SAE Paper 851674 (1985).

31. S. Matsushita, T. Inoue, K. Nakanishi and N. Kobayashi, "Development of the Toyota Lean Combustion System." SAE Paper 850044 (1985).

32. A. J. Scussel, A. O. Simko and W. R. Wade, "The Ford PROCO Engine Update." SAE Transactions 87, 2706-2725 (1978).

33. W. T. Tierney, E. Mitchell and M. Alperstein, "The Texaco Controlled-Combustion System - A Stratified Charge Engine Concept -- Review and Current Status." I. Mech. E. Paper C1/75, Conference on Power Plants and Future Fuels (1975).

34. R. R. Toepel, J. E. Bennethum and R. E. Heruth, "Development of Detroit Diesel Allison 6V-92TA Methanol-Fueled Coach Engine." SAE Transactions 92, 4.959-4.975 (1983).

35.  M. K. Martin, "Wankel Rotary Engine Development Status and Research
     Needs." Aerospace Report No. ATR-82(2869)-5ND (1982).
36.  R. R. Stephenson, "Should We Have a New Engine?"  Jet Propulsion
     Laboratory JPL SP 43-17 (1975).

# A REVIEW OF THE STRATIFIED CHARGE ENGINE CONCEPT

Duane Abata

Associate Dean
College of Engineering
Michigan Technological University
Houghton, Michigan U.S.A.

ABSTRACT

This paper presents an overview of stratified charge combustion engines and ties together recent advances with past achievements. A brief historical literature survey of the stratified charge engine concept is first presented. Selected stratified charge engine designs are then presented which represent the varied methods devised to achieve stratified charge combustion. Advantages and disadvantages of the stratified charge concept are then presented. Various topic areas pertinent to the subject are then discussed including hydrocarbon analysis, emissions, alternative fuel operation, modifications to conventional Diesel and rotary engines, and modeling of stratified charge combustion systems are also discussed. This overview summarizes with the current efforts underway to achieve successful stratification. This review is meant to augment existing reviews as well as focus on recent developments in the area.

## Introduction

Stratified engines have held an exciting place in the development of the internal combustion engine. Inventors, engineers, and research scientists have, repeatedly through this historical development, turned to the stratified engine concept as a solution to the many problems that have plagued conventional homogeneous charge engines since its inception. Problems of mechanical loading, fuel economy, engine knock, combustion emissions and alternative fuel utilization, to name a few, have all been investigated in the stratified charge engine with the thought that the stratified charge concept would offer unique solutions to these problems. Yet, despite optimism of engine developers, the stratified engine has suffered several setbacks. For instance, high hydrocarbon emissions at light load has been a continual difficulty. Although the stratified engine has had limited success in production, the concept is as viable today as it has been throughout its historical development. Axially stratification of today's electronically controlled engines, the two cycle rotary engine, and multifuel requirements of the military remain active areas of research.

The purpose of this paper is to review stratified charge engine concepts and tie together the historical development of the engine with present day achievements. A brief historical literature survey of the stratified charge engine concept is first presented. Selected stratified charge engine designs are then presented which represent the varied methods devised to achieve stratified charge combustion. Advantages and disadvantages of the stratified charge concept are then discussed. Hydrocarbon analysis, emissions, alcohol and other alternative fuel operation, modifications to conventional Diesels, the special application of stratified charge combustion to rotary engines and modeling of stratified charge combustion systems are also discussed. This overview summarizes with the current efforts underway to achieve successful stratification.

## Historical Literature Survey

Although stratified engines have received considerable attention in the past twenty years or so, the concept of mixture stratification in the internal combustion engine actually originated over a century ago. In 1867 the Hugon vertical engine, a decendent of the infamous non-compression Lenior engine, charge stratification was achieved to a limited extent where the mixture was locally rich for ignition and lean throughout the remainder of the combustion chamber. Soon after, in 1872 Nicholas Otto came upon the stratification concept and later, in 1876, capitalized on the idea proclaiming the quiet running of his newly introduced compression engine was not due to the working cycle but rather due to charge stratification (1).

Otto was concerned about shock loading on the piston. In an inspirational moment, while watching smoke ascend from a chimmey and gently disperse into the atmosphere, Otto came upon the thought of stratifying the charge, rich near the source of ignition and lean near the piston surface thereby cushioning the 'explosive' combustion. Otto actually constructed a hand cranked version model and observed the action of smoke entering during the intake stroke. Four years later, Otto introduced his compression engine and considered charge stratification one of the main advantages of the new four-stroke cycle claiming combustion was initiated in the rich portion of the charge with flame velocity decreasing as it progressed through the leaner mixture near the piston. Otto's main claim in his U.S. patent reads:

> *A gas motor engine wherein an intimate mixture of combustible gas or vapor and air is introduced into the cylinder, separate from a charge of air or other incombustible particles of combustible mixture will be close together at the point of ignition, but will be more and more dispersed in the charge of air forward of that point, whereby the development of heat and the expansion or increase of pressure produced by the combustion are rendered gradual ...(1).*

Throughout Otto's patents, the charge stratification concept was preserved for the sole purpose of reducing piston shock and contributing to a quieter engine. It was this concept that enabled him to defend the four stroke idea and keep his competitors out of the four stroke compression engine business. The subsequent success of his engines, produced by *Gasmotoren-Fabrik Deutz* credited Otto with the four stroke concept, displacing Alphanse Beau de Rochas, the original inventor of the four stroke concept.

Excepting the hand cranked model of Otto, the first actual experiments with stratification were carried out by Ricardo in 1915 (2). The concept of torch ignition was later introducted by Smirnov and others (3). Significant effort to stratify the charge was carried out during the war years 1940 through 1950. Charge stratification was thought to be an answer to the knock phenomena. Extensive work was conducted in the USSR where the main interest was to improve fuel consumption at part load and achieve values comparable to the Diesel engine.

Despite the effort during those years, variations of the stratified charge engine did not leave the experimental stage because the overall advantages of the conventional gasoline engine were always greater. The lack of successful development of the stratified charge engine at this time in history can be attributed to the inability of producing a varied and controlled fuel air mixture and to a general lack of technology. The introduction of the Clean Air Act in the United States gave new impetus to the development of the stratified charge engine and to engine combustion in general. The stringent limitation of NOx, the contributor to photochemical smog, was especially worrisome to engine designers. The importance of the air-fuel ratio to exhaust gas composition was established and soon after, the link between lower emissions of the three legally controlled pollutants, CO, HC and NOx to overall leaner fuel air mixtures was determined (4).

Engine manufacturers in the 1960's were forced to focus their work more than ever before on the combustion process to reduce the emission of the legally controlled noxious exhaust constituents. Stratified charge work received renewed attention for the simple fact that it was considered much more reasonable to try to reduce the amount of noxious exhaust constituents at their place of origin; in the cylinder itself, despite intensified use of emission control devices in the exhaust stream.

The first approach used to lower CO and HC emission levels in the SI engine was to lean the fuel-air mixture down to the stoichiometric ratio or even to slightly leaner mixtures.

It was soon realized however, that igntion and flame propagation deteriorated rapidly as the mixture was made leaner, thus disturbing the regularity of successive combustion cycles. Engines operated at stoichiometric or slightly leaner also showed excessive NOx emssions as well. Theoretical and experimental investigations on the origin of NOx (5,6,7) indicated that the formation of NOx was strongly dependent on the air/fuel ratio, the maximum combustion temperature and on cycle to cycle variations (8).

In 1970, Newhall and El-Messiri (9) were the first to demonstrate that multiphase or stratified combustion with overall operation of the working gas in the lean fuel air mixture region was necessary to achieve a noticeable reduction in  NOx generation. This work served as the motive for further investigations into the stratified engine concept (10,11,12). Reductions in NOx and CO were achieved but the problem of high HC emissions, particularly at light load and idle, arose. In most situations, overly high HC emissions, together with a decrease in fuel economy cast a general pall over stratified engine development. With the introduction of exhaust catalysts and technical advances in gasoline combustion (primarily lower compression ratios and exhaust gas recirculation), limitations of the conventional gasoline engine were stretched and acceptable emission levels were reached (13).

During this time period many of the original stratified charge concepts were re-investigated. Some of the more notable engines were studied quite thoroughly, although work remained primarily experimental in nature. The Texaco TCCS engine, the Ford Proco engine and the Honda CVCC engine recieved considerable attention. (These engines and others are discussed in the next main section.) Although one of these engines (the Honda CVCC engine) was placed into full production, advantages of the stratified engine concept were outweighed by disadvantages and the conventional gasoline engine with exhaust aftertreatment remained the inevitable solution to meet the needs of the automotive industry.

Experimental and theoretical work of the day did not provide all the answers needed to solve the difficulties of satisfactorily stratifying the charge while providing a repeatably ignitable mixture in the combustion chamber. Mathematical models, discussed in a later section in this paper, were beginning to evolve (14) to a point where results could be applied to engine development. The problems associated with fuel injection and associated fuel-air mixing leading to ignitable mixtures were found to be directly attributable to air motions within the chamber. The introduction of sophisticated two and three dimensional models, which could be used to help map out the complicated airflow patterns occuring in the cylinder, introduced yet another phase of stratified charge engine development. The Texaco TCSS engine, subsequently abandon by Texaco, was further developed by the United Parcel Service, UPS, with the aid of mathematical modeling to aid the  experimental effort (15,16). General Motors has continued with a program on stratified charge engine development at their research laboratories in Warren (17) with two and three dimensional mathematical models serving as the basis for further experimental work. Stratified combustion work on the rotary engine, first with Curtiss Wright and now recently acquired by John Deere, has been considerably augmented by mathematical modeling at NASA Lewis and elsewhere as interest has been renewed in the rotary engine for military applications (18,19,20,21,22,23).

Work on the stratified charge concept would have remained at this point to date with one notable exception - the development of the axially stratified engine. The advent and mass production of electronic fuel injectors, together with increased use of manifold port injection, introduced yet another concept of charge stratification. In 1976, Peters and Quader (24) demonstrated stratification of the charge in an axial direction on a pancake CFR engine equipped with appropriately timed port injection and a shrouded valve to induce intake swirl.  In later publications (25, 26), this concept was applied to a production engine and was shown to have reduced fuel consumption, reduced NOx emissions, and increased tolerance to low octane fuels when operated in the axially stratified mode. This engine remains under development at GM laboratories.  (This engine is discussed in further detail later in this paper.)

## Why Stratify the Charge?

In a report to the Environmental Protection Agency (27) Ricardo engineers presented this argument from a rather interesting and simplistic viewpoint. The thermal efficiency of the internal combustion engine, based upon cycle analysis, is a function of two principal variables: the compression ratio and the specific heat ratio of the working gas. Most practical engines can be approximated by the Otto cycle or some variation of the Otto cycle since combustion generally occurs at constant volume rather than constant pressure due to the slow speed of the piston at top dead center compared to the heat releasing reaction rate. The specific heat ratio is a direct function of mixture strength or better, equivalence ratio, of the working  gas. The lower the equivalence ratio of the working gas, the higher the theoretical efficiency. This can be easily shown through the well known efficiency relationship for the theoretical Otto cycle,

$$1 - (\frac{1}{\Gamma})^{\gamma-1}$$

From the above relationship, it can easily be shown that as equivalence ratio of a combusted gasoline/air mixture is decreased to zero (100% air) the efficency increases to a maximum value at constant compression ratio. This is shown in Figure 1.

Power output of the engine is controlled by varying the fuel delivery. The Diesel engine, because of inherent stratification of the fuel and nature of the ignition process, can successfully operate high compression ratios and, at light loads, low overall equivalence ratios. The homogeneous charge gasoline engine is limited to lower compression ratios because of the inherent knock phenomena, and near stoichiometric equivalence ratios because of ignition requirements. Throttling of both the fuel and air are necessary to achieve part load operation since decreasing the fuel strength of the mixture without limiting the air flow (i.e., lowering the equivalence ratio) will allow the mixture to pass through the lean limit of combustion and the spark will not be able to initiate combustion. The net effect is that theoretical efficiency of the homogeneous charge engine is somewhat independent of engine load and does not improve at light load as does the unthrottled Diesel engine. This simple analysis has disregarded pumping losses associated with air throttling which, of course, decrease efficiency even further. Unthrottled engines are thus more efficient at light loads.

The compression ratio of the engine also bears a direct relationship on engine efficiency. Increasing the compression ratio *ad infinitum* does not necessarily increase *engine* efficiency because of higher bearing loads and resulting frictional losses. Energy balances for full and part load operation for a typical Diesel engine (high compression ratio), a gasoline engine (low compression ratio) and a hypothetical stratified charge engine (moderate compression ratio) are shown in Figures 2 and 3. Engine Size was chosen so that total work produced by each engine was approximately the same. At full load, Figure 3, the overall efficiency of all three engines is about the same. Energy is lost to the coolant, lubricating oil and exhaust gases. While the Diesel engine looses energy due to frictional losses (high compression ratio) and smoke formation, the gasoline engine looses energy in the exhaust due to incomplete combustion (formation of carbon monoxide which contains unreleased chemical energy).

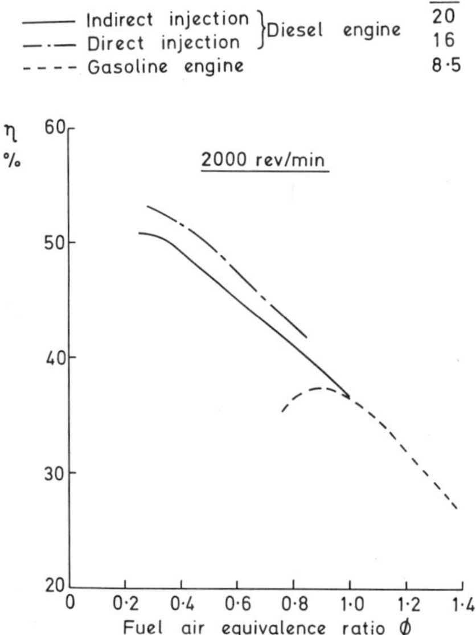


*Figure 1. Cycle Efficiency as a Function of Equivalence Ratio (27).*

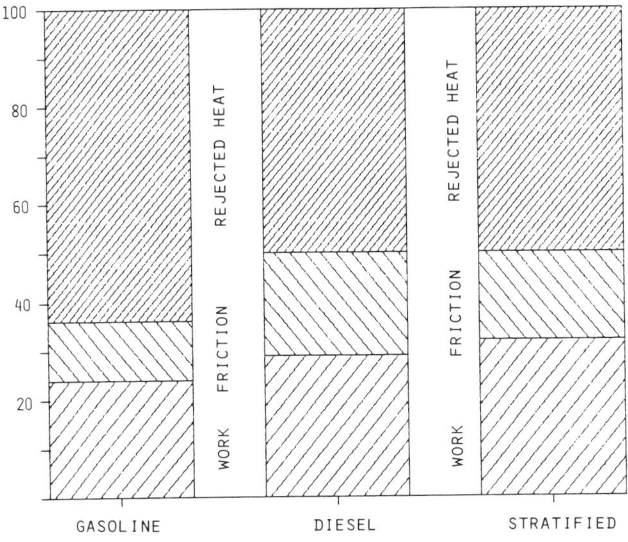



*Figure 2. Energy Balance - Part Load Operation.*

When the engines are operated at part load the efficiencies change. The Diesel engine combustion process is more efficent than the gasoline engine, but the high compression ratio contributes to a proportionately greater frictional load. While the frictional losses of the gasoline engine are smaller, pumping losses caused by the throttled intake air occur. The hypothetical unthrottled engine operating at a moderate compression ratio, with a relatively good thermal efficiency and moderate friction, has the highest overall operating efficiency. Frictional load remains the same as full load due to the high compression and is, therefore, proportionally greater at part load.

A combustion process that operates with a moderate compression ratio (about 12:1, suggested by Ricardo engineers) will, then, give the best fuel comsumption for light load automotive applications. Conventional Diesel engines requiring autoignition of the fuel will

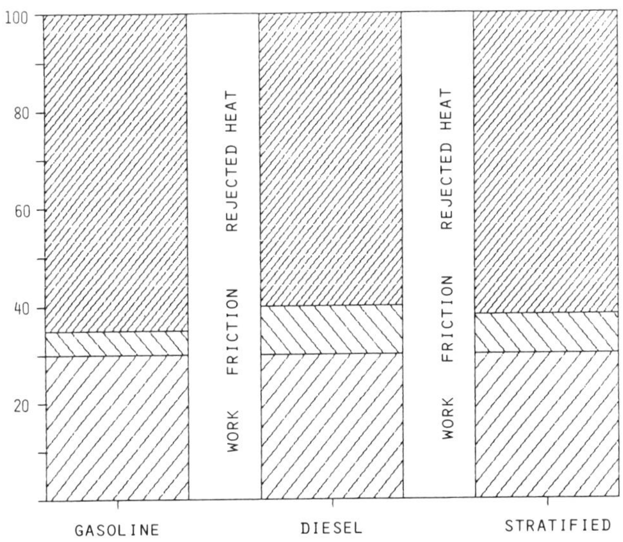



*Figure 3. Energy Balance - Full Load Operation.*

not operate at this compression ratio. Conventional homogeneous charge gasoline engines will tend to autoignite or knock and, as discussed earlier, will not operate at low equivalence ratios. A hybrid combustion process incorporating characteristics of the Diesel and gasoline engine is needed. This is the unthrottled stratified charge engine. If the combusting gas can be successfully confined from the working gas by either physically or aerodynamically stratifying the overall charge, resulting engine operation is theoretically more efficient.

## Definition and Categorization Scheme

Essentially, all engines that are not homogeneous are, by default, stratified in some way! Even the homogeneous charge engines are, more often the case, stratified in some fashion due to fresh charge gradients and exhaust gas residual/recirculation. Diesel engines are stratified charge. It is not the object of this review to cover this broad area of engines! Barnescu (28) prefers the acronym SADI (spark assisted stratified charge) to DISC (direct injection stratified charge) so as not to confuse Diesel engines with hybrid stratified charge engines. In any case, it is wise to present a reasonably clear definition of a stratified charge engine before discussing some of the more representative engines.

Hasslett et al. (29) developed a sound definition. It is repeated here with minor modification:

*A stratified charge engine is an engine where the combusting gases are not, by design, homogeneous throughout the combustion chamber but rather, separated either physically or aerodynamically into igniting, combusting and working regions having differing air to fuel ratios for the purpose of altering the combustion and emissions characteristics of an equivalent homogeneous charge. The igniting and combustng charge is then characteristically different from the working gas which may be air, recirculated exhaust gas or a combination thereof.*

To further clarify the issue, a few more thoughts are added. In the stratified charge engine, combustion is initiated by a ignition source, either a conventional spark plug, plasma injector or a glow plug to differentiate the engine from conventional, multipoint compression ignition typical of Diesel engine operation. Ignition generally occurs in a favorable fuel rich region with subsequent combustion moving into a fuel lean region, thereby employing the 'burn rich - burn lean' concept to reduce overall NOx generation. An overall lean air/fuel ratio is generally developed contributing (in theory anyway) to lower carbon monoxide and hydrocarbon emissions.

For many years, the development of stratified charge engines followed a path of creativity and inventiveness. Empirical design formed the basis of the approach to engine development. Only recently, within the past fifteen years, have scientific tools (such as computational modeling and laser velocimetry) been fully appreciated and utilized to refine the more promising designs. For this reason, stratified charge engines vary considerably in design and are unique in their own way. For the most part however, these engines fall into categories or groups of having similar methods of stratifying the charge. In a thorough review in 1975, Haslett et al. (29) placed the engines into seven categories or groups. Most of this categorization scheme has been preserved although some modification was necessary to include recent work the last ten years. The seven categories are:

1. Single combustion chamber, fuel injection before TDC, heat release controlled by reaction rate *(early injection)*.

2. Single main combustion chamber, fuel injection during combustion, heat release controlled by injection rate *(late injection)*.

3. Fuel lean main combustion chamber and fuel rich prechamber, fuel injection into the prechamber, heat release controlled by ignition and subseqent combustion of the main chamber by the prechamber.

4. Fuel lean main combustion chamber and fuel rich prechamber, carburetted air-fuel preparation charged with separate intake valves, heat release controlled by ignition and subseqent combustion of the main chamber by the prechamber *(three valve concept)*.

5. Single combustion chamber, manifold port injection electronically timed to produce charge stratification in the axial direction *(axially stratified engine)*.

6. Single combustion chamber, direct injection, two stroke cycle.

7. Miscellaneous engines of novel and promising design.

As with any classification scheme there are exceptions. Some engines are difficult to classify because they tend to straddle two groups or are unique and simply fall into the seventh 'catch-all' category. As development continues and new contributions are made, it may be again necessary to modify the groupings.

## Some Examples of Stratified Charge Engines

It would be lengthy to include all the stratified charge engines or variations in principal design here. As indicated earlier, many engines are very inventive and reflect the creativeness of engineering thought. Many are based upon empirical design rather than detailed scientific analysis (This is really true of most engine development. To some, it is this *art* of engine design that makes the field exciting and interesting!) Some of the more promising engines, along with a few others for the sake of interest, were choosen to be reasonably representative of the seven catagories mentioned earlier and are discussed here. Complete engine systems, e.g., injection systems, turbocharging, etc., are not discussed here but may be a pertinent component in many of these designs.

*Category 1. Single combustion chamber, early injection*

A well known stratified charge engine in Category 1 is the Ford Programmed Combustion engine, more often referred to as the Ford PROCO engine. This engine is reasonably representative of engines based upon early injection, where ignition occurs after injection is complete and heat release is controlled by the physical evaporation and mixing and chemical kinetics processes of combustion. This engine was one of two engines (the other being the Texaco TCCS engine discussed in Category 2) introduced in the late 1950's in response to an Army Tank Automotive Command (TACOM) program having an objective of improved fuel economy and multifuel capability in small transport vehicles (30). The Ford PROCO engine is a spark ignited, direct fuel injection stratified charge engine developed to operate on 91 Research Octane Number gasoline with a ten to fifteen percent better fuel efficiency than that of the conventional gasoline engine (31). This engine is shown below in Figure 4.

The PROCO engine is an *early injection* engine where fuel is injected during the compression stroke directly into the cylinder by a conical spray nozzle. The combustion

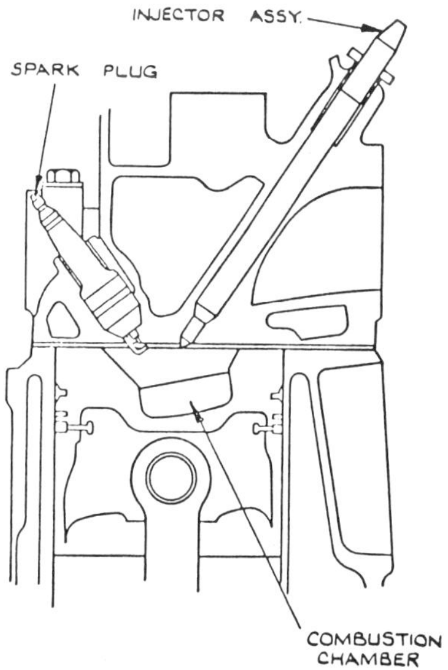


*Figure 4. The Ford PROCO Engine (27).*

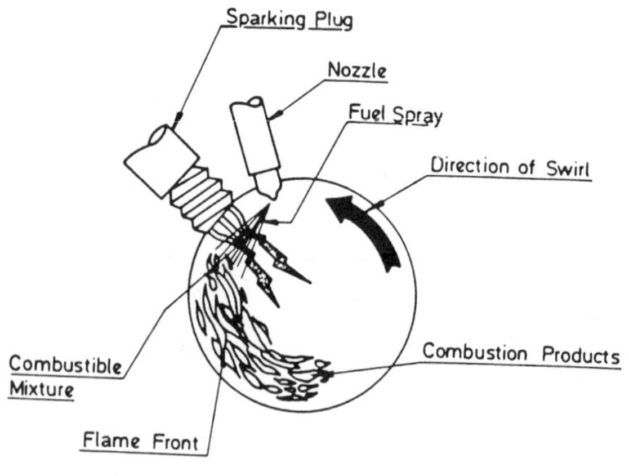


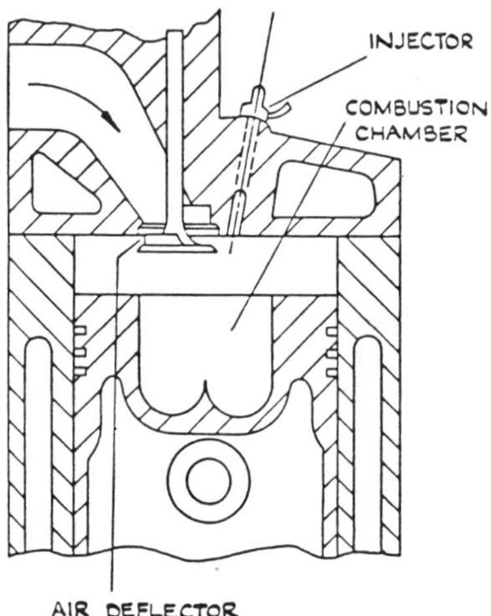


Figure 5. *The TCCS Engine (27).*

cavity, as shown above, is formed in the piston crown. The degree of stratification is controlled by timed fuel injection rather than a precombustion cavity (Categories 3 and 4). Injection timing allows some control on detonation through the stratification process. Early injection timing during the intake stroke is employed for improved mixing and air utilization at heavier loads, thus resulting in good power output. Later injection timing *still during the intake stroke* provides the high degree of stratification required for emission control and fuel efficiency at light loads. Combustion is initiated in the rich region with two spark plugs. Relatively fast combustion occurs due to turbulence caused by a purposely induced intake swirl and squish action. Degree of stratification is controlled by injection timing during the intake stroke with some control on detonation. For further information including emissions and fuel consumption data, the reader is referred to Reference 27.

*Category 2. Single combustion chamber, late injection*

The Texaco TCCS engine, is a well known stratified charge engine in Category 2 and is reasonably representative of DISC engines based upon late injection with heat release controlled by injection rate after the spray plumes have been ignited. This engine was the other of two engines as mentioned above in Category 1 introduced in the late 1950's in response to an Army Tank Automotive Command (TACOM) program having an objective of improved fuel economy and multifuel capability in small transport vehicles (30). The Texaco combustion concept has become one of the best known stratified charge engines due to the considerable development effort which has been expended over the last forty years by Texaco, White, UPS and other TACOM subcontractors (32,33,34,35,36,37,38,15,39,40,41). Other engines in this catagory include the Deutz combustion process (42) and the Curtiss-Wright combustion process (43,20), which are variations on the TCCS process, the latter being incorporated into the Wankel rotary engine. The TCCS engine is shown in Figure 5.

The TCCS engine achieves stratification by mixture formation. Port design and a shrouded inlet valve cause a circular swirl to the air on the intake stroke. The swirl persists throughout the combustion process and is increased by the conservation of momentum as the piston cup decreases the upper diameter of the combustion chamber. Fuel injection begins about thirty degrees before top dead center and continues for a duration of about 60 degrees (thirty degrees after TDC) at full load. Injection is shorter with correspondingly smaller quantities of fuel at part load operation. Fuel is injected both down and across the swirling airflow streamline (44).

The spark plug, located downstream of the injected fuel as physically close to the nozzle as possible but yet outside the envelope of the spray, is fired immediately after the start of injection. The flame thus initiates at the tip of the injector nozzle and propagates to engulf the plume which extends considerably downstream of the nozzle. The burning zone is, thus, very fuel rich with diluent air added to the spray zone by the swirling air until combustion is complete. Combustion is relatively slow and, as mentioned earlier, controlled through injection rate of the fuel as well as the physical processes of evaporation and mixing. The spark plug can be thought of as solving the ignition delay and octane/cetane sensitivity problem. For further information including emissions and fuel consumption data, the reader is referred to Reference 27.

*Category 3. Two combustion chambers, fuel injection*

This category is one of the more popular areas and includes several stratified engine systems produced by various engine manufacturers. This category was further subdivided into three smaller groupings by Ricardo Engineers (27) which were:

1) Prechambers which occupy less than 20% of the main chamber volume,
2) 20% to 40%, and,
3) More than 40% of the main chamber volume.

These subcategories covered a spectrum of combustion techniques from throttled engines with the prechamber charged rich and ignited to produce a jet of flame that would ignite a relatively homogeneous lean mixture in the main chamber (subcategory 1), to unthrottled engines with the prechamber charged very rich and ignited to produce a jet of flame and unburnt fuel completing oxidation in the air of the main chamber similar to an indirect prechamber Diesel engine (subcategory 3). As expected, unthrottled engines often produced higher hydrocarbon levels.

The Porsche stratified charge engine system (27) is reasonably representative of this category falling into subcetagory 1 as described above. This engine data is shown below in Figure 6.

In this specific engine, the prechamber occupies roughly 10-15% of the main chamber clearance volume. Fuel is injected into the prechamber and is ignited by the spark plug as shown in the figure. At medium loads, the prechamber air/fuel ratio is about 10:1 with an overall air/fuel ratio of 25:1. Throttling is used to medium loads after which load is controlled by the amount of fuel injected.

As mentioned earlier, this category was quite popular with other engine stratification systems that included:

    Freeman system - subcategory 1
    Clawson system - 1
    Broderson system - 2
    Volkswagen system - 2
    Schlamann system  - 2
    Pischinger system - 3
    Huber system - 3
    Newhall system - 2

Full descriptions of these engines, along with available performance and emissions data can be found in Reference 27.

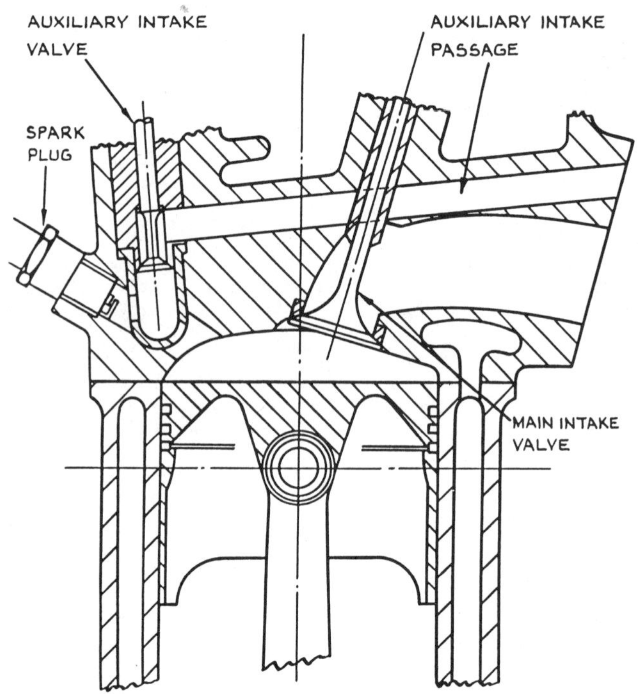


*Figure 6. The Porsche Engine (27).*

*Category 4. Divided Combustion Chamber with Separate Inlet Valve*

The category, referred more often as three valve scavenged prechamber engines (or simply 'three valve arrangements'), represents one of the first methods of achieving charge stratification (45,46,47,48,49). The concept has first appeared in the 1920's by Ricardo laboratories and reappeared in various forms throughout the years. It is interesting to note that in the 1960's and early 1970's Russian experimenters attempted to optimize this system for improved fuel economy over that of the conventional gasoline engine. Increases in efficiency at certain operating conditions were, however, not significant enough to warrant production until Honda showed that this concept could be used effectively to reduce combustion emissions.

The Honda CVCC engine, shown in Figure 7 below, met with enough reasonable success to place it into production for seven years. More recently, improvements in exhaust treatment and fuel economy of the conventional gasoline engine were sufficiently strong to once again overcome the advantages of this system (the Honda CVCC engine).

The main combustion chamber is operated fuel lean with the mixture blended in a carburetor (or low pressure manifold fuel injector) and induced through the main intake valve. The small prechamber is operated very fuel rich with the mixture blended in a separate carburetor (or manifold injector). During intake, these two mixtures are drawn into the chambers simultaneously. The intake process is a dynamic process rather than a simple one. Interaction between the two chambers occurs on intake, compression and expansion strokes which effect the combustion process. (An important design parameter is the amount of air entering the prechamber volume termed in the literature as 'prechamber fills'.) Stratification occurs not only between the prechamber and the main chamber but in the main chamber as well, since during the intake stroke some of the very rich mixture flows from the prechamber into the main chamber. The amount is dependent upon pressure gradients, valve timing and orfice size and shape of the prechamber. During compression, some lean mixture is forced into the prechamber diluting it to provide an ignitable mixture (A/F of about 9 to 12) near the spark plug. Following ignition, expansion of the combusting gas produces a torch like flame which enters the main chamber igniting the fuel lean main chamber charge. Expansion then occurs as usual. Exhaust stratification may also occur due to the complexities of the fluid flow between the two chambers and exhaust valve.

In most engines of this type, the prechamber is insulated to decrease heat transfer (surface area is larger than a conventional gasoline engine) and to increase the vaporization rate of any liquid fuel (especially during cold operation).

Similar to category 4 above, this category is also populated with many engine designs having the three valve configuration. Additionally, some engine manufacturers, in an attempt to meet emission standards of existing engines, developed retrofit devices - essentially add-on prechambers - that screwed into the existing spark plug hole. These retrofit devices generally met with failure. Engine designs include:

Ricardo (early 1920's)
GM  Scavenged Prechamber
Ford Scavenged Prechamber
Honda CVCC
Eaton Scavenged Prechamber
Volkswagen Scavenged Prechamber
Nilov, GAZ-52, ZIL-120 FK (Russian)
Philips (retrofit)
Teledyne (retrofit)
Walker Stratofire (retrofit)
Heintz

Descriptions of some of these engines, along with available performance and emissions data can be found in Reference 27.

*Category 5. Timed Port Injection, Axially Stratified Engines*

The introduction and mass production of low-cost, electronic mainifold port injection has introduced yet another possibility of stratifying the charge through sophisticated control of fuel injection timing during the intake process. This appears to be the least expensive

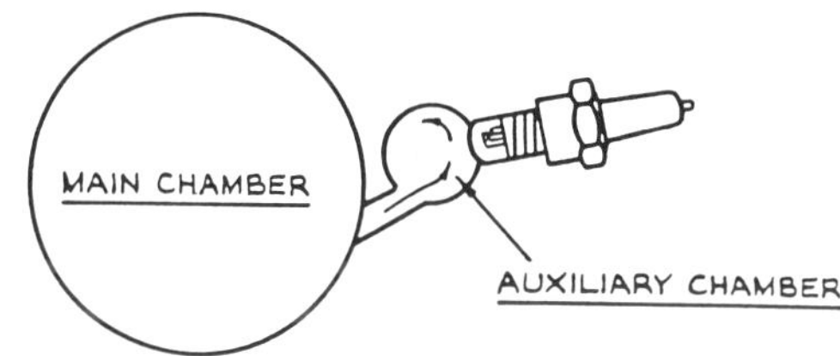


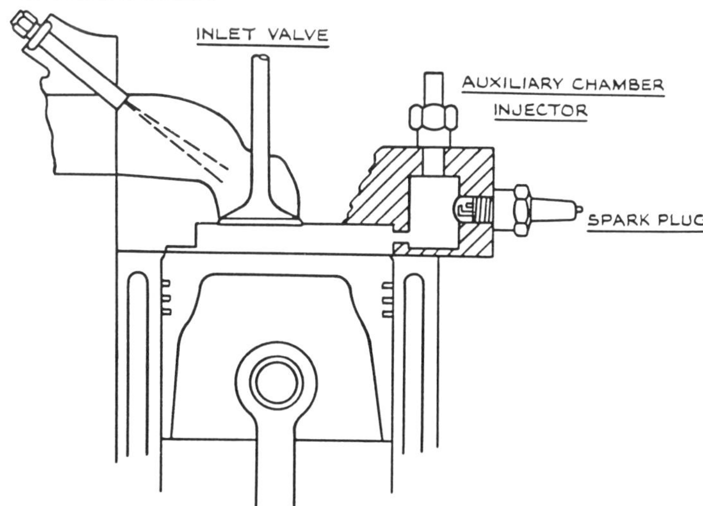


*Figure 7. The Honda CVCC Engine (27).*

approach, as most engines produced in the marketplace today are equipped with electronic engine control devices and electronic injection.

During the late 1970's and continuing to present day, studies by Peters and Quader (24,25,50,53) at General Motors Research Laboratories have shown that the use of timed inlet-port fuel injection in conjunction with gas swirl in the cylinder volume significantly changes the nature of combustion due to axial stratification. In an otherwise conventional gasoline engine, axial stratification has demonstrated reduced fuel consumption, reduced emissions and increased tolerance to exhaust dilution.

Axial stratification is produced by timing the port injector to begin fuel delivery during the intake stroke. The process is shown below in Figure 8.

Early in the intake stroke, air is drawn into the cylinder. Helical swirl is imparted to the intake air by either a shrouded valve or port design. Later in the intake stroke, fuel is injected and the layers of stratified air/fuel mixture are formed. The momentum of the air swirl and the lack of strong vertical flow components (tumbling) prevent mixing of the fuel/air charge along the cylinder axis. This stratification is maintained throughout the compression process resulting in a mixture that is rich at the top of the combustion chamber and lean near the piston surface. The spark plug, located at the top of the combustion chamber, ignites the rich mixture and the flame is propagated outward consuming the lean mixture near the piston surface as well. The axially stratified chrge concept allows control of the air/fuel ratio near the plug and extends the overall lean limit of the cylinder charge. The engine is essentially unthrottled with the exception of minor throttling at light load added to enhance idle and part load operation and reduce hydrocarbon emissions. Power output is regulated by the duration of injection (amount of fuel injected) and minor throttling. The engine can operate with an overall airfuel ratio as lean as 21.5:1 at part load. At full load, the stratification process does not occur as the fuel injector is open for the full duration of the intake stroke.

Conceptualization and development of this engine is credited to General Motors Research Laboratories. The GM Axially Stratified Charge (GM-ACS) engine concept was first developed on a conventional pancake piston CFR single cylinder engine in 1976 and later extended to a four cylinder engine. Development continues at GM with several prototype engines. Toyota recognized the significance of the concept and developed their own version, the Toyota Lean Combustion System (Toyota LCS), which is now in limited production in their Carina automobiles.

Toyota developed a split intake swirl port design that varies with engine speed and load and reduces the volumetric losses associated with swirl ports. This intake port design is shown in Figure 9.

The intake port is split into two sections with the flow into the cylinder controlled through a swirl control valve. The valve is closed during part load operation to induce a high swirl of the intake air. The valve opens at full load to allow more air to enter the cylinder thus increasing volumetric efficiency for maximum power. The engine is electronically controlled with a lean mixture sensor located in the exhaust to maintain stable combustion while operating in the stratified mode (part throttle).

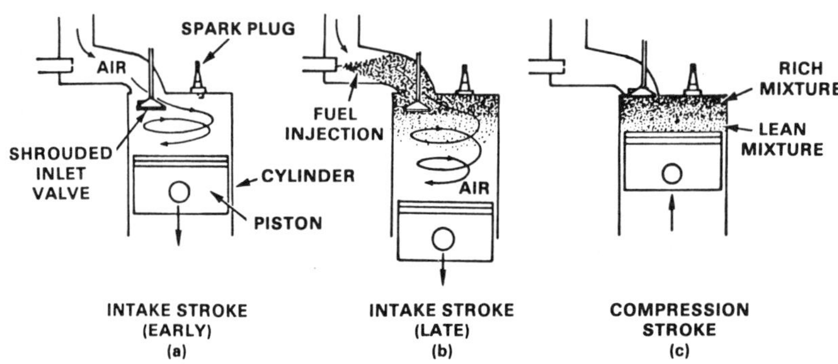


*Figure 8. The GM-ACS Engine (24).*

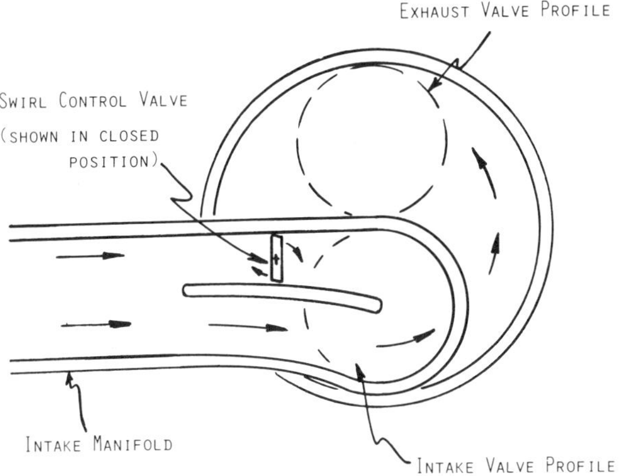



*Figure 9. Toyota Swirl Port Design.*

The viability of the GM-ACS (and Toyota LCS) concept to mass production in some form is strong, for the concept shows lower octane requirements, lower NOx emissions, lower CO emissions and improved combustion stability with today's electronically controlled engines. The concept is not without difficulties however. High swirl increases heat transfer and hydrocarbon emissions (most likely from the rich mixture entering the valve crevices in the cylinder head.) Misfire due to the overall lean airfuel ratio at some light load conditions is also a difficulty, thereby offsetting advantages in fuel economy and driving hydrocarbon emissions to unacceptable levels.

*Category 6. Two Stroke Engines*

Attempts at stratifying the charge in two-stroke engines are almost as numerous as that applied to the four-stroke engine. Ported intake and/or exhaust permits additional possibilities for stratification. Although numerous engines have been studied experimentally, only one engine, the Fairbanks Morse engine, was placed into production. This engine is shown in Figure 10.

The Fairbanks Morse engine is a four cylinder, opposed piston, two-stroke, stationary gas engine utilizing two small prechambers per cylinder. A spark plug and gas injector are located in each prechamber. The prechamber combusts rich and initiates combustion in the

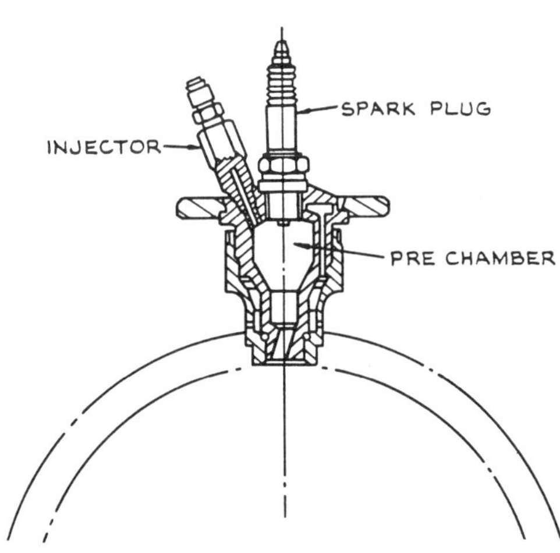



*Figure 10. The Fairbanks Morse Engine (27).*

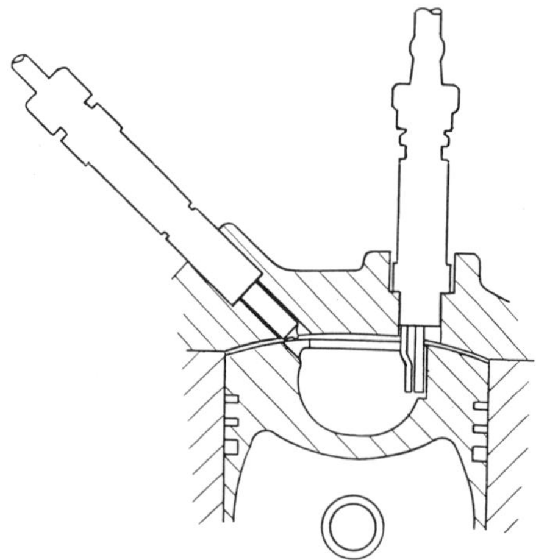

*Figure 11. The MAN-FM Engine (27).*

lean main chamber through torch ignition. The engine can operate reasonably at overall air/fuel ratios of 30 to 40:1 throughout the load range, thereby eliminating detonation problems frequently encountered in open chamber gas engines, particularly at the minimum fuel consumption point.

Two stroke engines working with varied concepts include:

Nippon Clean Engine
Yui and Ohnishi Combustion Process
Fairbanks Morse System (discussed above)
Kataoka and Hirako System
Clawson System
Heintz Ram Straticharge System
Richardo E65, APC-1 and E30 Engines
Kushul Engine

With the exception of the Fairbanks Morse engine and the Kushul engine, most two-cycle stratified engines have not been thoroughly evaluated. Fuel consumption of the two-stroke engine is usually higher than the four-stroke counterpart because of fresh charge loss during the scavenging process. Additionally, performance is generally irratic at light loads because of poor scavenging with subsequent poor ignition and combustion. The two-stroke stratified engine uses exhaust gas with a relatively low specific heat value as the bulk gas instead of fresh air and hence, the advantages of high specific heat (discussed earlier) contributing to fuel economy are lost. The added cost and complexities of the stratification devices offset the cost advantages and simplicity of the conventional two-stroke gasoline engine. Hydrocarbon emissions are characteristically high with the two-stroke engine. Stratification does not significantly offer improvement to this general shortcoming. For further information including emissions and fuel consumption data, the reader is referred to Reference 27.

*Catagory 7. Miscellaneous Engines*

Several engines have been included here because of the varied nature of the catagory. These are the MAN - FM engine, the IFP engine, and the Jessel exhaust dilutent engine.

*The MAN 'FM' Engine*

This process is fairly unique and is represented by only one engine, the MAN 'FM' (Fremdzundung) combustion system where fuel is injected onto the combustion chamber wall and, following ignition, subsequent combustion is controlled by fuel evaporation from the resulting fuel film. This engine was originally developed from the 'M' type diesel combustion system to allow multifuel capabilities. The high compression ratio was found to be impractical with gasoline due to the mechanical stresses from autoignition. Hence, a spark plug was adopted to initiate combustion early and avoid difficulties caused by later autoignition of the fuel.  This engine, along with fuel consumption and emissions data, is shown in Figure 11.

In this engine a helical inlet port imparts a very high swirl during the intake process. The swirl is further  increased due to the conservation of momentum as the airstream is confined to the spherical piston bowl late in compression.  During compression, fuel is sprayed from a single hole nozzle onto the wall of the combustion chamber.  High levels of swirl help spread the fuel over the chamber wall, thus maintaining a high degree of stratification.  Heat from the hot piston wall and compressed air cause the fuel to evaporate and mix with the swirling air which carries the rich mixture to the spark plug and subsequent flame front.

As with all stratification concepts, altering injector and spark plug location and orientation causes a marked difference in engine operation and emissions.  This is particularly true of this type of engine because of the wall evaporation factor which adds yet another parameter to the overall picture.  Another concern is vaporization and evaporation of the spray plume before wall contact.  This causes lean and rich pockets in the working gas, which are essentially controlled by the thermodyanmics and fluid mechanics of the swirling air as it penetrates the spray plume.

*The IFP Engine*

A rather simple solution to achieve stratification by minor modification to an existing gasoline engine was attempted by the Institut Francais du Petrole (IFP).  The most recent information on this engine is fairly out-of-date having been published in 1965, but is worthy of mention here because of the simplicity in design.  The concept was claimed to produce a heterogeneous mixture within the combustion chamber without resorting to fuel injection or air swirl.  This was achieved by directing a separately carburetted rich mixture into the combustion chamber via a tube placed in the intake port as shown below in Figure 12.

Pure air or a lean mixture was provided by the normal inlet port.  The engine was throttled at light loads and unthrottled at heavy loads.  Power output was controlled through use of the throttle at light loads and mixture strength at heavy loads.  The hemispherical combustion chamber was well suited in that it was relatively easy to direct the rich charge to the spark plug and the lack of squish did not cause much mixing during compression. Transient operation was difficult to control because of the inlet flow dynamics of the rich fuel mixture in the tube.  In later designs, the rich mixture flowed through a hollow intake valve and was timed by means of port alignment in the valve stem.  This modification was

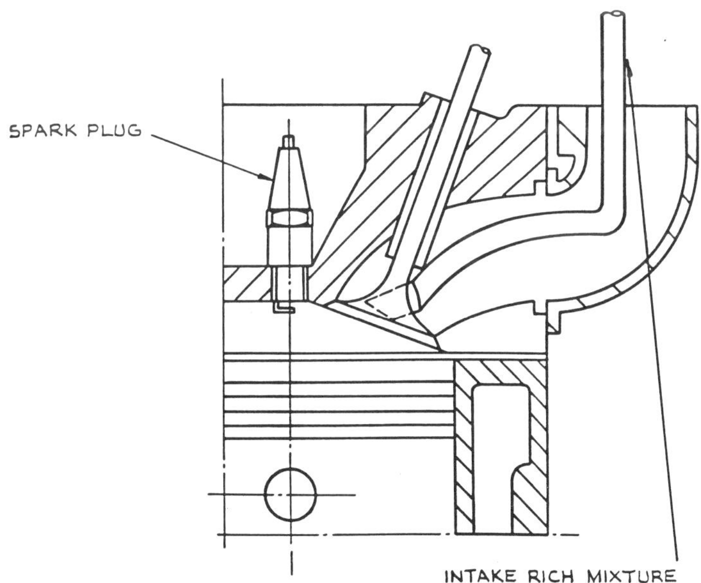


*Figure 12.  The IFP Engine (27).*

claimed to improve the transient operation of the engine significantly.  Data indicate the engine performed similar to a conventional gasoline engine at heavy loads (essentially operating with homogeneous combustion) with a 5% increase in fuel economy at lighter loads - a result of the stratification process.  Although hydrocarbon and CO emissions were similar to the conventional engine, levels were maintained over a wider A/F range.  NOx data was not available but were probably lower at lower loads due to the stratification process.

As mentioned earlier, information was very limited and one can only assume the stratification process was haphazard at best with location and size varying rich and lean zones throughout the speed and load range of this engine.  Once again, simplicity and ease of retrofit in an existing engine make the concept worthy mention.  A more optimum performance level could perhaps be found by systematically evaluating cylinder airflow with appropriate models and improving the airflow with appropriate port and cylinder chamber geometry to enhance stratification.  Like most other stratified charge concepts, advantages did not outweigh the increase in complexity and cost.

*The Jessel Exhaust Dilutent Engine*

The Jessel exhaust dilutent engine is unique in operating principle and is of interest because of its concept.  Information is very limited because the engine was not further developed beyond the single cylinder experimental phase.  The engine, developed by Jessel, Myers and Uyehara at the University of Wisconsin (the latter two individuals granted a U.S. patent on the idea) achieves stratification by returning a proportion of the exhaust gas back into the cylinder, via a recirculation valve, followed by a fresh combustible mixture.  Power output is controlled by the exhaust recirculating valve and hence, the amount of recirculated exhaust gas that displaces the fresh charge.  The engine is shown in Figure 13 below.

During intake, exhaust gas is allowed to enter the cylinder through the recirculation valve.  At some point during intake, depending on the power setting, the recirculation valve is closed and the unthrottled air/fuel mixture is drawn in from the carburetor.  This pocket of fresh charge remains near the spark plug during compression and, throughout combustion,

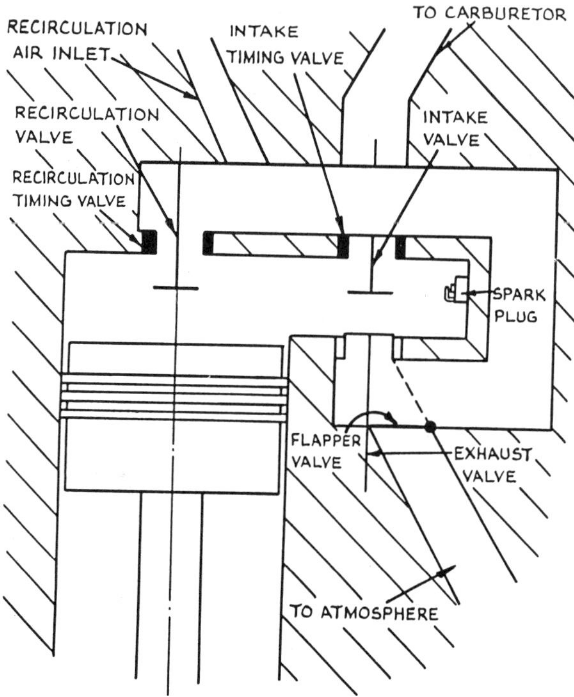


*Figure 13.  The Jessel-Myers-Uyehara Engine (27).*

52

these two zones remain stratified.  At the start of exhaust stroke, the once-combusted gas, nearest the exhaust valve, is directed into an intermediate connecting passage with a flapper valve.  During the exhaust stroke, the flapper valve diverts the remaining once- Once-recycled exhaust to waste, thereby allowing the exhaust to cycle only once.  Although only limited data is available, it was thought that this engine would have low HC and CO emissions because the exhaust is reheated through recirculation.  Low NOx emissions were also expected because of the use of stratified and controlled EGR at light to moderate loads.  At higher loads, stratification is limited or lost and the engine approachs that of a conventional gasoline engine with EGR.

## Overall Advantages and Disadvantages

1.  The unthrottled stratified charge engine cycle has a higher theoretical efficiency and hence, a predictable higher fuel economy than that of the conventional gasoline engine cycle.  This is generally not true of actual engine operation however, due to heat transfer and throttling effects of prechamber engines and misfire and incomplete combustion of single chamber engines.  Only the late injection stratified charge and the axially stratified charge engines have shown higher fuel ecomomy than the conventional gasoline engine when the charge ignited and combusted appropriately.

2.  Carbon monoxide emissions from most unthrottled stratified charge engines are usually low due to the excess air in the cylinder charge.

3.  Oxides of nitrogen emissions from unthrottled stratified charge engines are lower than those from conventional gasoline engines.  Of the stratified charge engine group, prechamber engines where the rich charge is physically separated from the lean main cylinder charge has the lowest NOx emissions of most engines examined and reported in the current literature.

4.  Some stratified charge engines have shown multifuel capability providing reasonable startup and operation through the speed and load range of the engine when emissions and efficiency (fuel economy) are disregarded.

5.  Stratified charge engines generally do not possess any starting difficulties when compared with the conventional gasoline engine.

*Overall disadvantages*

1.  Efficiency and hence, fuel economy of the engine, is generally lower than that of the conventional gasoline engine.

2.  In most cases, unthrottled stratified charge engines have high hydrocarbon emissions due to either incomplete combustion of the lean portion of the charge, or occasional misfire of the charge at the light load condition.

3.  Stratified charge engines generally have a higher noise level than that of the conventional gasoline engine.  Unthrottled stratified charge engines distinctively have  a higher noise level at idle and light load operation.

4.  Stratified charge engines generally are more complex, heavier and hence, costlier than the conventional gasoline engine.

5.  Stratified charge engines, despite claims, are not a panacea to the emissions problem.  All require exhaust treatment of some sort to meet U.S.  emission standards.

## The Rotary Engine - Another Look at Stratification

Although the rotary engine is a bonafide internal combustion engine it is often left to a class by itself.  Stratification has been reasonably successful and good operating characteristics have been demonstrated with distinct advantages as well (20).  Stratification has been achieved with a late injection design (Category 2 above) and with a prechamber design

(Category 3 above). The stratification of the rotary engine reduces the main disadvantage of the rotary engine, that is, the relatively large surface to volume ratio of the rotary combustion chamber when compared to that of the reciprocating piston engine. Indeed, recent advances in the stratification concept have been made with the rotary engine and this is one of the active areas of research in charge stratification today.

Rotary internal combustion engines, unlike their reciprocating piston counterparts, are a relatively recent development. The engine devloped from a rotary piston steam engine in the early 1900's but a practical design did not appear until 1943. This design was later refined by Wankel and built in cooperation with NSU Motorenwerke in West Germany in 1957. Shortly after this, Curtiss-Wright entered the license agreement and initiated a fairly significant research program (43,54-61). In 1967, several engines were designed and produced primarily for military applications with an injection system replacing the carburetted intake, thus beginning the evaluation of various stratified charge configurations.

In this engine, shown in Figure 14 below, fuel was directly injected into the combustion chamber by high pressure diesel type nozzles located adjacent to the spark plug. The engine was unthrottled and utilized the late injection concept in that all, or most, of the burning occured in a 'stationary' flame front as the fuel was injected. Combustion of various stratification configurations was systematically investigated along with multifuel capabilities and reported in the literature (43). Fuel consumption remained relatively high until later refinements improved this aspect as well. Low peak cycle temperatures contributed to lower NOx levels when compared against the conventional gasoline engine. Slightly throttling the intake air had a direct effect on increasing the exhaust temperature and, with exhaust catalysis, hydrocarbon emissions were reduced as well. Various problems (such as cost, and durability, to name a few) remained, which made the engine unacceptable for large scale production in light duty general applications.

Development continued, and in 1974 difficulties encountered between light loads and high power were overcome by the addition of a pilot injector (5% of fuel delivery at full load) working in conjunction with the main injector for improved ignition and subsequent combustion (43). Mathematical modeling was extensively used in the development process by researchers primarily at Princeton to better determine injector nozzle patterns (62,14) and improve combustion at various operating conditions. Full advantage of the pilot injection concept to ignite the multi-orfice main injection streams could be realized because of the shape of the combustion chamber. As a conseqence, the hydrocarbon emission problems associated with other stratified charge engines were not as severe, particularly at operating temperatures. The momentum of production of reciprocating engines, patent right questions and costs remained and as a consequence, offset advantages of the stratified rotary engine over the conventional piston engine which, by this time, had evolved to meet U.S. emission standards.

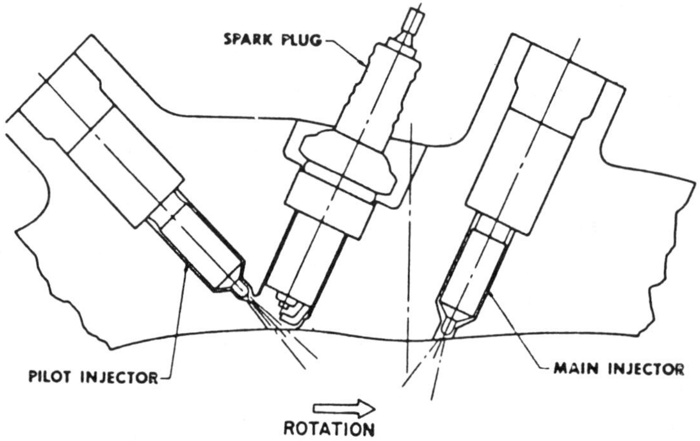


*Figure 14. Rotary Stratified Charge Injection System.*

Rotary engine development has continued with renewed vigor due to a recent transfer of patent rights from Curtiss Wright to John Deere Technologies International Inc. (division of Deere and Co.) in 1984. During the energy crisis of the 1970's, engineers at Deere and Co. became very much interested in engines with alternate fuel capabilities for agricultural applications. Deere and Co. was of the opinion that the stratified charge rotary engine had superior alternate fuel capability and power to weight ratio (leading to good fuel economy) when compared to other reciprocating engines and, successfully pursued acquisition of stratified charge engine patent rights from Curtiss Wright. All rights are now the concern of Deere and Co. with the exception of patent rights of the prechamber engines which now have expired. Several engines are now in development for military applications. All engines are turbocharged and have single orifice pilot injection with multiple orifice main injection as described above. A small 0.7 liter (150 - 160 hp) engine is being developed as a prototype for general duty aircraft applications. High speed electronic injection is now being evaluated to overcome speed limitation difficulties with conventional injection. This stratified rotary concept is also being applied to the development of a larger 400 hp powerplant for general aviation applications in joint work with Avco Lycoming. A larger, 5.8 liter, 750 hp engine is also has being developed for marine applications under contract with the U.S. Marine Corps (19,21). This development work is paralleled with three dimensional modeling and other analytical work at the Intermittent Combustion Engine Branch at NASA Lewis Research Center, at Princton University and elsewhere (63,64,65).

## Multifuel Capabilities

In some cases, stratified charge engines were not developed for multifuel capabilities. This is particularly true for those engines developed during the mid 1960's, primarily in response to the Clean Air Act and subsequent emission standards. Some stratified charge engines have been, however, either specifically designed for or, at least, have been tested for multifuel capability. This is particularly true for those engines developed fully, or in part, by the military. The TEXACO TCCS concept is a good example of this. It should be noted that in the past, multifuel capability usually implied that engine parameters (such as spark and injection timing) were not changed so that operating conditions would not be optimized for varied fuels. With the introduction of electronically controlled engine components, optimization could be achieved by simple movement of a switch.

The late injection (Category 2) engines do show multifuel capabilities. The Texaco TCCS engine has been operated on a wide range of fuels including common gasolines, diesel and jet aircraft fuels JP4 and JP 5. The Deutz engine concept has shown some multifuel capability. The John Deere stratified rotary engine (previously Curtiss-Wright) has also been operated quite successfully on several fuels. In general, Diesel fuel produces higher torque output and better fuel economy than gasoline at the same fuel delivery setting. Significant differences do occur with emissions however (27).

Wong (66,67) showed that in the TCCS engine, ignition of a fuel having high volatile components (40-300 c boiling range) occurs close to the spark plug, while for the same operating conditions ignition of diesel fuel occurs farther down stream of the spark plug. The distance between the injector tip and spark plug is critical for multifuel operation because a short distance, required by high volatility fuels, may be too short to allow proper mixing of a low volatility fuel leading to improper ignition. Clearly, viscosity, density and bulk elasticity are important parameters of the fuel. Evers et al. (68) showed that diesel fuel operation efficiency and emissions were not significantly affected by spark timing. He concluded that selfigniiton is the dominate ignition process in diesel fuel operation of the TCCS engine. Multifuel advantages were indeed demonstrated but the optimization process is complex because of straddling the autoignition curve.

Some of the difficulties of multifuel operation of the TCCS were described by Baranescu (69) including flame movement away from the spark source. Burning velocity is a function of the fuel and turbulence level, in this case swirl. If swirl is held constant for different fuels, one can expect different ignition quality and/or different heat release. Laminar burning velocity as a function of equivalence ratio for two fuels is shown in Figure 15 below. Although the flame surely will turbulent and not laminar, some inferences still can be made.

With regard to the TCCS system and most other 'late injection' systems, various engine parameters which are often considered 'mechanically fixed' (such as injection timing, injection rate, swirl) then may have to be changed to optimize the engine for multifuel operation.

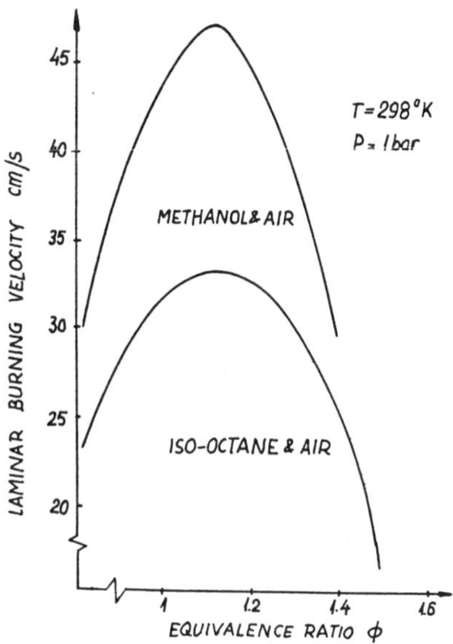


*Figure 15. Laminar Flamespeed as a Function of Equivalence ratio for Two Fuels.*

The MAN-FM system (Category 7, Miscellaneous Engines) has demonstrated good operating characteristics under several fuels including gasoline, diesel, JP-4 (kerosene) and alcohols (27, 70-73). This engine has been further developed to operate on methanol for city buses and has shown reliable operation on this fuel while meeting the stringent State of California emissions standards. The low lubricating qualities of methanol was offset by the development of an improved lubrication system and the use of a better lubricating oil. Ethanol can be used in the MAN system as well because of similar characteristics of this fuel.

Needham et al. (74) at Ricardo Consulting Engineers examined the interaction between several fuels and light duty engine combinations. Fuels primarily covered were D-2 diesel, naphtha and intermediate broad-cut blends. Engines studied included an IDI diesel, a DI diesel, a spark assisted ID diesel, the Texaco TCCS and the MAN-FM. All engines were unthrottled and classified as 'late' injection. These researchers concluded that, irrespective of fuel, the open-chamber stratified charge engine exhibits baseline hydrocarbon emission penalties when compared to other combustion systems. However, further development of these engines may improve the HC problem and allow the low NOx advantage to surface. Apparently, fuel and air mixing problems plague multifuel operation of these engines as well (refer to hydrocarbon analysis section of this paper).

Finally In some instances, modifications to conventional Diesel engines have been made to extend fuel tolerability or add multifuel capability to production engines. The concept of spark assisted diesel combustion is not new. Hesselman (75) began experimenting with spark plugs in Diesel engines in the early 1940's. Lange and Spindler (76) later studied local mixture strength and flame propagation of fuel oil and gasoline in a spark assisted diesel engine with a sampling technique. Wong (66) studied spark plug location in detail and reported the startability and operational characteristics of the spark assisted Diesel engine. Phatak and Komiyama (77) converted a single cylinder, low compression, open chamber, direct injection diesel engine for spark assist operation with alcohol based fuels. Several spark plug locations were studied over a range of alcohol/fuel oil blends and swirl combinations. More recently, Borman et al. (78) at the University of Wisconsin and Abata et al. (79) at Michigan Technological University have studied startability and operation of alternate fuels in diesel engines modified with spark assist under TACOM contract. High speed photographs (10,000 frames per second) of the combustion process of these fuels help characterize the combustion process and aid in determining optimum spark plug location and injection timing. An example of a schematic obtained from the high speed photographs is shown in Figure 16.

Results indicate that reasonable operation within a reasonably wide fuel range can be obtained for use in military applications. Work continues in an effort to optimize the process. Emission work has not been conducted.

56

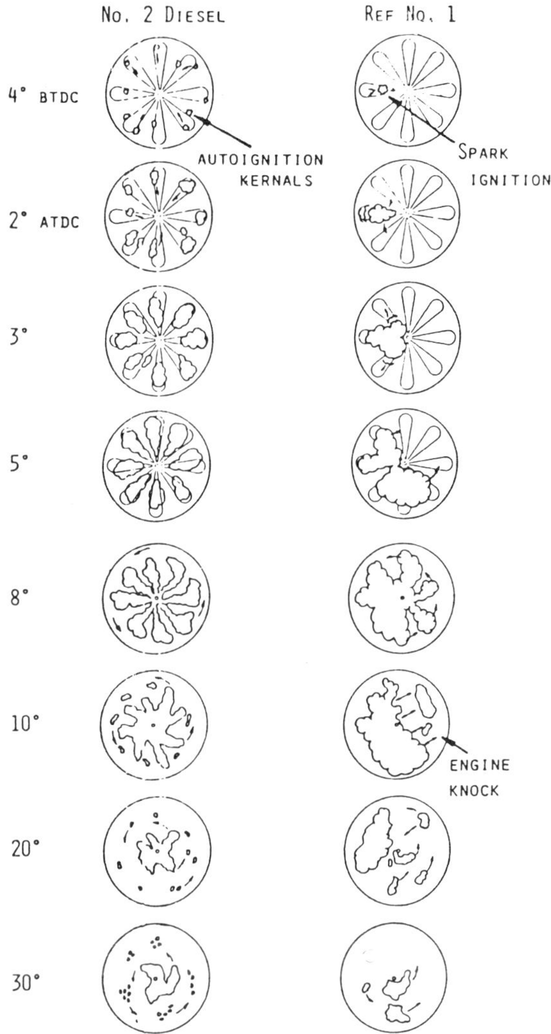


Figure 16. Schematic of SADI Engine Combustion from High Speed Photographs.

## Exhaust Emissions

It is well known that there are undefined tradeoffs between power, fuel economy and emissions. Trends can often be predicted. For instance, increased power output is generally accompanied by a decrease in fuel economy. Trends of the U.S. regulated pollutants, HC, CO and NOx are more difficult to predict. It is also well known that accurate comparisons of power, fuel economy and emissions between engines are difficult to obtain. Further, these pollutants can be, to a certain extent, varied in any given engine depending upon the application. While it may be inappropriate to compare exhaust emissions of selected stratified charge engines because of the many independent variables and uncertainties, this paper would perhaps be incomplete if at least some comparison were not presented. As a cautionary note, very minor design changes often change engine operation dramatically, particularly with the often irratic and fragile operation of some stratified engines. For instance, engineers at GM observed almost a two fold decrease in hydrocarbon emissions when the spark plug electrode was rotated 90 degrees in one of their tests with a particular stratified setup!

A comparison of emission data of various stratified charge engines appears in Figure 17 below. For the most part, this data was obtained from references 27, 20, and 53. Emissions

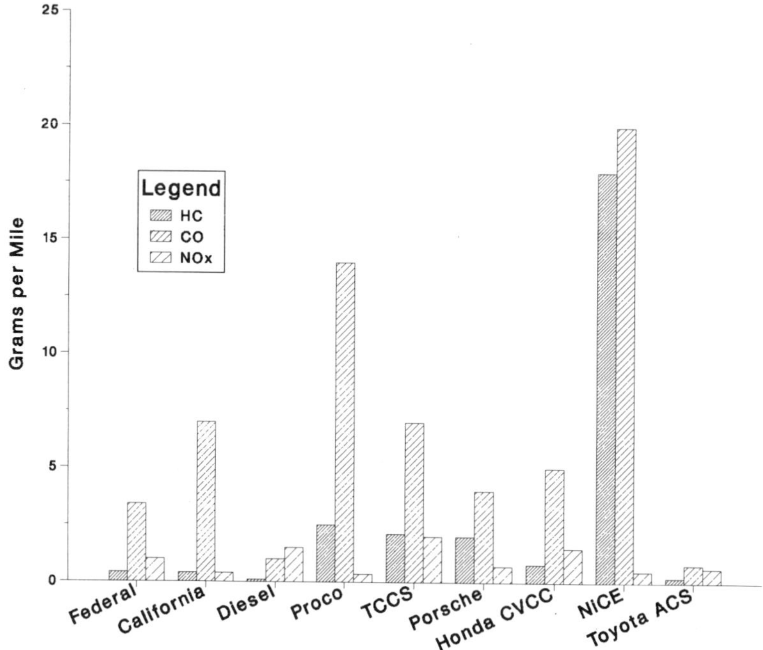

Figure 17. Emissions Data for Selected Stratified Charge Engines.

data were obtained with the relatively standardized CVS-CH (constant volume sampling, California highway) test cycle. Vehicles were not equipped with any exhaust aftertreatment devices and weight ranged from 2200 lbs to 2800 lbs.

In almost all cases, hydrocarbon emissions are very high and possibly beyond that which could be treated effectively with exhaust aftertreatment devices. Two exceptions to this observation is the GM ASC engine and the Deere stratified rotary engine. The GM ASC

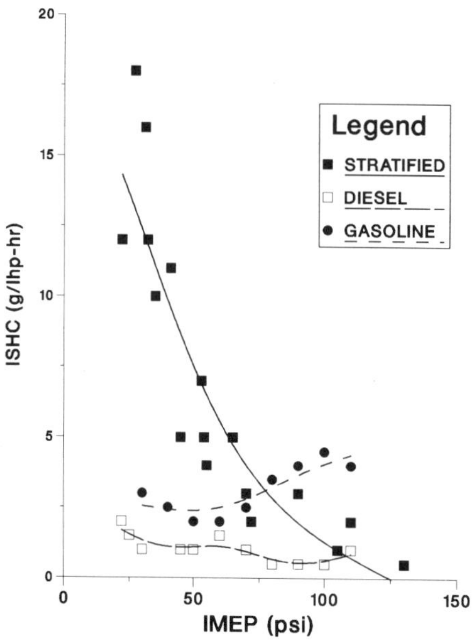

Figure 18. Hydrocarbon Emissions from Various Stratified Charge Engines.

engine appears to have good hydrocarbon emission control provided misfire does not occur. Data for the Deere engine obtained with the CVS-CH test cycle was not available and hence, is not presented above. Data presented in various publications by Deere, however, does indicate the hydrocarbon difficulty has been improved due to pilot injection and ignition design.

The hydrocarbon difficulty in the other cases can be improved. Ricardo Engineers (27) reported that hydrocarbon emissions decrease in all engines when throttling is used. Use of throttling is at the expense of fuel economy, however. Further, primary emissions limits could perhaps be attained with the use of EGR (exhaust gas recirculation) for control of NOx and exhaust aftertreatment devices for control of CO and HC.

Novel two stroke designs appear to have poor fuel consumption, exceptionally high hydrocarbon and carbon monoxide levels, but low oxides of nitrogen levels. This is most likely due to inherent EGR in the two cycle engine. Two cycle engines have not been installed in vehicles in the U.S. because of hydrocarbon levels. Recent work on the two cycle engine and, in particular, with pneumatic injection, may re-introduce this engine concept to vehicle applications.

## Hydrocarbon Analysis

One main problem limiting the utility of the stratified charge engine is the high level of hydrocarbon emissions, particularly at light loads. Indeed, the level of unburned hydrocarbon emissions from the direct injection stratified charge engine operating at light load is an order of magnitude greater than emissions from conventional gasoline engines under similar operating conditions. Since exhaust gas temperatures are also low at light load, catalytic treatment of the exhaust gases is difficult, which further aggravates the problem. Shown below are hydrocarbon emissions data from eight stratified charge engines plotted as a function of indicated mean effective pressure, IMEP (80).

This light load hydrocarbon emissions problem appears to be independent of combustion chamber geometry or fuel injection system as this behavior has been identified in almost all stratified combustion systems with both early and late injection schemes. Hydrocarbon emissions from diesel engines are low at all loads. Hydrocarbon emissions from conventional gasoline engines are also low over most of the load range, increasing slightly at heavy load. While hydrocarbon emissions from direct injection stratified charge engines are low at heavy load, light load emissions are greater by a factor of ten when compared with SI and diesel engines as shown in Figure 18. This is particularly interesting because, in some instances, injection systems for stratified charge engines have been retrofit from Diesel engines.

Hydrocarbon emissions have been found by a number of investigators, to be exponentially dependent on overall combustion chamber equivalence ratio in the stratified charge engine as shown in Figure 19 (81).

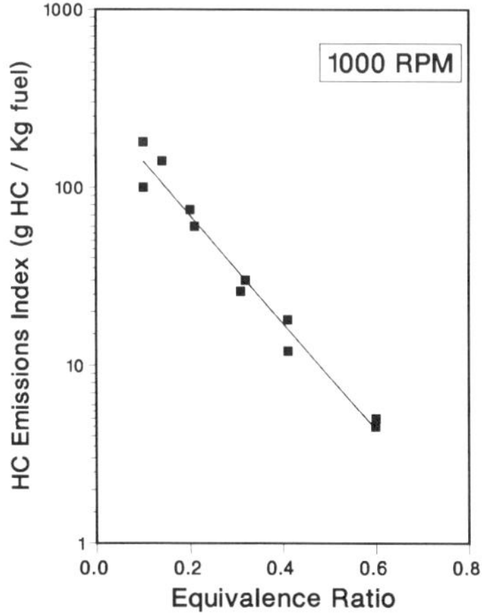


*Figure 19. Hydrocarbon Emissions vs. Overall Equivalence Ratio.*

Research into the the origins of this problem have focused on liquid fuel delivery, fuel-air mixing, cycle to cycle variation in cylinder air motion, and flame quenching in the bulk gases. In a sampling study, Wood (82) determined that the formation of mixtures too lean to burn is a major source of hydrocarbon emissions at light load, and that exhaust gas temperatures are too low to promote secondary oxidation in the exhaust. In model work, Johnston and Lancaster (83,84) predicted that volume flame quenching due to lean fuel-air regions mixture results in excessive unburnt hydrocarbons. Sinnamon et al. (85) concluded that cycle to cycle variations in cylinder air motion could cause ignition difficulties in stratified charge engines. In-cylinder air motion was found to distort the leading edge of the fuel spray causing separation from the main spray. In photographic work, Witze (86) observed that turbulence levels inside high swirl DISC engines determines whether or not the flame stays attached to the spark plug. Variations in turbulence levels between successive engine cycles cause different flame configurations and ultimately different burning rates contributing to incomplete combustion. In general, the fuel injection system has a strong influence on engine performance characteristics such as fuel economy, power, and emissions (81).

The injection process in a DISC engine is very critical. The injection system must consistently place an ignitable charge at the spark gap over a broad range of engine speed and load. Complications arise because the injection process is mechanically dependent on the speed of the engine. In addition, in-cylinder air motions are also dependent on the speed and load of the engine. The same strategy that is successful in producing the correct charge stratification during full load operation may lead to ignition problems during ligh load operation (81,69).

Giovanetti et al. (80) further investigated this hydrocarbon emissions problem by examining the injection parameters of fuel delivery pressure to determine the possible presence of a correlation between hydrocarbon emissions and fuel delivery. Based upon this experimental data, an estimate was made of the velocities of the fuel delivery and, in particular, the mass of fuel entering the chamber at velocities below that which would promote proper mixing of the fuel with the air. Two pressure time histories of injection pressure from a pencil type fuel injector taken at two engine loads are shown in Figure 20.

From Figure 20 it can be observed that as load decreases, secondary injection increases. This is particularly true at the light load condition. This 'needle bounce' has also been observed by several other investigators (79). Mass flow rate was determined from the equation:

$$dmf/dt = pf\ Cd\ A(2(Pinj - Pcyl)/pf)1/2$$

where the discharge coefficient, Cd, was taken at 0.6 based on square edged orvice exper-

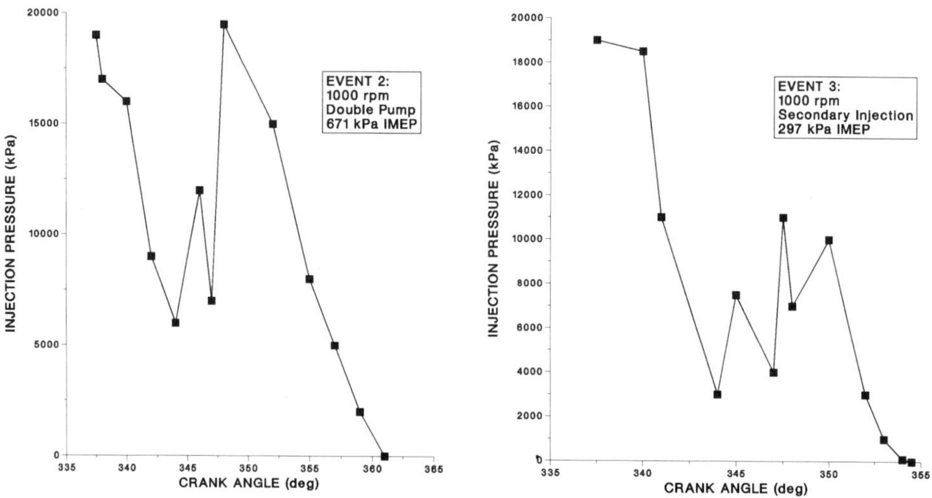



*Figure 20. Pressure-time Histories of Injection Pressure at Two Loads.*

60

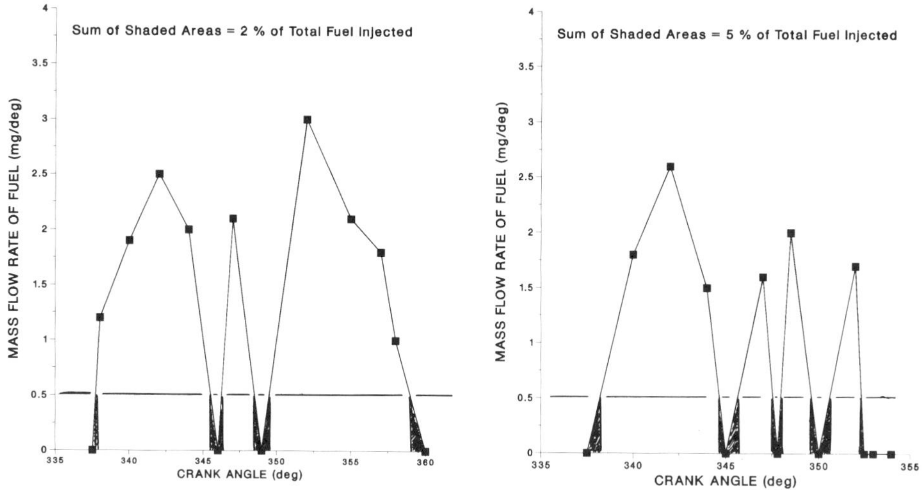

*Figure 21. Instantaneous Fuel Flow Rate Calculated from Data in Figure 20.*

iments (80). The instantaneous fuel flow rates predicted by the above equation from data obtained from Figure 20 are shown in Figure 21.

Here, the total mass of fuel delivered is equal to the area under the curve. The horizontal line in Figure 21 above indicates that fuel which entered the chamber at zero relative velocity, i.e., at a velocity approximately equal to the swirl velocity of the air. Giovanetti

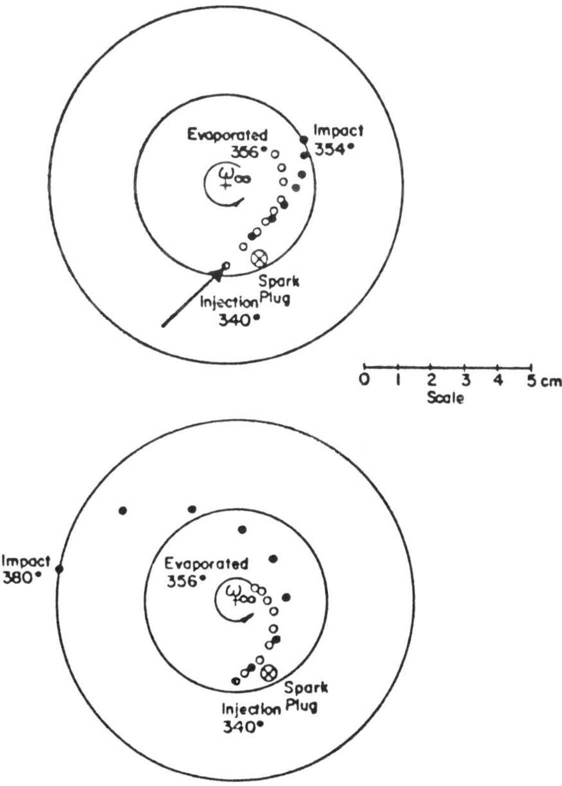


*Figure 22. Simulated Droplet Trajectories.*

et. al. observed that this 'zero zelocity' fuel was the same order of magnitude as that discharged into the exhaust.

Giovanetti then examined the effect of low and high velocities on individual droplets of selected sizes with a trajectory and evaporation model. These results are shown in Figure 22.

Results lead to the postulation that there are two ways the fuel-air Mixture could escape direct contact with the ignition source and flame. In the case of low velocity fuel injection, liquid fuel and combustible fuel-air mixture remains too close to the cylinder head and does not penetrate into the piston cup where most of the primary combustion occurs. In the case of high velocity fuel injection, liquid fuel impinges on the cup and diffuses into the bulk gases during expansion. If temperatures are high enough, as in heavy load operation, fuel in both cases will oxidize due to autoignition and secondary combustion. If temperatures are low, as in light load operation, both of these cases will contribute to hydrocarbon emissions. The relatively simplistic model did not address the development of local air/fuel mixtures reasonably placed near the flame zone but beyond the lean limit of combustion. This, too, could also be a contributing factor, although calculations of this effect by Giovanetti were not as conclusive. Hydrocarbon emissions at ligh load, then, could be summarized by the following mechanisms shown in Figure 23.

The results of this study and others indicate that hydrocarbon emissions in direct injection stratified charge engines could possibly be reduced by 1) an improved light load fuel delivery system and 2) increased cylinder gas pressure and gas temperature at light load. The latter could be achieved by increasing the compression ratio and/or utilizing ceramic components which would increase surface temperatures through decreased heat transfer.

## Modeling

While engine design and optimization have been around since the beginning of the internal combustion engine in the mid 1800's, mathematical modeling of the many fluid and

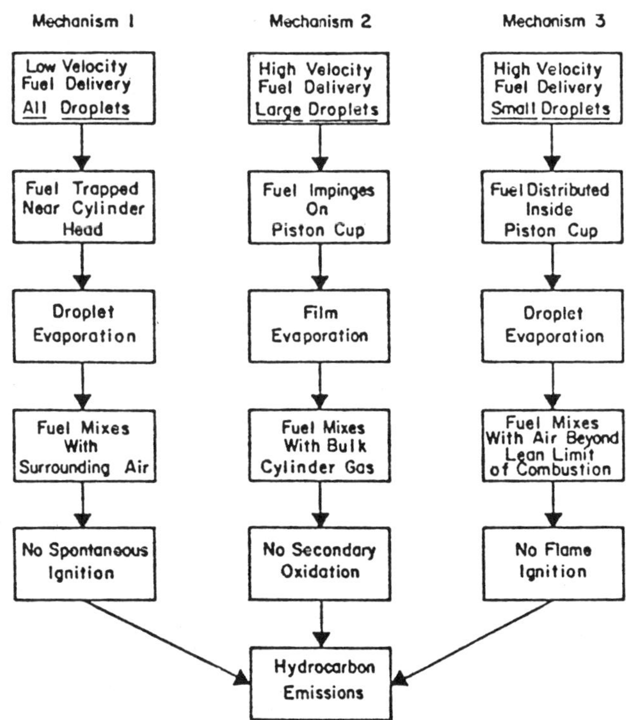


*Figure 23. Hydrocarbon Emissions Mechanism.*

chemical processes in the engine are a relatively recent contribution to engine research. Within the last twenty years or so mathematical modeling of the complex phenomena occuring in the engine has given the research scientist considerable insight into development and optimization of the internal combustion engine. Mathematical modeling of engine processes was limited to relatively simplistic analyses until, of course, the widespread use of modern high speed digital computers and accompanying numerical techniques. Today, the cost of producing and using a detailed model is negligible in comparison to the total development cost of the engine itself.

Models can be either *macroscopic* yielding information which is of a general nature where the cylinder charge is treated as a whole and only time is the independent variable or *microscopic* where the cylinder charge is broken into discrete finite elements and where time and space are the independent variables. Macroscopic models, often referred to as *zero-dimensional* models based primarily on equilibrium thermodynamics, can vary a great deal in complexity and some are very clever yielding valuable results. Some of the more popular zero dimensional models have been summarized by Foster (87). A zero dimensional model has been recently used at MIT to deduce the rate at which fuel is burned to study the causes of hydrocarbon emission mechanisms in a lightly loaded stratified charge engine (88). Microsopic models can be *one, two or three dimensional* in space. Some of the more simpler multidimensional models have yielded very informative results (89-93). The physical processes of air motion, turbulence level, fuel mixing and evaporation, combustion and heat transfer, are of obvious importance. Conditions at the intake and exhaust valves are of particular importance since macroscopic flow patterns are often the result of inlet and exhaust geometry. A fine grid system is often necessary to adequately resolve the fluid flow detail. A three dimensional simulation of the in-cylinder flowfield would necessarily include submodels such as a turbulence model, detailed fuel spray and mixing models, multi-step kinetics, and a grid size small enough to resolve important flow details near the surfaces and of the fuel spray. The extreme case of three dimensional, two phase, turbulent, unsteady, reacting fluid flow with detailed multi-step chemical kinetics is currently beyond the scope of present day computers. Because of the magnitude of calculations, results from such a predictive model would certainly be open to question. Indeed, the assumptions made with present day codes, together with the often inadequate experimentally-determined boundary conditions and associated machine round-off errors leave many to ponder the validity of the results. Nevertheless, seemingly complex computer codes that model engine processes (homogeneous or stratified) represent only the current limits of computer hardware availability and numerical technique which are continually evolving with time.

In a review paper, Brocco (94) summarized engine modeling work of investigators at Princeton and others elsewhere. Bracco begins with the generalized equations for two phase reactive flows which, in theory, completely describe the physical and chemical processes of air flow, fuel injection and atomization, fuel and air mixing, and combustion occuring in the engine. As mentioned earlier however, current understanding of numerical analysis and availability of computer hardware do not allow a complete solution of these general equations. Several constraints (or limitations) either must be or are often made. These are:

1. The process of fuel breakup and atomization must be approximated due to our inability to solve difficult equations describing this process.

2. The phenomena of gas turbulence must be approximated due to our inability to rapidly solve difficult equations with current numerical techniques and with a 'reasonable' grid size.

3. The processes of vaporization and evaporation of spray droplets within the droplet cloud must be simplified with thick and thin spray approximations once again due to our inability to solve difficult equations that fully describe this process.

4. The lack of kinetic mechanisms that fully describe the reactive chemistry within the combustion process of complex hydrocarbons and of precise kinetic data lead to approximations in the kinetic scheme.

5. The complexity of kinetic schemes involving many unstable and intermediate chemical species are often approximated by one-step and two step schemes.

6. The computer time and computer memory required to solve three dimensional processes lead many researchers to approximate three dimensions with two dimensions (axi-symmetric models).

*Table 1.  General Equations*

$$\partial\rho/\partial t + \nabla\cdot(\rho\underline{u}) = 0 \tag{1}$$

$$\partial(\rho\underline{u})/\partial t + \nabla\cdot(\rho\underline{u}\,\underline{u} + \underline{\underline{p}}) = 0 \tag{2}$$

$$\partial(\rho\underline{u}^2/2 + \rho e)/\partial t + \nabla\cdot(\rho\underline{u}^2\underline{u}/2 + \rho e\underline{u} + \underline{u}\cdot\underline{\underline{p}} + \underline{q}) = 0 \tag{3}$$

$$\partial(\rho Y_i)/\partial t + \nabla\cdot[\rho Y_i(\underline{u} + \underline{u}_i)] = w_i \tag{4}$$

where $i = 1\ldots N$

$$\underline{\underline{p}} = p\underline{\underline{U}} - [(k - 2\mu/3)(\nabla\cdot\underline{u})\underline{\underline{U}} + \mu(\nabla\underline{u} + \nabla\underline{u}^T)] \equiv p\underline{\underline{U}} + \underline{\underline{\dot{\sigma}}} \tag{5}$$

$$\underline{q} = -\lambda\nabla T + \rho\sum_{i=1}^{N} h_i Y_i \underline{u}_i \tag{6}$$

$$\nabla X_i = \sum_{j=1}^{N} \left(\frac{X_i X_j}{D_{ij}}\right)(\underline{u}_j - \underline{u}_i) \tag{7}$$

where $i = 1\ldots N$

$$\sum_{i=1}^{N} \nu'_{i,k} M_i \rightarrow \sum_{i=1}^{N} \nu''_{i,k} M_i \tag{8}$$

where $i = 1\ldots N$ and $k = 1\ldots K$

$$w_i = W_i \sum_{k=1}^{K} (\nu''_{i,k} - \nu'_{i,k})$$

$$\left[B_k T^{\alpha_k} \exp(-E_k/RT)\right] \prod_{j=1}^{N}\left(\frac{X_j\, p}{RT}\right)^{\nu'_{j,k}} \tag{9}$$

where $i = 1\ldots N$

$$p = \rho RT \sum_{i=1}^{N} (Y_i/W_i) \tag{10}$$

$$e = \sum_{i=1}^{N} h_i Y_i - p/\rho = \sum_{i=1}^{N}\left(h_i^{\circ} + \int_{T^{\circ}}^{T} c_{p_i}\, dT\right) Y_i - p/\rho \tag{11}$$

$$X_i = (Y_i/W_i) \sum_{j=1}^{N} (Y_j/W_j) \tag{12}$$

7. For the same reason above, one dimension (spherical or cylindrical coordinates) is often used as a further approximation with additional distortion.

8. In the simplest of cases, zero dimensional models (where time is the only independent variable) are often used to approximate the macroscopic behavior of the mixing and/or combustion process.

*The General Equations*

Bracco presented the general equations for two phase reactive flows for the case where liquid fuel is injected into the combustion chamber of an engine with no restrictions on engine type (rotary or reciprocating piston), engine geometry (open chamber, divided chamber, etc.) or fuel ignition (autoignition, spark assisted). The general equations are shown in Table 1.

The first four equations are the mass, momentum, energy and species conservation equations. Equation 5 gives the pressure tensor in terms of the hydrostatic pressure and viscosity shear tensor. Equation 6 shows the heat flux vector including heat transfer through a surface and heat transfer due to diffusion velocities. Equation 7 relates the diffusion velocities and the concentration gradients of each species. Equation 8 represents the elementary reaction steps. Equation 9 is an expression for the species source terms and represents the chemical kinetics phenomena of each chemical species. Equation 10 is the thermal equation of state (perfect gas model). Equation 11 is the caloric equation of state including the enthalpy of formation of each chemical species. Finally, Equation 12 relates the mole fractions to mass fractions of each chemical species.

Two sets of the equations listed in Table 1 are needed, one for the liquid and one for the gas. The set for the liquid phase would be simplified because of constant density and the fact that reaction does not take place in the liquid phase. Boundary conditions, including appropriate experimental data at the boundary, are also necessary. Finally, the equations would have to be step-solved with a suitable numerical technique. The reader can begin to get a feel for the complexity of the modeling problem.

There have been many modeling studies of the stratified charge engines over the past years. They are too numerous to detail here. Work at Princeton under the direction of F. Bracco (94,89,92,93), particularly with the rotary engine, is certainly worth mention, as is work at General Motors Research Laboratories under the direction of Ed Groff (17). Other laboratories have done equally important modeling work and have contributed significantly to stratified engine development. The Intermittent Combustion Laboratory at NASA Lewis has also been involved in modeling work for many years (23). Los Alamos Laboratory has developed sophisticated two and three dimensional computer codes 'packaged' for use by researchers in laboratories around the country (95-98).

In order to provide the reader with some insight into modeling work applied to the stratified charge engine, two examples have been chosen and are detailed below. The first is an example of a one-dimensional model applied to a planer combustion chamber for the purpose of determining the effect of volume flame quenching on hydrocarbon emissions. The second is an example of a axi-symmetric model (CONCHAS) applied to an early injection engine for the same purpose.

*Example of a One Dimensional Model*

Westbrook (99) used a one-dimensional model to follow fuel motion and pollutant formation in a stratified charge engine. He compared the homogeneous combustion case with that of the stratified charge case in an attempt to determine the effect of *volume flame quenching* as a source for unburned hydrocarbon emissions. He considered both global kinetic mechanisms (one step and two step schemes) and a detailed multistep kinetic scheme involving 66 reactions. The mathematical model was based on simplified one-dimensional equations for conservation of mass, momentum energy and each chemical species. The model was limited to one space dimension and assumed planar symmetry. With the simple geometry, the detailed chemical kinetics model was retained and interactions between the fluid flow and the chemical reactions on the calculation of flame structure, flame propagation and fuel motion. The model equations are shown in Table 2 below.

*Table 2. Equations Used in One-dimensional Model by Westbrook.*

**Model equations**

$$\frac{\partial \rho}{\partial t} + \frac{\partial \rho u}{\partial x} = 0$$

$$\frac{\partial \rho u}{\partial t} + \frac{\partial \rho u^2}{\partial x} = -\frac{\partial (P + Q)}{\partial x}$$

$$\frac{\partial \rho \mathcal{E}}{\partial t} + \frac{\partial \rho u h_s}{\partial x} = \frac{\partial}{\partial x}\left(\rho C_p \, \alpha \frac{\partial T}{\partial x}\right) + \frac{\partial}{\partial x}\left(\rho \sum h_i D_i \frac{\partial Y_i}{\partial x}\right) + \rho \dot{q}$$

$$\frac{\partial \rho Y_i}{\partial t} + \frac{\partial \rho u Y_i}{\partial x} = \frac{\partial}{\partial x}\left(\rho D_i \frac{\partial Y_i}{\partial x}\right) + (\rho \dot{Y}_i)_{kin}$$

$$K = AT^n e^{-E_a/RT}$$

where

$x$ = position
$t$ = time
$\partial$ = density
$u$ = velocity
$P$ = pressure
$Q$ = artificial viscosity
$E$ = total energy
$hs$ = stagnation enthalpy
$\xi$ = energy diffusivity
$T$ = temperature
$Cp$ = specific heat
$Di$ = diffusivity of species i
$hi$ = specific enthalphy of species i
$Yi$ = mass fraction of species i
$q$  = time dependent energy deposition rate used to simulate the ignition phase of combustion

The last term $(\rho \dot{Y}_i)_{kin}$ in the species conservation equation refers to the change of species i due to chemical reactions. The final equation in Table 2 is the general form of the Arrhenius equations used for all the chemical reaction rates. The Lewis number was assumed unity so that the sum of all the species conservation equations yield the overall mass conservation equation. A planar chamber was assumed with ignition at the left wall and flame propagating in the x+ direction as shown in Figure 24 below.

Although detailed kinetic mechanisms are desireable in flame studies of this nature, the availablity of these mechanisms are limited to a few simple hydrocarbon fuels. These me-

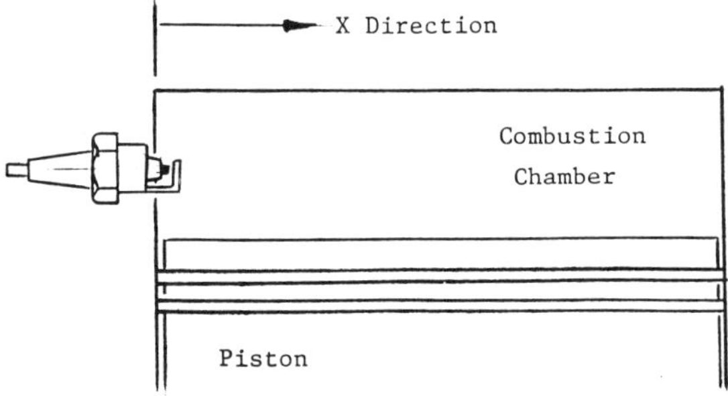


*Figure 24. One-dimensional Combustion Chamber of Westbrook.*

chanisms must often be simplified to facilitate numerical analysis. Westbrook used methane fuel in the model computations. Three situations were considered: 1) a single step mechanism shown in Table 3, 2) a two step mechanism shown in Table 4 and 3) the multistep mechanism shown in Table 5.

*Table 3. One Step Reaction Mechanism for Methane*

$$-\frac{d[CH_4]}{dt} = 1.5 \times 10^{13} \exp(-48400/RT) \times [CH_4]^{0.7} [O_2]^{0.8}$$

*Table 4. Two Step Reaction Mechanism for Methane*

$$-\frac{d[CH_4]}{dt} = 1.5 \times 10^{13} \exp(-48400/RT) \times [CH_4]^{0.7} [O_2]^{0.8}$$

$$\frac{d[CO_2]}{dt} = 5.62 \times 10^{14} \exp(-43000/RT) \times [CO]^{1.0} [H_2O]^{0.5} [O_2]^{0.25}$$

*Table 5. Multistep Reaction Mechanism for Methane*

**Reaction rate parameters, in mole-cm-sec-kcal units ($k = A\,T^n \exp(-E_a/RT)$)**

| # | Reaction | log A | n | $E_a$ | # | Reaction | log A | n | $E_a$ |
|---|---|---|---|---|---|---|---|---|---|
| 1 | $CH_4 + M = CH_3 + H + M$ | 17.30 | 0 | 88.40 | 34 | $O + H + M = OH + M$ | 15.90 | 0 | 0.00 |
| 2 | $CH_4 + H = CH_3 + H_2$ | 14.10 | 0 | 11.90 | 35 | $O + O + M = O_2 + M$ | 17.11 | -0.75 | 0.00 |
| 3 | $CH_4 + OH = CH_3 + H_2O$ | 12.50 | 0 | 3.77 | 36 | $H_2 + M = H + H + M$ | 15.49 | 0 | 104.00 |
| 4 | $CH_4 + O = CH_3 + OH$ | 13.30 | 0 | 9.20 | 37 | $O_2 + H_2 = OH + OH$ | 14.90 | 0 | 45.00 |
| 5 | $CH_4 + HO_2 = CH_3 + H_2O_2$ | 12.70 | 0 | 12.50 | 38 | $H_2O + M = H + OH + M$ | 21.69 | -1.0 | 123.09 |
| 6 | $CH_3 + OH = CH_2O + H_2$ | 12.60 | 0 | 0.00 | 39 | $H + O_2 + M = HO_2 + M$ | 15.22 | 0 | -1.00 |
| 7 | $CH_3 + O = CH_2O + H$ | 14.11 | 0 | 2.00 | 40 | $O + OH + M = HO_2 + M$ | 17.00 | 0 | 0.00 |
| 8 | $CH_3 + O_2 = CH_3O + O$ | 13.38 | 0 | 29.00 | 41 | $HO_2 + O = O_2 + OH$ | 13.70 | 0 | 1.00 |
| 9 | $CH_3 + CH_2O = CH_4 + HCO$ | 10.00 | 0.5 | 6.00 | 42 | $H + HO_2 = OH + OH$ | 14.40 | 0 | 1.90 |
| 10 | $CH_3 + HCO = CH_4 + CO$ | 11.48 | 0.5 | 0.00 | 43 | $H + HO_2 = H_2 + O_2$ | 13.40 | 0 | 0.70 |
| 11 | $CH_3 + HO_2 = CH_4 + O_2$ | 11.48 | 0.5 | 6.00 | 44 | $OH + HO_2 = H_2O + O_2$ | 13.70 | 0 | 1.00 |
| 12 | $CH_3 + HO_2 = CH_3O + OH$ | 13.30 | 0 | 0.00 | 45 | $HO_2 + HO_2 = H_2O_2 + O_2$ | 12.30 | 0 | 0.00 |
| 13 | $CH_3O + M = CH_2O + H + M$ | 13.70 | 0 | 21.00 | 46 | $H_2O_2 + M = OH + OH + M$ | 17.08 | 0 | 45.80 |
| 14 | $CH_3O + O_2 = CH_2O + HO_2$ | 12.00 | 0 | 0.60 | 47 | $H_2O_2 + H = HO_2 + H_2$ | 12.23 | 0 | 3.80 |
| 15 | $CH_2O + M = HCO + H + M$ | 16.70 | 0 | 72.00 | 48 | $H_2O_2 + OH = H_2O + HO_2$ | 13.00 | 0 | 1.80 |
| 16 | $CH_2O + OH = HCO + H_2O$ | 14.73 | 0 | 6.30 | 49 | $CH_3 + CH_3 = C_2H_6$ | 13.48 | 0 | 0.00 |
| 17 | $CH_2O + H = HCO + H_2$ | 13.13 | 0 | 3.76 | 50 | $C_2H_6 + CH_3 = C_2H_5 + CH_4$ | 11.54 | 0 | 11.30 |
| 18 | $CH_2O + O = HCO + OH$ | 13.70 | 0 | 4.60 | 51 | $C_2H_6 + H = C_2H_5 + H_2$ | 14.12 | 0 | 9.68 |
| 19 | $CH_2O + HO_2 = HCO + H_2O_2$ | 12.00 | 0 | 8.00 | 52 | $C_2H_6 + OH = C_2H_5 + H_2O$ | 13.05 | 0 | 2.45 |
| 20 | $HCO + OH = CO + H_2O$ | 14.00 | 0 | 0.00 | 53 | $C_2H_6 + O = C_2H_5 + OH$ | 13.30 | 0 | 0.00 |
| 21 | $HCO + H = CO + H_2$ | 14.30 | 0 | 0.00 | 54 | $C_2H_5 = C_2H_4 + H$ | 13.58 | 0 | 38.00 |
| 22 | $HCO + O = CO + OH$ | 14.00 | 0 | 0.00 | 55 | $C_2H_5 + H = C_2H_6$ | 13.50 | 0 | 0.00 |
| 23 | $HCO + HO_2 = CH_2O + O_2$ | 14.00 | 0 | 3.00 | 56 | $C_2H_5 + O_2 = C_2H_4 + HO_2$ | 12.00 | 0 | 5.00 |
| 24 | $HCO + O_2 = CO + HO_2$ | 12.52 | 0 | 7.00 | 57 | $C_2H_4 + O = CH_3 + HCO$ | 12.91 | 0 | 1.94 |
| 25 | $HCO + M = H + CO + M$ | 14.16 | 0 | 19.00 | 58 | $C_2H_4 + OH = CH_3 + CH_2O$ | 13.00 | 0 | 0.96 |
| 26 | $CO + OH = CO_2 + H$ | 7.18 | 1.3 | -0.77 | 59 | $O + NO = N + O_2$ | 9.37 | 1.0 | 38.64 |
| 27 | $CO + O + M = CO_2 + M$ | 15.77 | 0 | 4.09 | 60 | $O + N_2 = N + NO$ | 14.15 | 0 | 75.20 |
| 28 | $CO + O_2 = CO_2 + O$ | 12.33 | 0 | 47.97 | 61 | $NO + M = N + O + M$ | 20.60 | -1.5 | 151.00 |
| 29 | $CO + HO_2 = CO_2 + OH$ | 14.00 | 0 | 23.00 | 62 | $N_2O + M = N_2 + O + M$ | 14.70 | 0 | 58.00 |
| 30 | $H + O_2 = O + OH$ | 14.34 | 0 | 16.79 | 63 | $NO + NO = N_2O + O$ | 14.00 | 0 | 76.00 |
| 31 | $H_2 + O = H + OH$ | 10.26 | 1.0 | 8.90 | 64 | $NO_2 + M = NO + O + M$ | 16.00 | 0 | 65.00 |
| 32 | $H_2O + O = OH + OH$ | 13.83 | 0 | 18.35 | 65 | $NO + O_2 = NO_2 + O$ | 12.00 | 0.5 | 45.50 |
| 33 | $H_2O + H = OH + H_2$ | 13.98 | 0 | 20.30 | 66 | $N + OH = NO + H$ | 13.60 | 0 | 0.00 |

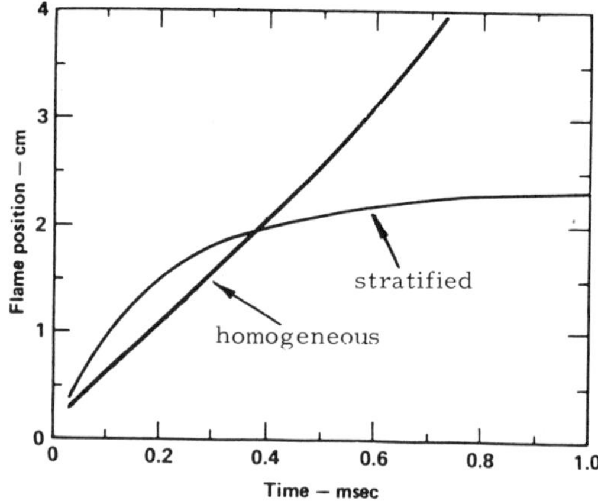

Figure 25. *Flame Position as a Function of Time.*

Specific heat data for each of the chemical species as a function of temperature were fit to a fifth order polynomial. The general equations were solved using an implicit, coupled finite difference scheme developed specifically for combustion applications (100,101). Reflective boundary conditions were assumed at both walls resulting in zero mass flux and heat flux through the walls. Turbulent diffusivity was assumed to be constant (500 cm2/s) for simplicity even though, in the actual engine environment, turbulence intensity is a complex function of both space and time, dependent on the intake, compression, and flame propagation processes.

In the stratified charge case, the initial fuel distribution consisted of an air fuel mixture with an equivalence ratio of 1.2 extending from $x = 0$ to $x = 1.7$ with no fuel beyond $x = 1.7$. Resulting computation of equivalence ratio as a function of position for various times is shown in Figure 25.

Flame was assumed at a position where the computed temperature exceeded 1500 K. The position of the flame as a function of time for both the homogeneous case and the stratified charge case is shown in Figure 25. With stratified charge, the flame is propagating into an exceedingly lean mixture after about 2.0 ms and as a consequence, flame speed (shown

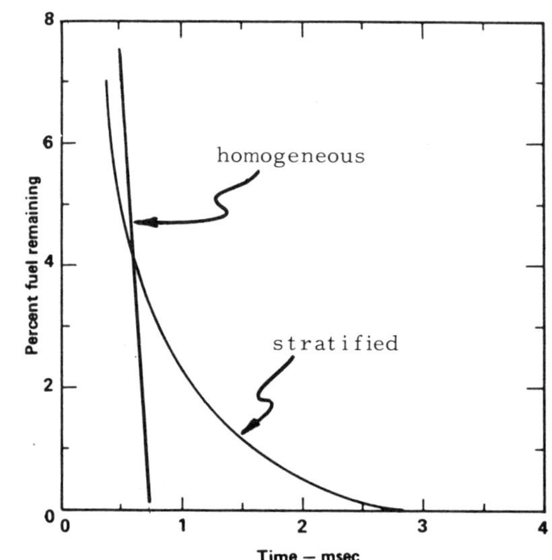

Figure 26. *Percent Fuel Remaining as a Function of Time.*

by the slope of the curve) drops off rapidly. This is an example of mid-volume quenching where the flame is gradually quenched as the fuel-air mixture becomes progressively more lean.

The amount of fuel remaining also gives a clue to the phenomena of volume flame quenching. Percent fuel remaining as a function of time is shown in Figure 26 for both the homogeneous case and the stratified charge case.

In the homogeneous case, all fuel is consumed before 0.8 ms. In the stratified charge case, a good portion of the fuel remains unburned after 1 ms and is not completely consumed until after 3 ms. This difference leads to the basic problem that with stratified charge, there may be insufficient time in a practical environment to permit the complete consumption of the fuel before the partially burned mixture is exhausted (leading to light hydrocarbon emissions).

Several points can be inferred from Westbrook's calculations. First, a simplified single step mechanism can give quite reasonable results for flame speed and energy release rates in a conventional environment. In marginal cases however, such as accurate heat release predictions or wall quenching problems, single step mechanisms should not be used unless corrections are applied. Predicted fuel consumption as a function of position is not entirely accurate because of the assumption that fuel consumption and energy release occur simultaneously. The model did, however, predict a reasonable trend. Finally, although turbulent diffusivity was held constant, the calculated trends suggest that the form of turbulent diffusivity may not be particularly important.

Results of the computations indicate that, with the stratified charge model assumed in this example, the fuel can be kept away from the outer wall of the combustion chamber to avoid wall quenching phenomena. Unfortunately it also appears that so much time is required with reasonable diffusivity (turbulence level) to combust the fuel, a good portion of the fuel would end up in the exhaust in a practical engine where only limited time is available for the event. It must be noted to the reader that this model is very simplistic in nature and does not consider the effects of air swirl, spray geometry or any other mechanisms which have a profound effect on fuel confinement and movement.

*Example of a Two Dimensional Model (Los Alamos Code)*

In recent years, highly developed computer codes have become available that are applicable for engine modeling work primarily from Los Alamos Scientific Laboratories. The RICE (95), CONCHAS (96), CONCHAS SPRAY (97) and KIVA (98) codes were developed to handle arbitrary multicomponent reacting flows two dimensional geometries and are espe-

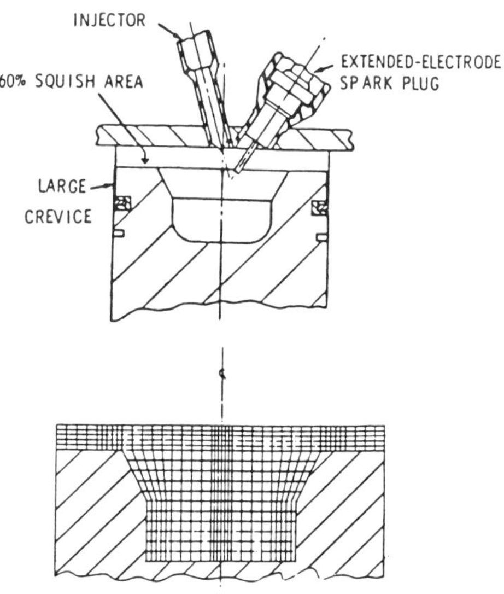


*Figure 27. Grid Cross Section Applied to an Early Injection Engine by Diwalker (102).*

cially applicable to stratified charge engine modeling research. These sophisticated codes have been used by many different researchers to examine, for example, piston/cylinder geometry and in-cylinder airflow on combustion. These codes have, to a certain extent, freed the researcher from developing in-house codes, and allowed him/her to focus on computational results. It should be mentioned however, it is wise for the researcher to develop a working knowledge of these codes before blindly applying them to the modeling situation.

R. Diwalker (102) applied the CONCHAS code to a direct injection stratified charge engine and compared the results to measurements obtained from engine sampling experiments by Landcaster (83). Specifically, CONCHAS is a time-marching code that solves the finite difference approximations to the partial differential equations governing the conservation of mass, momentum (Navier-Stokes) energy and chemical species in a two-dimensional axi-symmetric chamber with angular, solid-body swirl (the third dimension) and with a moving boundary (the piston). Provisions in the CONCHAS code are made for a gaseous fuel spray. (An appropriate fuel spray submodel was later added to the CONCHAS SPRAY code allowing a liquid fuel spray.) Diwalker developed the grid pattern shown in Figure 27.

The off-axis ignition simulated by the code is actually an ignition ring because of the axi-symmetric assumption. This assumption may not be far off because of the duration of the ignition source and, in most cases, the presence of air swirl.

Landcaster (103) had experimentally found that the principal source of hydrocarbon emissions was in the peripheral portion of the central fuel cloud due to a wave of unburned hydrocarbon fuel traveling outward into the squish volume at the start of the expansion stroke. In modeling work with CONCHAS, Diwalker determined the squish volume unburned fuel mass at selected time steps and compared the computation to measured values. Although absolute values were not directly comparable, trends between computations and experiment were in agreement. After the data was normalized to the MBT (max. brake torque) spark setting, values were comparable. Results are shown in Figure 28.

Diwalker concluded that, in spite of the many assumptions, qualitative trends of HC and NO emissions calculated from the model agreed with that from experiment when spark timing was varied at constant speed, injection timing and overall equivalence ratio. Fuel burned more rapidly in the model (higher absolute concentrations of NO and lower absolute concentrations of hydrocarbons) due to direct consequences of some of the assumptions made in the model. Finally, the model supported Landcaster's postulation that exhaust HC emissions result from bulk quenching of a lean-mixture wave eminating from the piston bowl and traveling outward into the squish volume during expansion.

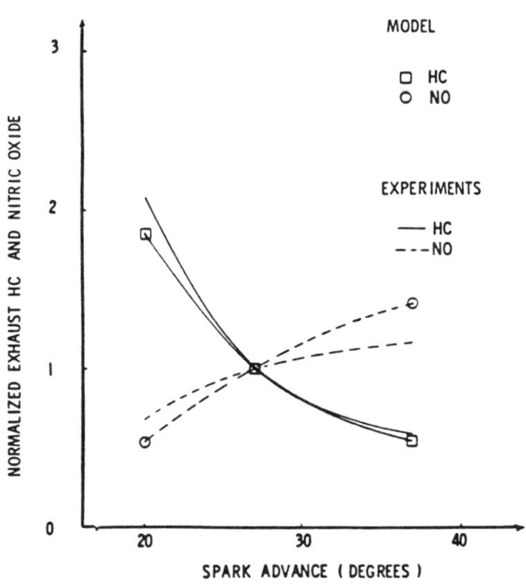


*Figure 28. Normalized HC and NO Emission Comparison (102).*

## Summary

The evolution and development of the stratified charge concept has paralleled that of the conventional homogeneous charge almost from the inception of the internal combustion engine. Nicholas Otto proclaimed that stratification of the cylinder charge helped ease the perceived problem of piston shock loading. Later, many other investigators looked upon stratification as a possible solution to the phenomena of engine knock which evolved to become a major factor limiting the power output of high performance engines. Excitement once again surfaced with the introduction of environmental standards in the United States and later throughout the world. Many innovative designs have surfaced and some have survived the rigorous test of time. In most instances, stratified charge concepts did not leave the prototype stage primarily due to a lack of detailed development and because of successful improvements in conventional gasoline engines and exhaust aftertreatment devices. Interest in stratified charge engines continued during the energy crisis as utilization of poorer quality fuels and alternate fuels became an active area of research. The military was especially interested in this capability. Despite today's successful homogeneous charge engines and an oil surplus (temporary at most) interest remains at engine laboratories throughout the country. Novel concepts will still emerge from time to time but perhaps it may safely be said that experimental 'trial and error' designs introduced prior to and during the sixties have been replaced with sophisticated engine analysis techniques and computer modeling.

Theoretically, the efficiency of the stratified charge engine is equal to the diesel engine with the added capability of combusting broad-cut low cetane fuels. Low emission levels of carbon monoxide and oxides of nitrogen are strong advantages. Poor performance at light loads accompanied by high hydrocarbon emissions and lower overall efficiencies are distinct disadvantages but with the recent developments with pilot injection (as with the rotary engine) and axial stratification, this problematic area seems to be improving. The difficulty of hydrocarbon emissions appears to be due to the interface region between the combusting mixture and working gas (usually air) where the equivalence ratio moves beyond the lean limit of combustion. Rather than static, the interface region is dynamic and is influenced by inlet aerodynamics, piston movement and combustion induced air movement. Sophisticated analysis techniques are needed to design around this complex phenomena.

The stratified charge engine concept is clearly a logical step in the evolutionary process of powerplant development as engine design moves toward the solution of a high efficiency, low emission, multifuel engine. The concept of controlling combustion (and hence emissions, multifuel capability, etc.) 'in situ' in the chamber is, to say the least, academically pleasing. Reduction or elimination of exhaust aftertreatment devices with the added capability of multifuel utilization is most certainly a worthwhile goal. Advances in electronic control of fuel injection, new ignition techniques and use of ceramics in engine design will significantly contribute to the development of the stratified charge concept.

## Acknowledgements

I would like to thank the many individuals who assisted me with information for the preparation of the manuscript of this review paper. Special thanks goes to Karen Brown who typeset the manuscript and to my graduate students for their comments and suggestions.

## References

1. Cummins. C. Lyle Jr., Internal Fire, Carnot Press, Oregon, 1975.

2. Stratified Charge Engines, Ricardo News, 1975, No. 4.

3. Voinoff, A. N., "Combustion Processes in High-Speed Piston- Engines," (Russ.), Moscow, 1965.

4. Gruden, D., U. Markovac and H. Lorcher, "Development of the Porsche SKS Engine," IMechE, C243/76, pp. 21-28, 1976.

5. Eberan-Eberhorst, R., D. Gruden and G. Schwarzbauer, "Statistische Analyse der Schwankungen von Arbeitszyklen im Ottomotor," Automobil-Industrie, 1973, 4.

6. Schwarzbauer, G., "Verbrennungsablauf und Stickoxidbildung im Ottor," MTZ 34, 1973, 3.

7. Schwarzbauer, G. and D. Gruden, "Die Stickoxidbidlung im Ottomotor," FISITA-Krongreb, London, 1972.

8. Gruden, D., Th. F. Brachert and W. Wurster, "Porsche One, Six and Eight Cylinder Stratified Charge Engine with Divided Combustion Chamber (SKS-Engine)," IMechE, C396/80, pp. 55-62, 1980.

9. Newhall, H. K. and El-Messiri, "A Combustion Chamber Concept for Control of Engine Exhaust Air Pollutant Emissions," Combustion and Flame, Vol. 14, No. 1, 1970.

10. Date, T., S. Yagi, A. Ishizuya and I. Fujii, "Research and Development of the Honda CVCC-Engine," SAE Paper 740605, 1974.

11. Decker, G. and W. Brandstatter, "Results from the VW-Stratified- Charge-Process," MTZ 34, 1973, No. 10.

12. Gussak, L. A., "High Chemical Activity of Incomplete Combustion Products and a Method of Pre-chamber Torch Ignition for Avalanche Activation of Combustion in Internal Combustion Engines," SAE Paper No. 750890, 1975.

13. Rhodes, K. H., "Project Stratofire," SAE Paper 660094, 1966.

14. Bracco F.V., "Stratified Charge Engine," Collection of papers published in Combustion Science and Technology. (Volume 8. Nos. 1 and 2) 1973.

15. Lewis, J.M., "United Parcel Service Applies Texaco Stratified Charge Engine Technology to Power Parcel Delivery Vans, Progress Report," SAE Transaction Vol. 89, SAE Paper No. 801429, 1980.

16. Lewis, J. M., "UPS Multifuel Stratified Charge Engine Development Program - Field Test," SAE Paper No. 860067, 1986.

17. General Motors Public Interest Reports 1975-1986, General Motors Corporation, Warren, Michigan.

18. Jones, C., "A Progress Report on Curtiss-Wright's Rotary Stratified Charge Engine Development," SAE Paper No. 741206, 1974.

19. Jones, C., "Advanced Development of Rotary Stratified Charge 750 and 1500 hp Military Multi-fuel Engines at Curtiss-Wright," SAE paper No. 840460, 1984.

20. Jones, C., "An Update of Applicable Automotive Engine Rotary Stratified Charge Development," SAE Paper No. 820347, International Congress and Exposition, Detroit, MI, February 1982.

21. Jones, C., J.R. Mack, and M.J. Griffith, "Advanced Rotary Engine Developments for Naval Applications," SAE Paper No. 851243, 1985.

22. Shih, T. I-P., S.L. Yang, and H.J. Schock, "A Two-Dimensional Numerical Study of the Flow Inside the Combustion Chamber of a Motored Rotary Engine," SAE Paper No. 860615, 1986.

23. Schock, Harold J., "A Review of Internal Combustion Engine Combustion Chamber Process Studies at NASA Lewis Research Center," Paper No. AIAA-84-1316.

24. Peters, B. D. and A. A. Quader, "Wetting the Appetite of Spark Ignition Engines for Lean Combustion," General Motors Research Laboratories Research Publication GMR-2586R, F&L-645, February 27, 1978.

25. Quader, A.A., "The Axially-Stratified-Engine," SAE Transaction Vol 91, SAE Paper No. 82131, 1982.

26. Peters, B.D., "Fuel Droplets Inside the Cylinder of a Spark Ignition Engine with Axial Stratification," SAE Transaction Vol. 91, SAE Paper No. 820132, 1982.

27. U.S. Environmental Protection Agency, " A Study of Stratified Charge for Light Duty Power Plants," EPA-460/3-74-011A, 1975.

28. Baranescu, G. S., "Some characteristics of Spark Assisted Direct Injection Engine," SAE Paper No. 830589, 1983.

29. Haslett, R.A., M.L. Monaghan, and J.J. McFadden, "Stratified Charge Engines," SAE Paper No. 760755, 1976.

30. Uyehara, O. A., P. S. Myers, E. E. Marsh and G. E. Cheklich, "A Classification of Reciprocating Engine Combustion Systems," SAE Paper 741156, 1974.

31. Choma, M. A., P. H. Havstad, A. O. Simko and W. F. Stockhausen, "Fuel Tolerance Tests with the Ford PROCO Engine," IMechE, C398/80, pp. 41-48, 1980.

32. Barber E.M., Reynolds B. and Tierney W., "The Elimination of Combustion Knock - Texaco Combustion Process," SAE preprint 473. June 1950.

33. Davis C.W., Barber E.M. and Mitchell E., "Fuel Injection and Positive Ignition: A Basis for Improved Efficiency and Economy," SAE preprint 190A. June 1960.

34. Barber E.M., Reynolds B. and Tierney W.T., "Texaco Combustion Process Gives Knock-Free Operation," SAE Journal, September 1950.

35. Tierney W.T., Mitchell E. and Alperstein M., "The Texaco Controlled Combustion System. A Stratified Charge Engine Concept Review and Current Status," I. Mech.E. London Conference "Power plants and future fuels," January 1975.

36. Mitchell, E., J.M. Cobb, and R.A. Frost, "Design and Evaluation of a Stratified Charge Multifuel Military Engine," SAE Transaction Vol. 77, SAE Paper No. 680042, 1968.

37. Mitchell, E., M. Alperstein, J.M. Cobb, and C.H. Faist, "Stratified Charge Multifuel Military Engine - A Progress Report," SAE Paper No. 720051, 1972.

38. Alperstein, M., G.H. Schafer, and F.J. Villforth, III, "Texaco's Stratified Charge Engine - Multifuel, Efficient, Clean, and Practical," SAE Paper No. 740563, 1974.

39. Witze, P. O., "Influence of Air Motion on the Performance of a Direct-Injection Stratified-Charge Engine," Stratified Charge Automotive Engines Conference, London, 25-26 November, 1980.

40. Witze, P.O., "Influence of Air Motion Variation on the Performance of a Direct Injection Stratified Charge Engine," Sandia Laboratories Report No, SAND79-8756, September 1980.

41. Jain, B.C., Rife, J.M., and Keck, J.C., "A Performance Model for the Texaco Controlled Combustion Stratified Charge Engine," SAE Transactions Vol. 85 (1976), SAE Paper No. 760116, Automotive Engineering Congress, February 1976.

42. Finsterwalder, G., "A New Deutz Multi-fueled System," SAE Paper No. 720103.

43. Jones, C., H.D. Lamping, D.M. Myers, and R.W. Loyd, "An Update of the Direct Injected Stratified Charge Rotary Combustion Engine Developments at Curtiss-Wright," SAE Transaction Vol. 86, SAE Paper No. 770044, 1977.

44. Obert, E. F., Internal Combustion Engines.

45. Purins, E.A., "Pre-chamber Stratified Charge Engine Combustion Studies," SAE Transaction Vol. 83, SAE Paper No. 741159, 1974.

46. Sakai, Y., K. Kunii, S. Tsutsumi, and Y. Nakagawa, "Combustion Characteristics of the Torch Ignited Engine," SAE Transaction Vol. 83, SAE Paper No. 741167, 1974.

47. Davis, G.C., R.B. Krieger, and R.J. Tabaczynski, "Analysis of the Flow and Combustion Processes of a Three-valve Stratified Charge Engine with a Small Prechamber," SAE Transaction Vol. 83, SAE Paper No. 741170, 1974.

48. Date, T., and S. Yagi, "Research and Development of Honda CVCC Engine," SAE Paper No. 740605, 1974.

49. Yagi S., Date T., and Inoue K., "NOx Emission and Fuel Economy of the Honda CVCC Engine," SAE 741158.

50. Peters, B. D., "Fuel Droplets Inside the Cylinder of a Spark Ignition Engine with Axial Stratification," SAE Paper 820132, 1982.

53. Chang, M., M.P. Nolan, J.H. Rillings, and A.A. Quader, "The Axially Stratified-charge Engine: Control, Calibration, and Vehicle Implementation," SAE Paper No. 851674, 1985.

54. Jones, C., "A Progress Report on Curtiss-Wright's Rotary Stratified Charge Engine Development," SAE Paper No. 741206, Presented at SAE International Stratified Charge Engine Conference, October, 1974.

55. Jones, C. and H. Lamping, "Curtiss-Wright's Development Status of the Stratified Charge Rotating Combustion Engine," SAE Transactions 1971, Vol. 80, Paper No. 710582.

56. Lloyd, R. W., "Curtiss-Wright Stratified Charge Rotary Combustion Engine Development," Combustion Science and Technology, 1976, Vol. 12.

57. Bentele, M., "Curtiss-Wright's Developments on Rotating Combustion Engine," SAE Transactions, Vol. 69 (1961), Paper 288-B.

58. Jones, C., "The Curtiss-Wright Rotating Combustion Engines Today," SAE Transactions 1965, Vol. 73, Paper No. 886d.

59. Jones, C., "New Rotating Combustion Powerplant Development," SAE Transactions 1966, Vol. 74, Paper No. 640723.

60. Jones, C., "The Rotating Combustion Engine - Compact, Lightweight Power for Aircraft," SAE Transactions 1966, Vol. 76, Paper No. 670194.

61. Cole, D. E. and C. Jones, "Reduction of Emissions from the Curtiss-Wright Rotating Combustion Engine with an Exhaust Reactor," SAE Transactions 1970, Vol. 79, Paper No. 70074.

62. Bracco F.V., "Theoretical Analysis of Stratified, Two-Phase Wankel Engine Combustion," Combustion Science and Technology, 1973, Vol. 8p. 69-84.

63. Willis, E. A., "Development Potential of Intermittent Combustion (IC) Aircraft Engines for Commuter Transport Applications," SAE Paper No. 820718, 1982.

64. Bracco, F.V., "Modeling of Engine Sprays," SAE Paper No. 850394, 1985.

65. Personal communication with representatives at John Deere International Technologies.

66. Wong, V.W., J.M. Rife, and M.K. Martin, "Experiments in Stratified Combustion with a Rapid Compression Machine," SAE Transaction Vol. 87, SAE Paper No. 780638.

67. Wong, V. W., "A Photographic Performance Study of Stratified Combustion Using a Rapid Compression Machine," Thesis, MIT, June, 1976 (see also reference 66).

68. Evers, L.W., R.D. Fleming, and R.W. Hurn, "Efficiency and Emissions of a Stratified Charge Engine Optimized for Various Fuels," SAE Paper No. 780236, 1978.

69. Baranescu, G. S. (see reference no. 28).

70. Chmela, F. G., "High Compression Stratified Charge Engines and Their Suitability for Conventional and Alternative Fuels," Conference on Stratified Charge Automotive Engines, Londonw, Nov. 1980, Paper C400/80.

71. Neitz, A. and F. Chmela, "M.A.N. FM Process to Enable Diesel Engines to Burn Methanol," First International Automotive Fuel Economy Research Conference, Arlington, Virginia, USA, Oct. 31 - Nov. 2, 1979.

72. Neitz, A. and F. Chmela, "Results of MAN FM Diesel Engines Operating on Straight Alcohol Fuels," IV International Symposium of Alcohol Fuels Technology, Guaruja-SP, Brazil, Oct. 5 - Oct. 8, 1980.

73. Nietz, A. and F. Chmela, "The M.A.N. Methanol Engine Powering City Buses," Fifth International Alcohol Fuel Technology Symposium, Auckland, New Zealand, May 13-18, 1982, Paper C2-23.

74. Needham, J. R., S. R. Norris-Jones and B. M. Cooper, "An Evaluation of Unthrottled Combustion System Options for Future Fuels," SAE Technical Paper No. 830374, 1983.

75. Heldt, P. M., "High-Speed Diesel Engines," P.M. Heldt, Nyack, NY, 1944.

76. Lange, K. and W. Spindler, "Investigation of Local Mixture Strength and Flame Propagation with Aided and Unaided Ignition," Proc. Instn. Mech. Engrs., Vol. 184, Paper 11, pp. 109-121, 1969-70.

77. Phatak, R. G. and K. Komiyama, "Investigation of a Spark-Assisted Diesel Engine," SAE Paper 830588, 1983.

78. Borman, G., J. Cramer, H. Y. Wang, R. Sowls and P. Myers, "An Investigation of the Combustion Characteristics and Performance of a Spark-Ignited Diesel Using Alternative Fuels," U.S. Army TACOM Report No. DAAE-07-82-C-4057, 1985.

79. Abata, D. L., S. Fritz, and B. Stroia, "A Photographic Study of Low Cetane Fuels in a Diesel Engine With Spark Assist," SAE Paper No. 860066, 1986.

80. Giovanetti, A. J., J. A. Ekchian, E. F. Fort and J. B. Heywood, "Analysis of Hydrocarbon Mechanisms in a Direct Injection Spark-Ignition Engine," SAE Paper No. 830587, SAE Transactions Vol. 2, p. 925 (1983).

81. Balles, E.N., J.A. Ekchain, and J.B. Heywood, "Fuel Injection Characteristics and Combustion Behavior of a Direct-injection Stratified-charge Engine," SAE Transaction Vol. 93, SAE Paper No. 841379, 1984.

82. Wood, C.D., "Unthrottled Open-chamber Stratified Charge Engines," SAE Paper No. 780341, 1978.

83. Lancaster, D. R., "Diagnostic Investigation of Hydrocarbon Emissions from a Direct-Injection Stratified-Charge Engine with Early Injection," General Motors Research Laboratories Report No. GMR-3344. November 1980.

84. Johnston, S. C., "Precombustion Fuel/Air Distribution in a Stratified Charge Engine Using Laser Raman Spectroscopy," SAE Paper No. 790433, Congress and Exposition, Detroit, MI, February 1979.

85. Sinnamon, J. F., Lancaster, D. R., and Stiener, J. C., "An Experimental and Analytical Study of Engine Fuel Spray Trajectories," SAE Paper No. 800135, Congress and Exposition, Detroit, MI, February 1980.

86. Witze, P. O., "Influence of Air Motion Variation on the Performance of a Direct Injection Stratified Charge Engine," Sandia Laboratories Report No. SAND79-8756, September 1980.

87. Foster, D., "An Overview of Zero Dimensional Thermodynamic Models for IC Engine Data Analysis," SAE Paper No. 852070, (1970).

88. Poulos, S. G. and Heywood, J. B., "The Effect of Chamber Geometry on Spark-Ignition Engine Combustion," SAE Paper 830334, 1983.

89. Bracco, F. V., editor, "Stratified Charge Engines," Combustion Science and Technology, Vol. 8, Nos. 1 and 2, (October 1973) pp. 1-100.

90. Sirignano, W. A. "One-Dimensional Analysis of Combustion in a Spark-Ignition Engine," Combustion Science and Technology, Vol. 7, No. 3 (May 1973), pp. 99-108.

91. Rosentweig-Bellan, J., "A Theory of Turbulent Combustion and Nitric Oxide Formation for Dual-Carbureted Stratified-Charge Engines," Ph.D. Thesis No. 1172-T, Dept. of Aerospace and Mechanical Sciences, Princeton University, Princeton, N.J., August 1974.

92. Bracco, F. V. and W. A. Sirignano, "Theoretical Analysis of Wankel Engine Combustion," Combustion Science and Technology Vol. 7, No. 3, pp. 109-123, May 1973. Originally presented at the 7th IECE Conference in Sept. 1972.

93. Bracco, F. V., "Theoretical Analysis of Stratified, Two-Phase Wankel Engine Combustion," Combustion Science and Technology, Vol. 8, Nos. 1 and 2 (October 1973), pp. 69-84.

94. Bracco, F.V., "Introducing a New Generation of More Detailed and Informative Combustion Models," SAE Paper No. 741174, 1974.

95. Rivard, W. C., O. A. Farmer, and T. D. Butler, "RICE: A Computer Program for Multi-Component Chemically Reactive Flows at All Speeds," Los Alamos Scientific Laboratories Report LA-5812, March 1975.

96. Butler, T. D., L. D. Cloutman, J. K. Dukowicz, and J. C. Ramshaw, "CONCHAS: An Arbitrary Lagrangian-Eulerian Computer Code for Multi-component Chemically Reactive Fluid Flow at All Speeds," Los Alamos Scientific Laboratories Report LA-8129-MS, November 1979.

97. Cloutman, L. D., J. K. Dukowicz, J. D. Ramshaw, and A. A. Amsden, "CONCHAS-Spray: A Computer Code for Reactive Flows with Fuel Sprays."

98. Los Alamos National Labs, Los Alamos, NM 87545, "KIVA - A Computer Program for Two and Three Dimensional Fluid Flows with Chemical Reactions and Fuel Sprays," Reprint LA-10245-MS (February 1985), update LA-10534-MS (October 1985).

99. Westbrook, C.K., "Fuel Motion and Pollutant Formation in Stratified Charge Combustion," SAE Paper No. 790248, 1979.

100. Westbrook, C, K., "A Generalized ICE Method for Chemically Reactive Flows in Combustion Systems," J. Computational Physics.

101. Westbrook, C. K. and L. L. Chase, "A One-Dimensional Combustion Model," University of California Lawrence Livermore Laboratory report UCRL-52297, July 1977.

102. Diwalkar, R., "Multidimensional Modeling Applied to the Direct- Injection Stratified-Charge Engine-Calculation Versus Experiment," SAE Paper No. 810225, 1981.

103. Tauschek, M. J., "Spark-ignition Engine of Tomorrow," SAE Paper No.650478 (SP-270), 1965.

104. Gay, E. J., "Commercial and Industrial Power-Plants-Future Prospects and Applications," SAE Paper NO.650631, 1965.

105. Witzky, J. E. and J. M. Clark, Jr., "Study of the Swirl Stratified Combustion Principle," SAE Transaction Vol. 75 SAE Paper No. 660092, 1966.

106. Willis, D. A., W. E. Meyer and C. Birnie, Jr., "Mapping of Air Flow Patterns in Engines with Induction Swirl," SAE Transaction Vol. 75, Paper No. 660093, 1966.

107. Rhodes, K. H., "Project Stratofire- Development of a Stratified Charge Combustion System for Automotive Engines," SAE Paper No. 660094, 1966.

108. Bascunana, J. L. and L. D. Conta, "Conversion of Propane-burning Stratified Charge Combustion System for Automotive Engines," SAE Transaction Vol. 75, SAE Paper No. 660095, 1966.

109. Adams, W. E. and R. V. Kerley, "Next Decade for Piston Engines, with Particular Reference to Air Pollution Problems," SAE Transaction Vol. 76, SAE Paper No. 670685, 1967.

110. Bishop, I. N. and A. Simko, "Ford Stratified-charge Engine has Fuel-injection System and Other Features That Provide High Thermal Efficiency Using Gasoline," SAE Transaction Vol. 77, SAE Paper NO. 680041, 1968.

111. Bolt, J. A., "Air Pollution and Future Automotive Powerplants," SAE Paper No. 680191, 1968.

112. Yui, S. and S. Ohnishi, "New Concept of Stratified-charge Two-stroke Engine Combustion Process," SAE Paper No. 690468, 1969.

113. Simko, A., M. A. Choma, and L. L. Repko, "Exhaust Emission Control by the Ford Programmed Combustion Process-PROCO," SAE Transaction Vol. 82, SAE Paper No. 720052, 1972.

114. Miyake, M., "Developing a New Stratified-charge Combustion System With Fuel Injection for Reducing Exhaust Emissions in Small Farm and Industrial Engines," SAE Transaction Vol. 81, SAE Paper No. 720196, 1972.

115. Jones, C., "Survey of Curtiss-Wright's 1958-1971 Rotating Combustion Engine Technological Developments," SAE Paper No. 720468, 1972.

116. Emery, D. B., "An Evaluation of the Performance and Emissions of a CFR Engine Equipped with a Prechamber," SAE Transaction Vol. 82, SAE Paper No. 730474, 1973.

117. Peart, J. R., and T. V. Huber, "Urban Vehicle Design Competition - A Practice in Design," SAE Paper No. 730508, 1973.

118. Jessel, A. J., O. A. Uyehara, and P. S. Myers, "A Spark Ignition Engine with an In-cylinder Thermal Reactor," SAE Transaction Vol. 82, SAE Paper No 730634, 1973.

119. Johnson, P. R., S. L. Genslak, and R. C. Nicholson, "Vehicle Emission Systems Utilizing a Stratified Charge Engine," SAE Paper No. 741157, 1974.

120. Varde, K. S., and M. J. Lubin, "The Roll of Connecting Nozzle and the Flame Initiation Point in the Performance of a Dual Chamber Stratified Charge Engine," SAE Paper No. 741161, 1974.

121. Pischinger, F. F. and K. J. Klocher, "Single-cylinder Study of Stratified Charge Process with Prechamber-injection," SAE Transaction Vol. 83, SAE Paper No. 741162, 1974.

122. Tabaczynski, R. J. and E. D. Klomp, "Calculated Nitric Oxide Emissions of an Unthrottled Spark Ignited, Stratified Charge Internal Combustion Engine," SAE Paper No. 741171, 1974.

123. Evers, L. W., P. S. Myers, and O. A. Uyehara, "A Search for a Low Nitric Oxide Engine," SAE Paper No. 741172, 1974.

124. Brandstetter, W. R., G. Decker, H. J. Schafer, and D. Steinke, The Volkswagen PCI Stratified Charge Concept. Results from the 1.6 Liter Air Cooled Engine," SAE Paper No. 741173, 1974.

125. Watfa, M., D. E. Fuller, and H. Daneshyar, "The Effects of Charge Stratification on Nitric Oxide Emission from Spark Ignition Engines," SAE Paper No. 741175, 1974.

126. Hurter, D. A. and W. D. Lee, "A Study of Technological Improvements in Automobile Fuel Consumption," SAE Paper No. 750005, 1975.

127. Ferguson, C. R., G. A. Danieli, J. B. Heywood, and J. C. Keck, "Time Resolved Measurements of Exhaust Composition and Flow Rate in a Wankel Engine," SAE Transaction Vol. 84, SAE Paper No. 750024, 1975.

128. Olikara, C. and G. L. Borman, "A Computer Program for Calculating Properties of Equilibrium Combustion Products with Some Applications to I.C. Engines," SAE Paper No. 750468, 1975.

129. Tierney, W. T., E. M. Johnson, and N. R. Crawford, "Energy Conservation: Optimization of the Vehicle-fuel-refinery System," SAE Transaction Vol. 84, SAE Paper No. 750673, 1975.

130. Brandstetter, W. R., G. Decker, and K. Reichel, "The Water-cooled Volkswagen PCI-stratified Charge Engine," SAE Transaction Vol. 84, SAE Paper No. 750869, 1975.

131. Gruden, D., "Combustion and Exhaust Emission of an Engine Using the Porsche-Stratified-Charge-Chamber-System," SAE Paper No. 750888, 1975.

132. Siewert, R. M. and S. R. Turns, "The Staged Combustion Compound Engine (SCCE); Exhaust Emissions and Fuel Economy Potential," SAE Transaction Vol. 84, SAE Paper No. 750889, 1975.

133. Newhall, H. K., "Combustion Process Fundamentals and Combustion Chamber Design for Low Emissions," SAE Transaction Vol. 84, SAE Paper No. 751001 (SP-396), 1975.

134. Turkish, M. C., "Prechamber and Valve Gear Design for 3-valve Stratified Charge Engines," SAE Transaction Vol. 84, SAE Paper No. 751004 (SP-396), 1975.

135. Hires, S. D., A. Ekchian, J. B. Heywood, R. J. Tabaczynski and J.C. Wall, "Performance and NOx Emissions Modeling of a Jet Ignition Prechamber Stratified Charge Engine," SAE Transaction Vol. 85, SAE Paper No. 760161, 1976.

136. Boekhaus, K. L. and L. C. Copeland, "Performance Characteristics of Stratified Charge Vehicles with Conventional Fuels and Gasoline Blended with Alcohol and Water," SAE Paper No. 760197 (SP-403), 1976.

137. Johnson, R. T., R. K. Riley, and M. D. Dalen, "Performance of Methanol-Gasoline Blends in a Stratified Charge Engine Vehicle," SAE Paper No. 760546, 1976.

138. Heywood, J. B., and R. J. Tabaczyniski, "Current Developments in Spark-ignition Engines," SAE Paper No. 760606 (SP-409), 1976.

139. Ciccarone, A., C. Antonini, and U. Virgilio, "Fuel Consumption in European Passenger Cars Powered by Gasoline, Diesel, and Direct Injection Stratifed Charge Engines," SAE Paper No. 760796, 1976

140. Evers, L. W., P. W. Myers, and O. A. Uyehara, An Experimental Study of the Delayed Mixing Stratified Charge Engine Concept," SAE Paper No. 770042, 1977.

141. Ekchian, A., J. B. Heywood, and J. M. Rife, "Time Resolved Measurements of the Exhaust from a Jet Ignition Prechamber Stratified Charge Engine," SAE Transaction Vol. 86, SAE Paper No. 770043, 1977.

142. Gabele, P. A., J. N. Braddock, and R. L. Bradow, "A Characterization of Exhaust Emissions from Lean Burn, Rotary, and Stratified Charge Engines," SAE Paper No. 770301, 1977.

143. O'Neill, E. B., and D. W. Taylor, "Landing Vehicle Assault (LVA)," SAE Paper No. 770340.

144. Ciccarone, A., "Possible Advances in European Passenger Cars Fuel Economy," SAE Paper No. 770846, 1977.

145. Novak, J. M., "Simulation of the Breathing Processes and Air-fuel Distribution Characteristics of Three-valve, Stratified Charge Engines," SAE Paper No. 770881, 1977.

146. Haselman, L. C., and C. K. Westbrook, "A Theoretical Model for Two-Phase Fuel Injection in Stratified Charge Enginers," SAE Paper No. 780318, 1978.

147. Asanuma, T., M.K.G. Babu, and S. Yagi, "Simulation of Thermodynamic Cycle of Three-valve Stratified Charge Engine," SAE Transaction Vol, 87, SAE Paper No. 780319, 1978.

148. Wall, J. C., J. B. Heywood, and W. A. Woods, "Parametric Studies of Performance and NOx Emissions of the Three-value Stratified Charge Engine Using a Cycle Simulation,: SAE Transaction Vol. 87, SAE Paper No. 780320, 1978.

149. Kerimov, N. A. and R. I. Mektiev, "Engines with Stratified Charge," SAE Paper No. 780342, 1978.

150. Hull, W. L. and S. C. Sorenson, "Research on a Dual-chamber Stratified Charge Engine," SAE Paper No. 780488, 1978.

151. McKee, D. E., F. C. Ferris, and R. E. Goebora, "Unregulated Emissions from a PROCO Engine Powered Vehicle," SAE Paper No. 780592 (SP-431), 1978.

152. Hillyer, B. J. and W. R. Wade, "Single-cylinder PROCO Engine Studies - Fuel and Engine Calibration Effects on Emissions, Fuel Economy and Octane Number Requirements," SAE Paper No. 780593 (SP-431), 1978.

153. Baudino, J. H. and L. C. Copeland, "Atypical Fuel Volatility Effects on Driveability, Emissions, and Fuel Economy of Stratified Charge and Conventionally Powered Vehicles," SAE Paper No. 780610 (SP431), 1978.

154. Bechtold, R. L., "Performance, Emissions, and Fuel Consumption of the White L-163-S Stratified-charge Engine Using Various Fuels," SAE Paper No. 780641.

155. Scussel, J., A. O. Simko, and W. R. Wade, "The Ford PROCO Engine Update," SAE Transaction Vol. 87, SAE Paper No. 780699, 1978.

156. Hayden, A.C.S., "The Effects of Technology on Automobile Fuel Economy Under Canadian Conditions," SAE Paper No. 780935, 1978.

157. Wall, J. C. and J. B. Heywood, "The Influence of Operating Variables and Prechamber Size on Combustion in a Prechamber Stratified-charge Engine," SAE paper 780966, 1978.

158. Johnston, S. C., C. W. Robinson, W. S. Rorke, J. R. Smith, and P. O. Witze, "Application of Laser Diagnostics to an Injected Engine," SAE Transaction Vol. 88, SAE Paper No.790092, 1979.

159. Syed, S. A. and F. V. Bracco, "Further Comparisons of Computed and Measured Divided-chamber Engine Combustion," SAE Paper No. 790247, 1979.

160. Davis, G. C. and J. C. Kent, "Comparison of Model Calculations and Experimental Measurements of the Bulk Cylinder Flow Processes in a Motored PROCO Engine," SAE Paper No. 790290, 1979.

161. Sorenson, S. C. and S. S. Pan, "A One-dimensional Combustion Model for a Dual Chamber Stratified Charge Spark Ignition Engine," SAE Paper No. 790355, 1979.

162. Johnston, S. C., "Precombustion Fuel/Air Distribution in a Stratified Charge Engine Using Laser Raman Spectroscopy," SAE Transaction Vol. 88, SAE Paper No. 790433, 1979.

163. Wood, C. D., "Performance of a Stratified Charge Engine," SAE Paper No. 790434, 1979.

164. Dimick, D. L., S. L. Genslak, R. E. Greig, and M. J. Malik, "Emissions and Economy Potential of Prechamber Stratified Charge Engines," SAE Transaction Vol. 88, SAE Paper No. 790436, 1979.

165. Sinnamon, J. F. and D. E. Cole, "The Influence of Overall Equivalence Ratio and Degree of Stratification on the Fuel Consumption and Emissions of a Prechamber, Stratified Charge Engine," SAE Transaction Vol. 88, SAE Paper No. 790438, 1979.

166. Yagi, S., I. Fugii, M. Nishikawa, and H. Shirai, "A New Combustion System in the Three-valve Stratified Charge Engine," SAE Paper No. 790439, 1979.

167. Reitz, R. D. and F. B. Bracco, "On the Dependence of Spray Angle and Other Spray Parameters on Nozzle Design and Operating Conditions," SAE Paper No. 790494, 1979.

168. Gussak, L. A., V. P. Karpov, and Y. V. Tikhonov, "The Application of Lag-process in Prechamber Engines," SAE Transaction Vol. 88, SAE Paper No. 790692, 1979.

169. Rysiewski, R. R., R. B. Katnik, and H. D. Albers, "An Interactive Computer Graphics Finite Element Modeling and Evaluation program," SAE Paper No. 790995 (P-83), 1979.

170. Hiraki, H. and J. M. Rife, "Performance and NOx Model of a Direct Injection Stratified Charge Engine," SAE Transaction Vol. 89, SAE Paper No. 800050, 1980.

171. Sorenson, S. C., J. J. Bruckhauer, and G. R. Gehrke, "Mixing and Charge Preparation Effects in a Dual Chamber Stratifed Charge Spark Ignition Engine," SAE Transaction Vol. 89, SAE Paper No. 800107, 1980.

172. Sinnamon, J. F., D. R. Lancaster, and J. C. Steiner, "An Experimental and Analytical Study of Engine Fuel Spray Trajectories," SAE Transaction Vol. 89, SAE Paper No. 800135, 1980.

173. Johnston, S. C., "Raman Spectroscopy and Flow Visualization Study of Stratified Charge Engine Combustion," SAE Transaction Vol. 89, SAE Paper No. 800136, 1980.

174. Sod, G. A., "Automotive Engine Modeling with a Hybrid Random Choice Method, II," SAE Paper No. 800288, 1980.

175. Yagi, S., I. Fujii, M. Nishikawa, and H. Shirai, "A Newly- Developed 1.5L CVCC Engine for Some 1980 Models," SAE paper No. 800321, 1980.

176. Komiyama, K. and I. Hashimoto, "Spark-assisted Diesel for Multifuel Capability," SAE Transaction Vol. 90, SAE Paper No. 810072, 1981.

177. Furukawa, J. and T. Gomi, "On the Propagation of Turbulent Jet- flame in a Closed Vessel," SAE Paper No. 810777, 1981.

178. Kuo, T. and F. V. Bracco, "On the Scaling of Transient Laminar, Turbulent, and Spray Jets," SAE Paper No. 820038, 1982.

179. Diwalker, R., "Direct-injection Stratified-charge Engine Computations with Improved Submodels for Turbulence and Wall Heat Transfer," SAE Paper No. 820039, 1982.

180. Kuo, T., and F. V. Bracco, "Computations of Drop Sizes in Pulsating Sprays and of Liquid-core Length in Vaporizing Sprays," SAE Transaction Vol. 91, SAE Paper No. 820133, 1982.

181. Hugelman, R. D., "Recent Developments in Swirl Induced Turbulent Mixing for 4-stroke Cycle Engines," SAE Paper No. 820157, 1982.

182. Ingham, M.C., P.S. Myers and O.A. Uyehara, "In-cylinder Sampling of Hydrocarbons in a Texaco L-141 TCP Engine," SAE Transaction Vol. 91, SAE Paper No. 820361, 1982.

183. Freeman, L.E., R.J. Roby, and G.K. Chui, "Performance and Emissions of Non-petroleum Fuels in a Direct-injection Stratified Charge SI Engine," SAE paper No. 821198 (SP-527), 1982.

184. Lewis, J.M., and T.K. McBride, "UPS Multifuel Stratified Charge Engine Development-Progress Report," SAE Transaction Vol. 92, SAE Paper No. 831782, 1983.

185. Duggal, V.K., and T. Kuo, and F. Lux, "Review of Multi-fuel Engine Concepts and Numerical Modeling of In-cylinder Flow Processes in Direct Injection Engines," SAE Transaction Vol. 93, SAE Paper No. 840005, 1984.

186. Onishi, S., S. Hong Jo, P.Do Jo, and S. Kato, "Multi-layer Stratified Scavening (MULS)- A New Scavenging Method for Two-Stroke Engine," SAE Transaction Vol. 93, SAE Paper No. 840420, 1984.

187. Clarke, B.C., and T. Canup, "EPIC - An Ignition System for Tomorrow's Engines," SAE Paper No. 840913 (SP-556), 1984.

188. Kuzak, D.M., R.C. Belaire, S. Le, and D.R. Brigham, "Parametric Simulation of the Fuel Consumption Effects of Engine Design Variation with Advanced Transmission Powertrains," SAE Transaction Vol. 93, SAE paper No. 841243, 1984.

189. Gatowski, J.A., EnN. Balles, K.M. Chun, F.E. Nelson, J.A. Ekchian, and J.B. Heywood, "Heat Release Analysis of Engine Pressure Data," SAE Transaction Vol. 93, SAE Paper No. 841359, 1984.

190. Evers, L.W. and O. Baasch, "Quantum Combustion Chamber for the Digital Engine," SAE Paper No. 850033, 1985.

191. Sierens, R., and W. Verdonck, "Development of a Natural Gas Stratified Charge Rotary Engine," SAE Paper No. 850034, 1985.

192. Primus, R.J. and V.W. Wong, "Performance and Combustion Modeling of Heterogeneous Charge Engines," SAE Paper No. 850343, 1985.

193. Amsden, A.A., Butler, T.D., P.J. O'Rourke, and J.D. Ramshaw, "KIVA-A Comprehensive Model for 2-D and 3-D Engine Simulations," SAE Paper No. 850554, 1985.

194. Kimbara, Y., K. Shinoda, H. Koide, and N. Kobayashi, "NOx Reduction is Compatible with Fuel Economy Through Toyota's Lean Combustion System," SAE Paper No. 851210, 1985.

195. Foster, D.E., "An Overview of Zero-dimensional Thermodynamic Models for IC Engine Data Analysis," SAE paper No. 852070, 1985.

196. Kim, C., and D.E. Foster, "Aldehyde and Unburned Fuel Emission Measurements from a Methanol-fueled Texaco Stratified Charge Engine," SAE Paper No. 852120 (SP-638), 1985.

197. Witsky J.E., "Stratification and Air Pollution," Institution of Mechanical Engineers. 21st June 1971.

198. Witsky J.E., "Stratified Charge Engines," A.S.M.E. 1st March 1964.

199. Anonymous, " The Hesselman Spark Ignited Oil Engine with Fuel Spray and Low Compression," Diesel Power, September 1930.

200. Heller A., "The Hesselman Oil Engine for Motor Vehicles," Mechanical Engineering, Vol. 52, No. 12. December 1930.

201. The Hesselman Spark-ignited Oil Engine with Fuel Spray and Low Compression. Diesel Power Vol. 8. No. 9. September 1930.

202. Starr A.M., "Fuel Injection Engine with Spark Ignition," SAE Journal, September 1947.

203. Collins D., "Windsor-Smith Stratified Charge Combustion System," Ricardo Report SN. 17994.

204. Eisele E., Hiereth H. and Charzinski P., "A Study of a Charge Stratification System for Heavy Commercial Vehicle Engines," I.Mech. E. C91/75.

205. Meurer J.S., "Towards Stratified Charge," Automobile Engineer, February 1967.

206. Meurer J.S. and Urlaub A., "Development and Operational Results of the MAN-FM Combustion System," SAE 690255.

207. Urblaub A. and Chmela F.G., "High Speed, Multi-Fuel Engine: L9204 FMV," SAE 740122.

208. Conta L.D. and Durbetaki P., "A Method of Change Stratification for Four Stroke-Cycle Spark Ignition Engines," ASME No. 58-OGP- 5, 1958.

209. Conta L.D. and Durbetaki P., "Research on Charge Stratification of Spark Ignition Engines," ASME 60-WA-314, 1961.

210. Baudry J. et al., "Stratified Charge Systems for Spark Ignition Engines," SAE Journal Vol. 69 No. 9, September 1961.

211. Conta L.D. et al., "Stratified Charge Operation of Spark Ignition Engines," SAE 3758B, 1961.

212. Newhall H.K. and El Missiri I.A., "A Combustion Chamber Designed for Minimum Engine Exhaust Emissions," SAE 700491.

213. Newhall H.K. and El Missiri I.A., "A Combustion Chamber Concept for Control of Engine Exhaust Air Pollutant Emissions," Combustion Flame. February 1970. Vol. 14 No. 1.

214. Breisacher P., Nichols R.J., and Hicks W.A., "Exhaust Emission Reduction Through Two-Stage Combustion," Combustion science and technology 1972 Vol. 6.

215. Date T. Yagi S. Ishizuya A. and Fujii I., "Research and Development of the Honda CVCC Engine, " SAE 740605.

216. Kuck H.A. and Brandstetter W.R., "Investigations on a Single Cylinder Stratified Charge Engine with a Scavenged Pre-chamber," I.M.E. C92/75.

217. Turkish M.C., "3-valve Stratified Charge Engines: Evolvement, Analysis and Progression", SAE 741163.

218. Rhodes K.H., "Project Stratofire," SAE 660094.

219. Yamagishi G., Satio T. and Iwasa H., "A Study of Two-Stroke Cycle Fuel Injection Engines for Exhaust Gas Purification," SAE 720195.

220. Kataoka K. and Hirako Y., "Improvements of Combustion and Clean Exhaust Gas of a Two-cycle Gasoline Engine with Charge Stratification," I.Mech.E. C93/75.

221. Beale N.R. and Hodgetts D., "The Cranfield - Kushul Engine," I.Mech.E. C90/75.

222. Hussman A.W., Kahoun F. and Taylor R.A., "Charge Stratification by Fuel Injection into Swirling Air," SAE Transactions, Vol. 71, 1963, p. 421-444.

223. Uyehara, O. A., "A Classification of Reciprocating Engine Combustion Systems", SAE 741156.

224. Heywood J.B. et. al., " Automotive Spark Ignition Engine Emission Control Systems to Meet the Requirements of the 1970 Clean Air Amendments," National academy of Sciences. May 1973.

225. Bellan, J. R. and W. A. Sirignano, "A Theory of Turbulent Flame Development and Nitric Oxide Formation in Stratified Charge Internal Combustion Engines," Combustion Science and Technology Vol. 8. 1973.

226. Witsky, J. E. and J. M. Clark Jr., "Stratification and Combustion in Reciprocating Engines," A.S.M.E. 68-DGP-4.

227. Schweitzer, P. H. and L. J. Grunder, Hybrid Engines, SAE Translation 1963.

228. Newhall, H. K., "Combustion Process Fundamentals and Combustion Chamber Design for Low Emissions," SAE Paper 751001.

229. Eisele, Hiereth, and Charzinski, "Stratification System for Heavy Commercial Vehicle Engines," Institution of Mechanical Engineers Publication No. C91/75, pp. 97-103, March 1975.

230. Delichatsios, M.D., "A Model of Fuel Spray Formation and Evaporation in the Ford PROCO Stratified Charge Engine-- A Photographic and Analytic Study," Ph.D. Thesis, Department of Mechanical Engineering, M.I.T., Cambridge, MA, February 1976.

231. Miyake, M., Okada, S., Dawahara, Y., and Asai, K., "A New Stratified Charge Combustion System (MCP) for Reducing Exhaust Emissions," Combustion Science and Technology, Vol.12, pp. 29- 46, 1976.

232. Miyake, M., "Recent Development of Mitsubishi's Stratified Charge Engine MCP," Institution of Mechanical Engineers Conference Publication No. C259/76, Automobile Division and the Combustion Engine Group, November 1976.

THE DUAL FUEL ENGINE


Ghazi A. Karim

Department of Mechanical Engineering
The University of Calgary
Calgary, Alberta, Canada


INTRODUCTION

Interest in the efficient utilization of gaseous fuel resources for
the production of power using conventional internal combustion engines, has
been increasing worldwide in view of the inevitable declining resources of
petroleum and projected limitations on the availability of refined liquid
fuels, particularly those of the right quality.  This is in contrast to
statements made that the proven reserves of natural gas are increasing[1] and
that there are potentially enormous reserves of natural gas[2] that can be
utilized as fuels in engine applications.  Moreover, there is increased
perceived availability of other forms of gaseous fuels such as liquid
petroleum gases, gas fuel mixtures produced from the processing and upgra-
ding of various fossil fuel resources such as coal, oil sands and shales
and from the processing of organic and vegetable wastes in the form of bio-
gas.

Operation of diesel engines on gaseous fuels is neither new nor recent.
It is possible to trace its origin back to the beginning of the century
when Dr. Rudolph Dierel patented a compression ignition engine to run on a
gaseous fuel, coal gas[3].  Subsequently, more successful commercial applications appear to have been made, mostly for stationary applications, prior
to the war.  During the second world war, some efforts were directed
towards using coal gas mixtures, sewage gas or methane as well as stocks of
poor quality gasoline in the form of gasified vapour to run conventional
diesel engines for a variety of applications.  After the war interest in
these applications fluctuated depending on the relative cost of such fuels
and the extent of competition from other conventional fuels.

It is fair to state that until relatively recently most of these efforts
can be described as having been made on ad hoc bases with relatively little
and inadequate basic understanding of the complex processes involved.  The
combustion processes in a typical dual fuel engine tend to be complex
showing combinations of the problems encountered both in diesel and spark
ignition engines.  There is much room and need for further development to
bring about more effective and trouble free conversions of engines to operate
on a wide range of gaseous fuels.  The recent aberration in the world energy
pricing picture should not deter from the long range task and the associated
perceived priorities in the whole energy field in general and alternative
gaseous fuels in particular.

THE DUAL FUEL ENGINE

The dual fuel engine is a conventional diesel engine of the compression ignition type in which some of the energy release by combustion comes about from the combustion of a gaseous fuel while the diesel liquid fuel continues to provide throughout, through timed cylinder injection, the remaining part of the energy release. The term "dual fuel" should not be confused with bi-fuel applications of spark ignition engines where the liquid fuel is not combusted simultaneously with the gaseous fuel.

Dual fuel applications can involve one of the following two main procedures:

i. The injection of a small quantity of diesel liquid fuel in a conventional diesel engine in the normal way so as to provide mainly means for igniting the engine charge, which is normally a lean mixture of a gaseous fuel in the air. The objective is to maximize the use of the gaseous fuel component and economize in the relative use of the liquid fuel, which is termed as the 'pilot'. This approach which has been the normal practice in most dual fuel engines involving stationary applications, tends to represent primarily a fuel substitution strategy.

ii. The addition of some gaseous fuel to the incoming air of a fully operational diesel engine so as to provide extra fuel loading and hence produce additional power. This supplementary fuelling which is applied to the diesel engine with little or no alterations to existing injection equipment, tends to have the advantage of being flexible. This has been the practice in most applications of the dual fuel engine in the transport field. No supplementary gaseous fuel is usually employed at light load but at above some prescribed load level, an increasing amount of the supplementary fuel is introduced with the air. Normally, the maximum allowable fraction of the energy release arising from this is limited eventually by the onset of "knock".

Ideally, there is a need for optimum variation in the liquid fuel quantity used anytime in relation to the gaseous fuel supply so as to provide for any specific engine the best in performance over the whole load range desired. Usually, the main aim, largely due to economic reasons, is to minimize the use of the diesel fuel and maximize its replacement by the cheaper gaseous fuel throughout the whole load range.

The dual fuel engine is an ideal multifuel engine that can operate effectively on a wide range of different fuels while maintaining the capacity for operation as a conventional diesel engine. Normally, the change over from dual fuel to diesel operation and vice versa, can be made automatically even under load.

Over the years the dual fuel engine has been employed in a very wide range of applications. Numerous stationary installations were and are being used for power production, co-generation, compression of gases and pumping duties. In transport, limited examples of conversions can be seen in trucks both for long as well as short haul duties, in buses both municipal and for schools, in commercial delivery vans and in taxis. Other successful applications can be found for marine transport in cargo ships, ferries and fishing vessels and in some limited traction duties. Some notable agricultural applications on the farm, involving the operation of machinery and tractors, have also been made. Though stationary engine applications can utilize the gas supplies conveniently at practically any pressure, the problem of the portability of the gaseous fuel and the provision of compact storage facility in mobile applications remain a field of urgent long term

research that can have the potential for opening widely the market for the
dual fuel engine and the increased exploitation of gaseous fuel resources,
particularly in the transport sector.


CONVERSIONS TO DUAL FUEL OPERATION

The dual fuel combustion system in compression ignition engines
features essentially the rapid compression of the gas-air mixture to below
its autoignition conditions.  The charge is then ignited during the compression stroke at some point near the top dead centre position by the ignition
of diesel liquid fuel in the usual way.  The engine retains the ability to
run on either diesel liquid fuel only or in association with a gaseous fuel-
air mixture over a wide range of concentration.  The diesel liquid fuel is
injected through the conventional diesel fuel system and the quantity of the
pilot charge per injection can be either fixed regardless of engine output
or made to vary in relation to the gaseous fuel supplied in a prescribed
fashion.

The motivation to run diesel engines on gaseous fuels is largely
economic.  The dual fuel engine enables the utilization of relatively cheap
gaseous fuel resources while saving on the consumption of good quality
diesel fuel.  Much of the superior qualities and advantages associated with
the diesel engine can be maintained while using existing standard diesel
engine installations with little modifications, yet without underming diesel
performance, nor losing the option to run entirely on diesel or convert
instantaneously to diesel operation, whenever needed.  For example, diesel
engines are highly efficient, have superior torque-speed characteristics,
high power outputs, amenable to turbocharging and have good emission characteristics.  They tend to be highly reliable, have long operational life, low
on maintenance and of robust construction that can make diesel engines
particularly suited to withstanding occasional shock and knock loading.

It can be shown that the dual fuel engine with appropriate conversion
methods, has indeed in principle superior characteristics to those of the
straight-diesel engine operation.  For example, dual fuel operation can
provide higher output with better specific energy consumption than the corresponding diesel operation, as shown typically in Fig. 1.[4]  This can be
achieved while displaying superior emission characteristics, quieter and
smoother operation, improved low ambient operation and reduced thermal
loading in comparison to diesel operation.

The dual fuel engine can also have far superior operational characteristics to the corresponding spark ignited gas engine.  For example, it offers
superior control and operational safety characteristics while accepting a
much wider spectrum of fuels without requiring the various spark ignition
timing changes needed with changes in load or fuel.  Dual fuel engines display far much lesser cyclic variations even at very light load arising
mainly from the employment of deliberate and reliable ignition by the liquid
pilot.  Moreover, its ability to operate over a wide range of charge mixture
strength permits quality control through changes in the concentration of the
fuel without resorting to throttling and its attendant deficiencies.  Of
course, meanwhile throughout dual fuel operation, the quality of the diesel
liquid fuel cannot be sacrificed so as to ensure deliberate and reliable
ignition of the entire charge.  Moreover, whenever comparison is to be made
of the relative performance of an engine as a dual fuel with its corresponding performance either as a diesel or a spark ignition engine, care is
needed so that the comparison is made for the same engine setup and under
the same operating conditions.

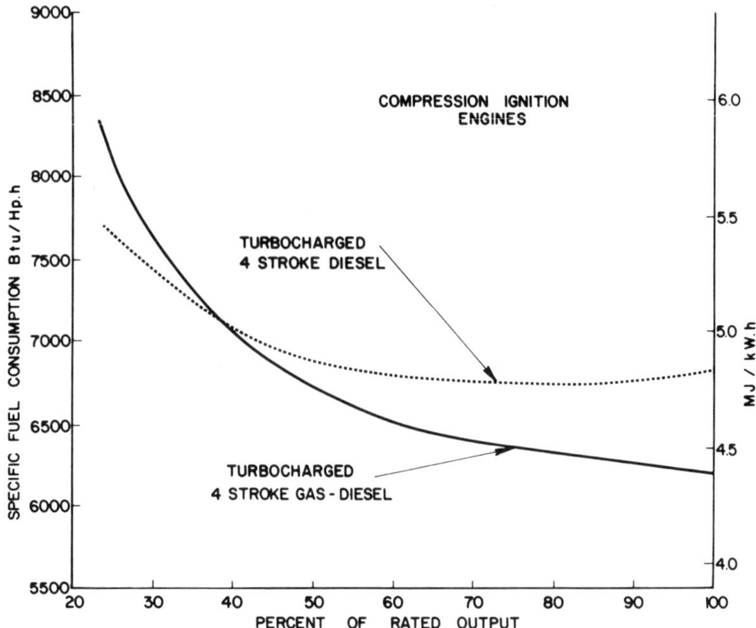


Fig. 1   Typical variation of the specific fuel consumption
with load of a turbocharged diesel engine and when
fuelled with methane.[4]

Currently, there is a number of commercially available diesel conversion
systems developed to operate on a range of gaseous fuels as dual fuel engines.
All of these systems though work, tend not to optimize the operation of the
engine as much as they should whether in terms of the overall efficiency or
in terms of the amount of diesel fuel replaced or in terms of the power output or even in terms of exhaust emissions.  For dual fuel engines to be
viable alternatives to diesel engines, considerable research and development
work is required for engine control systems and engine design itself.  In
general, the control system has to be matched both to the characteristics of
the engine to be converted and the fuel to be consumed which impose serious
limitations on the universality of such conversion equipment.

The gaseous fuel is usually introduced into the air outside the cylinder,
either from a high or low pressure supply, in a fumigated fashion continuously or through timed gas introduction.  Direct gas injection inside the
cylinder is being increasingly used, particularly for two stroke or highly
turbocharged engine applications.  The injection of the gaseous fuel well
into the compression stroke tends to be a more complex matter that needs
careful evaluation and optimization.  Apart from the need for adequate compression of the gaseous fuel, serious problems associated with control and
mixing can arise.

Adequate controls need to be provided to dual fuel engines adding to
the complexity and cost of the installation.  Such controls must cater simultaneously for diesel operation, pilot injection, gaseous fuel introduction
and providing protection against operational hazards such as engine overspeed,
gaseous fuel leakage and accumulation during starting, etc.  There is much
room for innovation and improvements in this field to provide economic, reliable and simple controls.  Accordingly, the question of the cost of conversion of typical diesel engines to operate on the dual fuel principle

remains too complex to be answered adequately even in general terms.  The
cost of conversion apart from depending on how many conversions are made
and by whom, will depend on the following factors:

   i. The type of engine being converted, e.g. whether turbocharged,
two or four stroke, size and number of cylinders, direct or indirect injection, new or existing engines, etc.
  ii. The gaseous fuel to be used, its composition and whether multi-
fuel applications are to be planned for.
 iii. The ratio of gas to diesel injection planned, i.e. whether pilot
diesel application or diesel operation to be topped up with
gaseous fuel introduction are to be employed and to what extent
optimization to be desired for such operation.
  iv. Field of application, e.g. stationary, mobile or marine.
   v. Whether engine operation is to be fully automated or some manual
controls, such as in transport applications, are to be retained.

There is always of course the question whether in trying to achieve
effective dual fuel operation, all features of diesel performance are to be
maintained, whenever required, intact or some of the modifications introduced permitted to undermine somewhat such a performance.  In any case, as
a long term aim, there is always the need to develop and design a totally
and optimally dedicated dual fuel engine, whether for operation on one
specific gaseous fuel or permit some tolerance to changes in the composition
of such a fuel.

The high compression ratio diesel engine is imminently suitable for
dual fuel operation with methane and much of the operational experience reported in the literature involves dual fuelling with methane.  However,
largely by virtue of the provision of an adequate and consistently timed
pilot ignition, almost any gaseous fuel or vapour can be utilized to a
varying degree of success in such engines.  Accordingly, gaseous fuels such
as ethane, propane, butane, hydrogen, ethylene, acetylene, and ammonia have
been employed.  Moreover, various gaseous fuel mixtures such natural gases,
biogases, liquefied petroleum gases, "low btu" gases, etc. have been successfully employed.

Some of the additional distinct advantages associated with dual fuel
operation include cleaner and longer lasting lubricants, with fewer filter
changes, potentially cleaner operation and longer engine life.  However, it
should not be assumed that dual fuel operation is without limitations nor
problems.  Some of these may be so serious as to set on occasions, with our
present state of knowledge, a limit to the effective utilization of dual
fuel engines.  Active research and development are needed so as to alleviate
the consequences of these undesirable features.  For example, to start with
it can be suggested that since dual fuel operation requires the simultaneous
availability of two or more fuels, this would constitute in principle an unattractive feature that can bring about increased complexity in controls,
additional cost, the need for specialized storage facilities and undesirable
operational features arising from the possible chemical interaction between
the gaseous fuel being used and the preignition processes and subsequent
combustion of the pilot.  Moreover, a serious problem associated with the
dual fuel engine is the relatively poor light load and idling performance
associated with low efficiency and inferior emission characteristics.  The
extent of this deterioration in performance depends largely on the gaseous
fuel being used and the engine employed.  Furthermore, when very high outputs are desired, the problem of knock may be encountered with most gaseous
fuels including methane.  Thus, a serious practical barrier is set for the
maximum load that can be achieved for any engine with any gaseous fuel.

The injection of a small quantity of diesel fuel in the form of a pilot is normally done on the basis that the diesel fuel vapour will autoignite and provide a multitude of ignition centres from which turbulent flames travel subsequently to consume the lean homogeneous gaseous fuel-air mixture. Changes in the magnitude of the pilot, particularly at light load when very lean homogeneous mixtures are involved, should bring about proportionally changes in the size of the initial combustion zone, the number of ignition centres and the associated thermal energy released. However, the processes involved then are found in reality to be much more complex than this perceived picture, affecting the performance of dual fuel engines at light load adversely. The introduction of a gaseous fuel with the engine air, even in very small quantities, can have a significant effect on the cylinder charge during compression affecting markedly the processes of preignition and subsequent combustion of the pilot and the cylinder charge.

Accordingly, a serious problem associated with dual fuel engines is the relatively poor light load and idling performance. The extent of this deterioration in performance depends largely on the extent of pilot quantity employed, the gaseous fuel being used, operating conditions and the engine employed. With certain fuels and engines, there are occasions when idling or even light load operation becomes totally impaired. The ignition delay of the pilot fuel increases considerably with the introduction of the gaseous fuel but decreases later with further gas addition as shown in Fig. 2.[5] This increase in the delay period is far in excess of that caused by the slight reduction of the partial pressure of oxygen by the addition of the gaseous fuel or by the reduction in the temperature of the charge at around the top dead centre position as a result of the higher overall specific heat of the charge. The gaseous fuel, including the not so reactive methane, can undergo significant reactions during the relatively long compression stroke. Accordingly, the homogeneously dispersed gaseous fuel and its partial oxidation products within the core of the pilot region can participate actively in the preignition chemical process of the pilot fuel to contribute significantly to these variations in the delay. The competition between the diesel fuel vapour and the gaseous fuel for active radicals can affect the preignition processes of the pilot adversely. Only when a considerable amount of the gaseous fuel is added that significant amounts of energy and species are produced during compression, aiding the ignition processes following pilot fuel injection and helping to reduce the ignition delay.

At very light load, as shown in Fig. 3[6], a significant proportion of the low concentrations of the gaseous fuel added to the intake will not burn completely. This is despite the presence of much excess air and pilot ignition. The small amount of gaseous fuel added, even at the very high temperatures encountered near the peak of compression and even after the ignition of the diesel pilot releasing quickly energy and raising the effective temperature, is too low for flames starting from these pilot ignition centres to propagate throughout the lean mixture. Thus, the flame will be quenched and some of the gaseous fuel can remain unreacted and survive to the exhaust stage. For the same pilot quantity, only by using greater gas concentrations that the effective flammability limit within the mixture environment will be achieved. Normally, associated with this low gaseous fuel utilization at lean mixtures is a significant increase in the carbon monoxide concentration well beyond that normally observed in the corresponding diesel operation, at the same fuelling rate. Thus, the tendency in diesel conversion to dual fuel application is to retain diesel operation for idling and low load operation.

If the cylinder charge is too lean, the flame fronts propagating from the various ignition centres do not extend to all regions of the cylinder

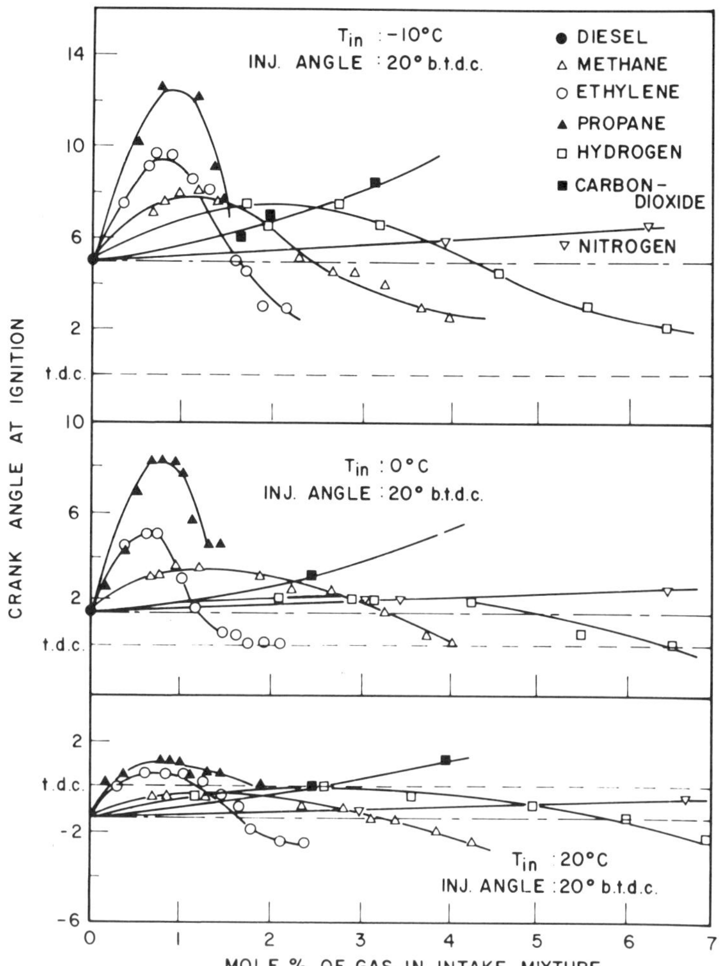


Fig. 2  Variation of the point of ignition with the extend
of gaseous fuel addition in the intake mixture by
volume for a number of common gaseous fuels as well
as for carbon dioxide and nitrogen additions at
constant injection timing, normally aspirated, direct
injection single cylinder engine operating at 1000
rev/min.  Results from three intake temperature are
shown.[5]

but leave some of the homogeneously dispersed gaseous fuel unchanged.  Some
partial oxidation products will be formed in regions at the fringes of the
main combustion zone as a result of the quenching of the combustion.  Thus,
only the fuel within these combustion zones will be oxidized.  When the
gaseous fuel concentration, however, exceeds some limiting range these flame
fronts will sweep throughout, providing a faster energy release rate, higher
rates of pressure rise and improved power output and efficiency.[7]  As a result of this inadequate fuel utilization at very lean mixture concentrations,
the specific energy consumption based on the total fuel charge, that is the
pilot diesel fuel and the gaseous fuel supplied and the associated power
output are inferior to that of the "straight" diesel, Fig. 4.[5] This situation

becomes particularly severe at low intake temperature operation, such as
those associated with cold ambient temperature or when using liquefied
natural gas as the main fuel.[5] Moreover, light load operation under ambient
conditions can become equally unacceptable when using low quality gaseous
fuels having excessive amount of diluents, such as in certain natural or
industrially produced gases containing significant concentrations of nitrogen
or carbon dioxide, Fig. 5.[8]

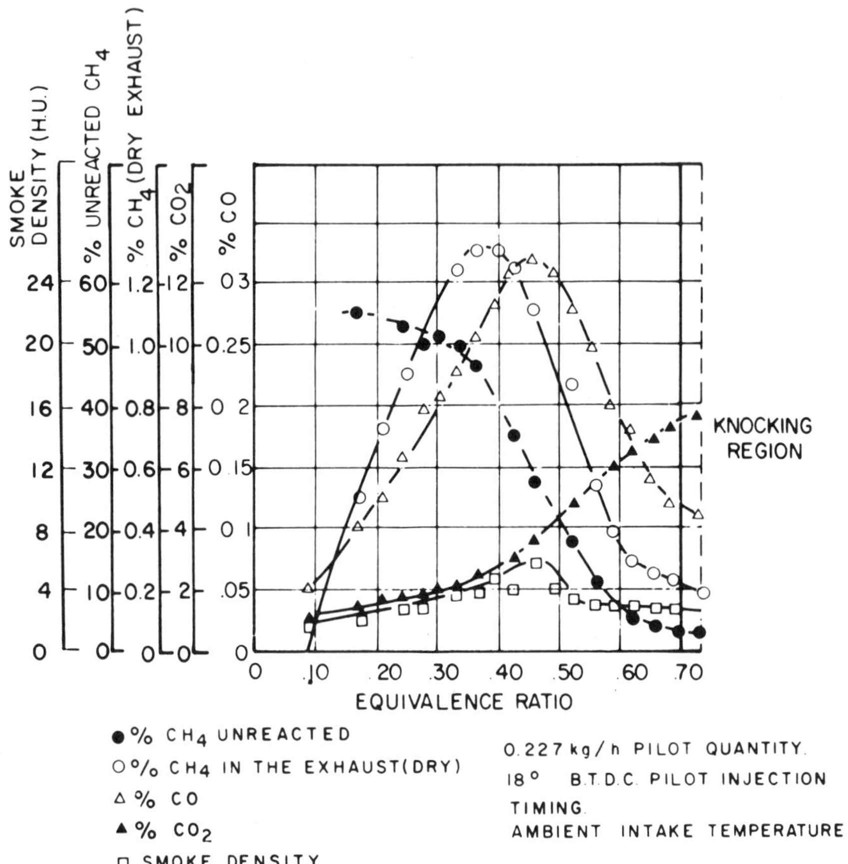


Fig. 3 Typical variation of the exhaust gas composition of
a dual fuel engine with the total charge equivalence
ratio (f/A relative to the stoichiometric f/A) at
1000 rev/min.[9]

Normal dual fuel operation at light load can be improved by extending
the lower operational limit boundary to lower gaseous fuel concentration in
the air charge through the employment of a number of procedures. The most
common approach of course, is to use a relatively large pilot fuel quantity
which on combustion, will increase the overall charge temperature, providing
a greater multiplicity of ignition centres and a larger combustion zone,
Fig. 6.[9] Moreover, the quantity of larger mass of mixture can be increasingly modified through mixing with diesel fuel vapour. As can be shown in
Fig. 7, based on performance data such as those of Fig. 6, the limiting

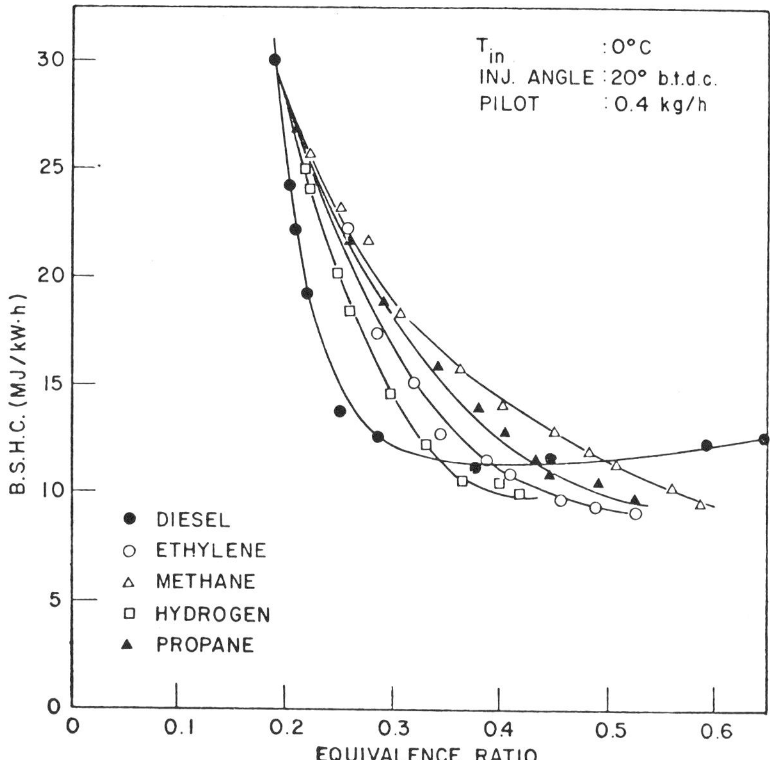


Fig. 4   Typical variation of the total brake specific energy
consumption (B.S.H.C.) of a dual fuel engine fuelled
in turn with a number of common gaseous fuels with
total equivalence ratio at constant pilot quantity
and a speed of 1000 rev/min.[5]

conversion of the methane to products is essentially linearly related to the
quantity of the pilot employed.

Similarly, the slight preheating of the lean intake gas-air mixtures
provides a higher mixture temperature at the end of compression than under
ambient intake temperature conditions.  Thus, the percentage of the unreacted gaseous fuel surviving the combustion process decreases, as shown
in Fig. 8.[9]  The limiting extent of conversion of methane to products,
essentially, is linearly related to the initial temperature employed, Fig.
9.  This is consistent with the tendency for the lean flammability limit of
all common fuels in air to widen approximately linearly with temperature.[10]
Of course, preheating of the charge is unnecessary at higher loads and when
deliberately employed may lead to the undesirable onset of knock.

Dual fuel performance at light load when using relatively small pilot
quantities can be further improved by incorporating the following changes:

i.   The use of the lowest injectar nozzle opening pressure associated
     with satisfactory engine operation as a diesel.  As shown in Fig.

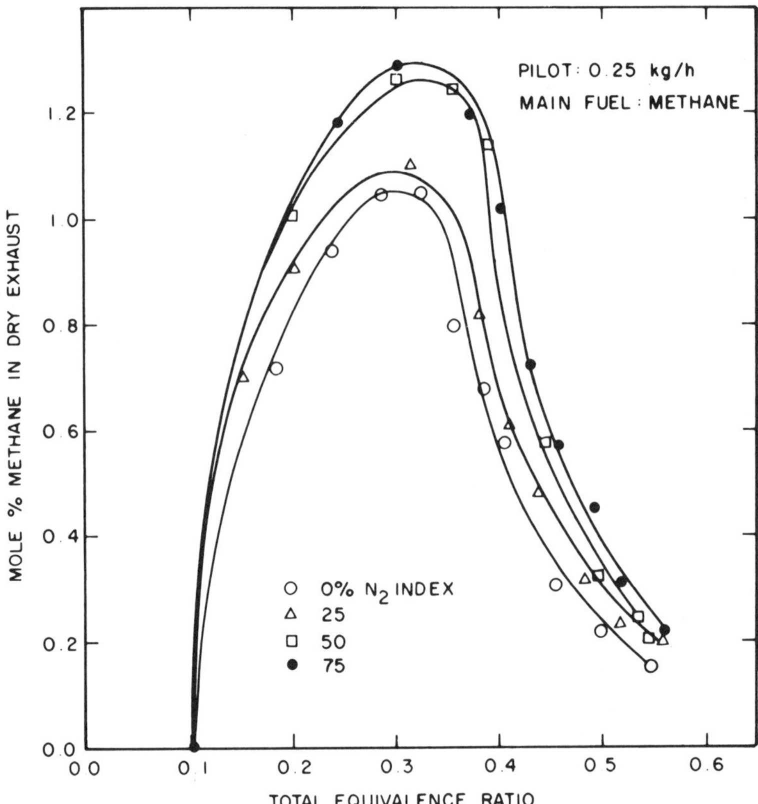



Fig. 5   Variation of the concentration of methane in the dry
exhaust with total equivalence ratio for various
"methane nitrogen index", $N_2/(CH_4 + N_2)$ by volume at
constant pilot and speed of 1000 rev/min, direct
injection single cylinder engine.

$10^9$, the employment of high injectar needle lift fuel line
pressure does lead to lower conversion of the gaseous fuel
when the same pilot liquid quantities are being injected.
This is probably due to the earlier dispersal of the small
amount of fuel injected during the prolonged delay period,
reducing significantly the number of the ignition centres and
the proportion of the gaseous fuel that will burn.

ii.   Advancing the injection timing of the pilot fuel somewhat, in
creases the residence time and the activity of the partial
oxidation reactions and widens the lower limit boundary.  This
modification, of course will be limited such that it will not
undermine diesel operation.  Moreover, advancing the injection
timing excessively can undermine severely dual fuel operation.

iii.  Partial restriction of the air component of the charge so as to
produce an effectively richer mixture at the end of compression.
For the same amount of gaseous fuel addition, this throttling
tends to improve part load performance up to a point.  However,
this approach, which brings about also complexity of controls,

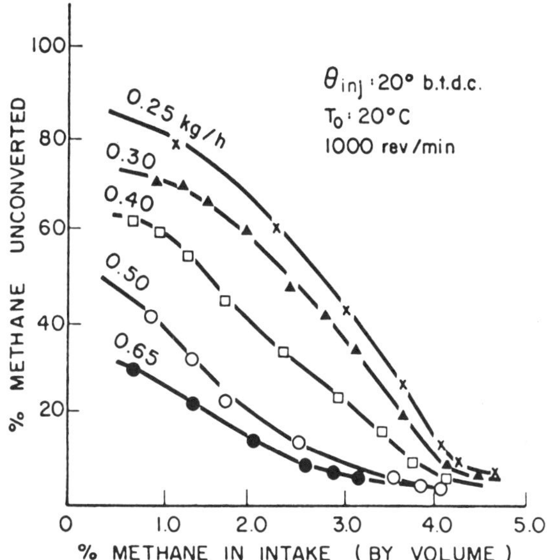


Fig. 6 Variation of the extent of methane unconverted with
extent of methane addition to the intake air of a
direct injection diesel engine for a range of pilot
diesel quantities at constant injection timing.

should be employed with care so that pilot ignition is not under-
mined by this throttling.

iv. Some stratification of the gaseous fuel component in relation to
the air, if can be achieved effectively in practice, may help to
improve light load operation through providing slightly richer
mixture in the vicinity of the pilot fuel. Indirect injection
engines may be more suitable for such stratification than direct
injection engines.

v. The selective addition of small amounts of various auxiliary fuels
to the main gaseous fuel supply, such as hydrogen, higher hydrocarbons, or gasoline vapour, as shown in Fig. 11, [11] can improve
light load operation significantly. This, however, is achieved
at the expense of increased complexity and the use of yet another
fuel system. Beyond the light load range, such a practice apart
from producing higher specific energy consumption, may have to be
avoided as it can lead to the onset of knock prematurely.[11]

vi. Resort to exhaust gas recirculation without much cooling can also
help in improving light load performance through increased initial

change temperature and seeding with active products.[12]

vii. Taking active measures to reduce heat transfer from the charge
during compression through operating with higher water jacket
temperature or judicial use of insulated parts of the engine.
However, when operating beyond the light load region this procedure may lead to knock prematurely.

viii. In turbocharged dual fuel engine applications at light load,
excess air may be ducted back into the compressor intake, heating
the charge and improving performance. The excess air may also be
made to bypass the engine altogether. Of course, the air manifold
pressure cannot be reduced below atmospheric pressure, which may
be required, particularly with methane operation.

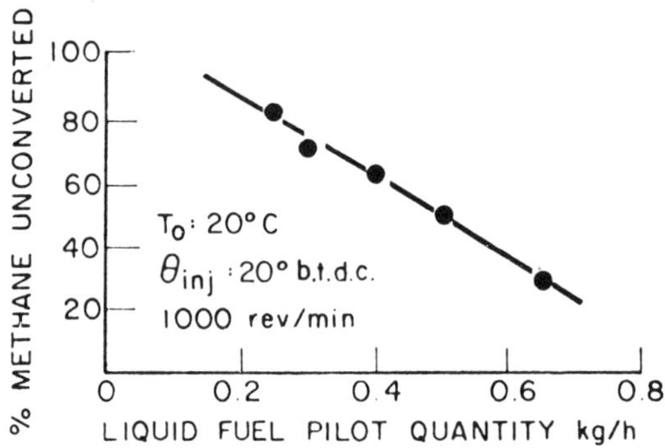


Fig. 7 Typical estimated limiting percentage of feed methane
unconverted with the extent of liquid pilot quantity
employed under constant operating conditions.

Most probably, resorting to a judicial combination of the foregoing
factors, particularly using a larger pilot quantity and preheating the
charge with some partial air restriction can be effective in improving the
light load performance of the dual fuel compression ignition engine.

In some applications, where there may be very cheap and abundant
supplies of gaseous fuels, such as natural gas, the main objective may well
be the reduction of the diesel fuel quantity being used, even at the expense
of the inefficient utilization of the gaseous fuel. For transport applications, the need to reduce fuel wastage tends to be more urgent than in
stationary engine applications.

Designers and engine manufacturers need to investigate the feasibility
of such approaches, so as to minimize the use of unnecessarily large pilots

and improve the conversion of the gaseous fuel at light load.  Thus, the utilization of significant proportions of the gaseous fuel can be made quite effectively, even under idling conditions.

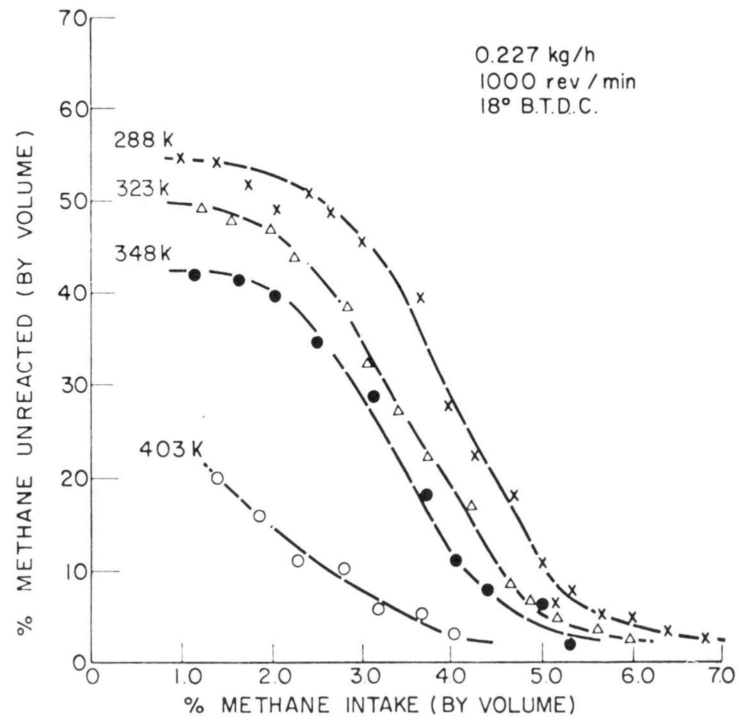


Fig. 8  Variation of the extent of methane unconverted with the extent of methane addition to the intake air of a direct injection diesel engine for a range of heated intake mixture temperatures when using a fixed pilot quantity.[9]

DUAL FUEL ENGINE PERFORMANCE AT THE HIGH LOAD REGION

At very light load, diesel operation is superior to that of the dual fuel producing slightly more power output for the same amount of total fuel induced.  When relatively higher fuelling rates are employed, the improvement in diesel operation is reduced and may even begin to deteriorate because of the longer injection and combustion times relative to gaseous fuel applications where very significant increases in power can be obtained. As shown typically in Fig. 12, [5], the output then can be well beyond that of the conventional diesel.  Methane may tend to produce slightly less power than propane which produces less power than hydrogen because of the faster burning rates of the latter.  However, it would be possible to produce ultimately higher power outputs with methane because of the lower likelihood of the onset of knock with methane.[13,14]

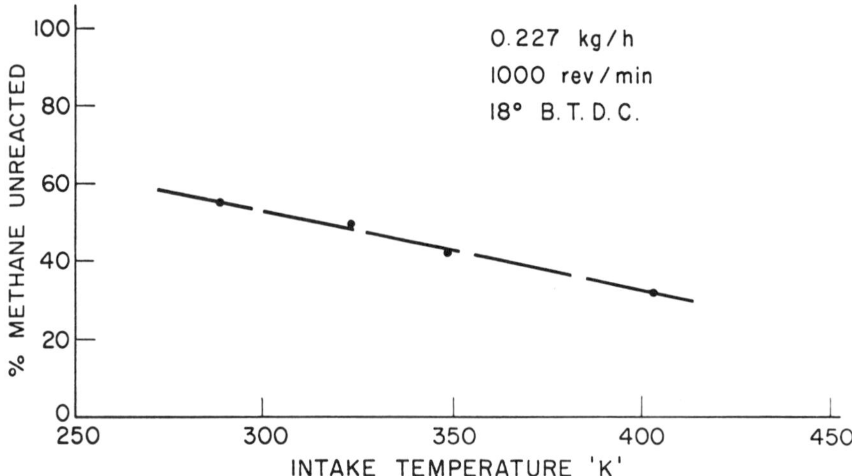


Fig. 9  Typical estimates of the limiting extent of methane
unconverted as a function of the mixture intake
temperature for a constant pilot quantity and timing.

The brake specific energy consumption, being a measure of how much total
energy is required to produce one unit of useful work, is shown typically in
Figs. 1 and 4, for a dual fuel engine.  It can be seen that beyond half load,
the efficiency of dual fuel operation improves sufficiently to surpass later
that of the corresponding diesel.  It can be clearly seen then, that operation
with methane is superior to that of the diesel at high load.  Moreover, as
far as exhaust emissions are concerned, oxides of nitrogen from diesel engines
tend to increase almost linearly with the amount of fuel being used.  However,
with the addition of natural gas, the production of oxides of nitrogen can
be delayed significantly.  Increasing the relative amount of diesel fuel injection to that of methane, more oxides of nitrogen will be produced indicating the need for using a very small amount of pilot fuel to reduce the
oxides of nitrogen emissions.  The presence of diluents with the natural gas
though will reduce relatively the power output, it will suppress significantly
the production of oxides of nitrogen and may influence adversely the specific
energy consumption.  Moreover, the presence of some hydrocarbon fuels such as
gasoline vapour with the methane will improve the power output as well as the
specific energy consumption if such fuels are added in relatively small
quantities.  Should significant proportions of the fuel induced be made up
of gasoline vapour, then an increase in the specific energy consumption will
be encountered as well as the maximum power that can be produced without
mechanical problems and the onset of knock will be severely limited.  This
is mainly due to the fact that the gasoline vapour, being more reactive than
methane, will undergo various pre-ignition reactions during the compression
process so that the energy released cannot be utilized usefully, as well as
making the mixture more prone to autoignition prematurely bringing a limit to
the ultimate useful output of the engine then.  This is a common problem
associated with certain natural gases that contain significant concentrations
of higher hydrocarbons.  The peak cylinder pressure is reduced considerably
with the gaseous fuel addition, contributing towards quieter engine running
than that of the diesel.  Similarly, the corresponding rate of pressure rise
is much lower.  This is a very definite advantage for dual fuel operation in
relation to that of the diesel, particularly where low temperatures are
concerned.

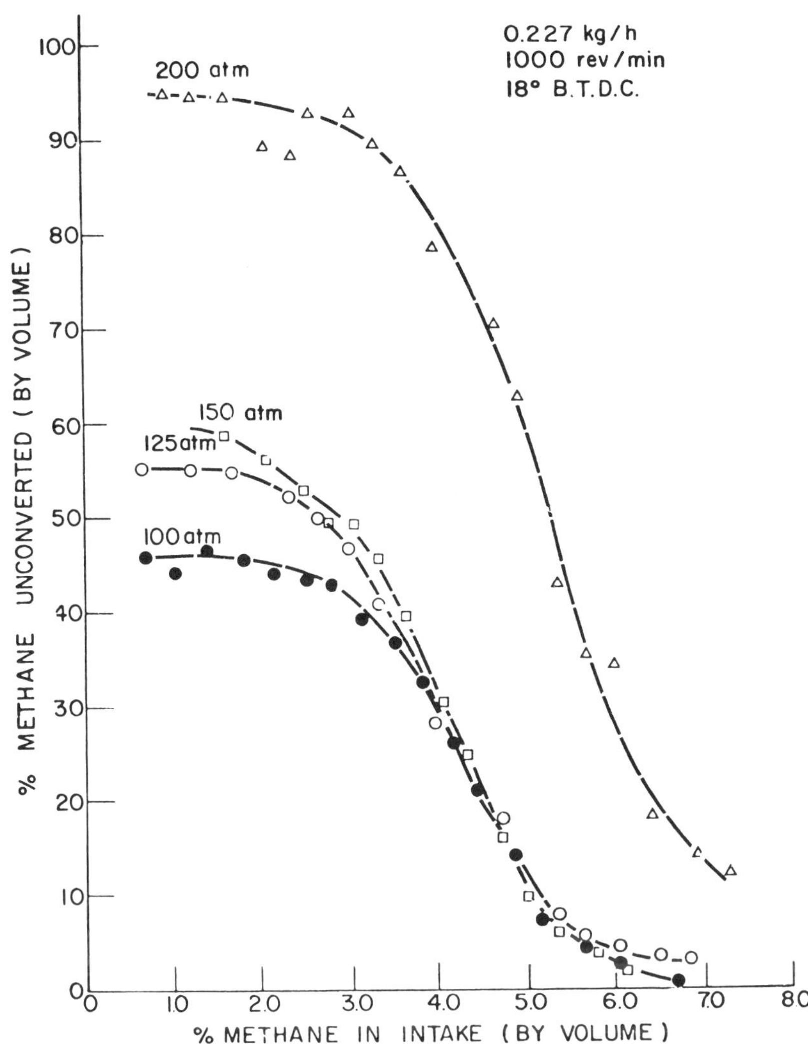


Fig. 10   The extent of unconverted feed methane with methane
          addition to a dual fuel engine for a number of
          situations when the injectar needle lift fuel line
          pressure varied, at constant pilot quantity and
          injection timing.[9]

Normal operation with various gaseous fuels can yield very satisfactory
operation with most common diesel engines.  However, it is only when very
high power outputs or very high intake temperatures and pressures are involved that the problem of knock, even with the very knock resistant fuel
such as methane, may be encountered.[15]  Knock is very strongly dependent on
the type of gaseous fuel being employed.  The knock ignition characteristics
of a gaseous fuel such as natural gas can be modified significantly through

the presence of small amounts of higher hydrocarbon vapour.[16,17] Knock will be associated with high rates of pressure rise, increase in heat transfer to the walls and consequent loss in thermal efficiency. In most cases persistent knock is highly objectionable and may lead to mechanical problems that may eventually lead to failures. Fortunately, the knocking region is normally out of most common operations with methane, unless highly supercharged, large bore diesel engines or large pilot quantities are employed. However, for some other gaseous fuel mixtures involving higher hydrocarbons the occurrence of knock in unmodified diesel engines is often encountered.

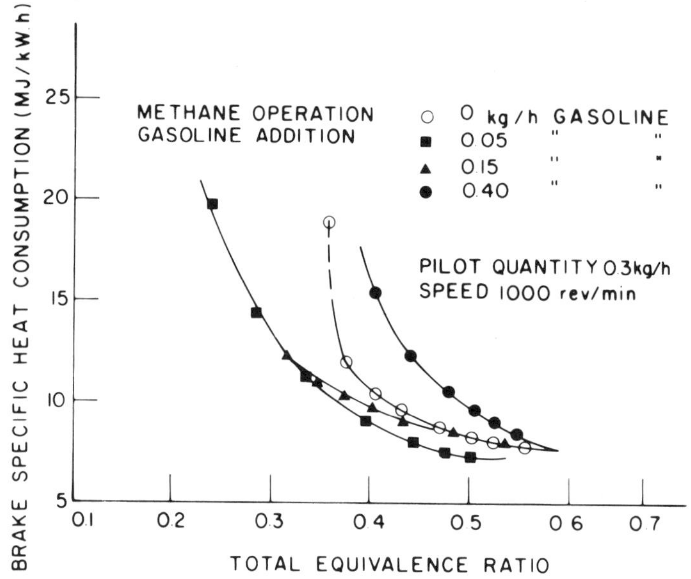


Fig. 11   Variation of the overall brake specific thermal consumption of a dual fuel engine fuelled with methane with total equivalence ratio at a constant pilot injection quantity for a range of gasoline aspirations into the intake charge at a constant speed of 1000 rev/min.[11]

It is acknowledged that the knock phenomenon observed in the dual fuel engine is of an autoignition nature, most probably of the gaseous mixture in the neighbourhood of the ignition centres.[15,18] The onset of knock can be delayed somewhat through the lowering of induction temperatures and pressures, water jacket temperatures and pilot quantities. Lower compression ratios and slightly later fuel injection can also be employed, subject to these actions not undermining seriously the performance of the engine as a diesel. Enhancing the quality of the diesel fuel by increasing its cetane number will have a relatively minor effect.[15] However, the quality of the gaseous fuel employed will have a very significant influence on the onset of knock, particularly through the presence of various hydrocarbon impurities with methane in some natural gases or through the presence of some hydrogen gas. The presence of diluents in the methane such as carbon dioxide, steam or nitrogen will suppress the onset of knock. Moreover, the use of smaller pilot quantities will aid in suppressing the onset of knock.

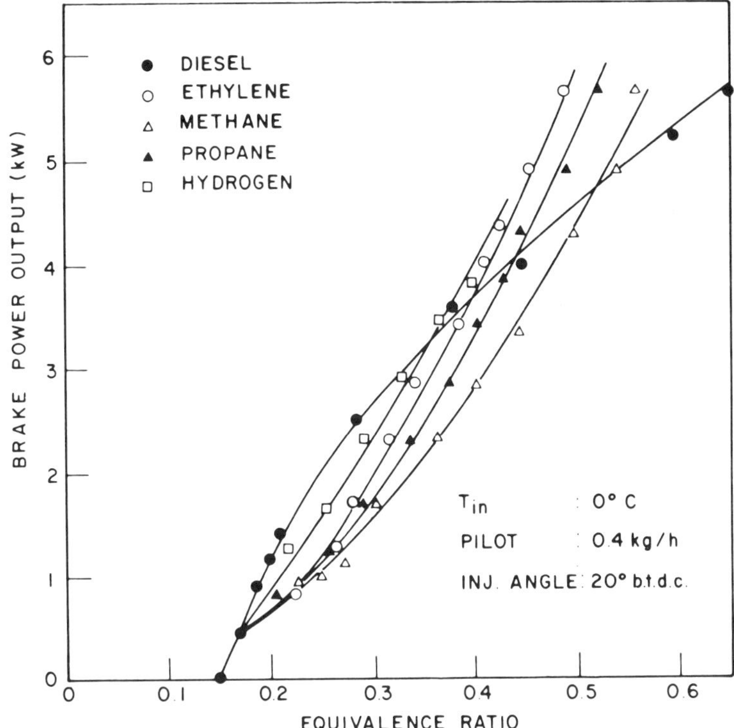


Fig. 12   Variation of the brake power output with total equivalence ratio in a single cylinder dual fuel engine,
at constant pilot quantity, when operating with various
gaseous fuels.  Corresponding normal diesel operation
is shown for comparison.[5]

Some prediction attempts to determine the knock limited power output
for a dual fuel engine when operating on gaseous fuels have been made in
terms of the pilot fuel quantity and the effective activation energy of
the gaseous fuel being used.[19,20]   It can be shown that knock can be simulated approximately by the autoignition of the gaseous charge at the
temperature that corresponds to the end of constant volume combustion of
the pilot quantity.  Accordingly, it can be seen that the knock limited output will be reduced through the increase of the pilot quantity at constant
injection timing, as well as when using gaseous fuels that are prone to preignition reactions and hence lower effective octane number.

One important factor that needs to be considered is how big a liquid
pilot is to be used for optimum performance and economy.  Most current
applications involving transport engines indicate that relatively large
pilots are being used.  This is done primarily for convenience and due to
the practical limitations of the control equipment being used that were
designed primarily for diesel operation.  This approach, in principle, is
unnecessarily restrictive and potentially costly in certain applications
that need to maximize the relative use of the gaseous fuel.  The amount of
diesel fuel being injected can be cut down to extremely small amounts,

particularly at high load.  Beyond the light load region, there is no need
to use large pilots and there appears to be a definite improvement associated with using a very small pilot, such as in terms of reduced tendency
to knock and the improvement in exhaust and noise emissions.  This would
clearly point out to the desirability of having a variable diesel pilot to
fuel gas mass ratio over the whole load range.  At light load then, a relatively large pilot is used.  At higher loads, the pilot quantity can be
reduced, yet still producing very good performance, providing that adequate
cooling is maintained of the fuel injection system that is handling a very
small amount of liquid fuel in a relatively hotter environment.


SOME FURTHER DEVELOPMENT WORK NEEDED

Despite the significant experience gained and research conducted in
relation to dual fuel engines when operating on a variety of gaseous fuels,
there is still a number of areas that need further attention in terms of
providing additional research and development.  These areas may include the
following:

i.    A dual fuel engine in principle, may be considered as an excellent
      median for the effective utilization of a wide range of gaseous
      fuels for the production of power.  There is a need to assess
      accurately for any engine the consequences of the changes in the
      quality and composition of the fuels being used, particularly when
      complex mixtures are employed and the corresponding necessary
      changes needed to bring about satisfactory performance of the
      engine under a wide range of operating conditions.

ii.   There is a need to overcome some of the practical problems asso-
      ciated with the exploitation of the significant potential for the
      use of the dual fuel engine in applications involving a wide range
      of fuels such as liquefied natural gas (LNG), liquid hydrogen
      ($LH_2$), low quality gasoline, oxygenated hydrocarbons such as
      alcohol, low heating value fuels, etc.

iii.  There is a need to develop more flexible and effective controls
      that will allow optimum performance with dual fuel operation over
      a wide range of fuels.  This would enable the optimum use of the
      diesel fuel in relation to the gaseous fuel component.  Thus, the
      pilot quantity can be made variable and minimized for efficient,
      reliable and acceptable engine performance, particularly for
      transport and automotive applications.

iv.   Additional efforts need to be directed towards the satisfactory
      employment of two-stroke engines for dual fuel engine application
      without much increase in complexity and cost or deterioration in
      performance.

v.    Although exhaust driven turbocharged dual fuel engines are rela-
      tively common, there is a need to optimize their operation in dual
      fuel applications without undermining the corresponding diesel
      performance nor engine cost or life.

vi.   The potential capacity of most dual fuel engines to produce signi-
      ficant overload in relation to diesel engines needs to be actively
      developed and exploited with most fuels and applications.

vii.  The exhaust emissions characteristics of the dual fuel engine need
      further refinement, particularly in terms of the reduction of the
      emissions of unburnt hydrocarbons and carbon monoxide under light
      load operation and oxides of nitrogen under high loads and when
      using large pilot quantities.

viii. The problem of knock in dual fuel engines, though may be fairly
      well understood, further research is still needed to contribute
      towards reducing the tendency of engines to knock particularly with
      some gaseous fuels and in turbocharged engine applications.

ix.  There is a need for fundamental research into the role of turbulence
     and gas motion within the cylinder during the pilot injection period
     and subsequent flame development.  It is not certain whether the
     experience in this area gained either from entirely diesel engine
     or spark ignition operations can be applied without modification to
     dual fuel operation.


ACKNOWLEDGEMENT

    The author wishes to acknowledge, with much appreciation and pleasure,
the major contribution made to this review by the many former students and
colleagues towards dual fuel engine research, over the years, that allowed
him to draw from it so very freely.  The financial assistance of The
Canadian National Science and Engineering Council is also acknowledged.


REFERENCES

1.  R. E. Wyman,"World Abundance of Methane from Unconventional Sources",
        Proc. Int. Congress on the use of Methane in the Transport Sector,
        Bologna, Italy (1982).
2.  T. Gold, "Earth Outgassing of Methane", A Chapter in "Methane - Fuel
        for the Future", p. 45, ed., P. McGeer and E. Durbin, Plenum Press,
        (1982).
3.  R. L. Boyer, "Status of Dual Fuel Engine Development", S.A.E. Journal,
        May (1949).
4.  C. L. Segaser, "Internal Combustion Piston Engines", Rep. No. ANL/CES/TE
        77-1, National Technical Information Service, U.S. Dept. of
        Commerce, Wash. D.C.
5.  G. A. Karim and K. S. Burn, "The Combustion of Gaseous Fuels in a Dual
        Fuel Engine of the Compression Ignition Type with Particular Refer-
        ence to Cold Intake Temperature Conditions", Proc. of S.A.E. Int.
        Congress, Detroit, paper No. 800263 (1980).
6.  G. A. Karim, "Some Considerations of the Use of Natural Gas in Diesel
        Engines", Proc. of the Third Symposium on Nonpetroleum Vehicular
        Fuels (III), Arlington, VA, (1982).  Published by The Institute
        of Gas Technology, (1983).
7.  G. A. Karim, "The Dual Fuel Engine of the Compression Ignition Type -
        Prospects, Problems and Solutions - A Review".  Transactions of
        the S.A.E., Vol. 92, p. 3.569, (1984).
8.  G. A. Karim and N. Amoozegar, "Examination of the Performance of a Dual-
        Fuel Diesel Engine with Particular Reference to the Presence of
        Some Inert Diluents in the Engine Intake Charge", presented at the
        S.A.E. Fuels and Lubricants Meeting, Toronto, S.A.E. Paper No.
        821222, (1982).
9.  D. Azzouz, "Some Studies of Combustion Processes in Dual Fuel Engines",
        M.Sc. Thesis in Mech. Engg., Imperial College, London University,
        (1966)
10. H. F. Coward and G. W. Jones, "Limits of Flammability of Gases and
        Vapours", U.S. Bureau of Mines Bulletin 503 (1952).
11. G. A. Karim and N. Amoozegar, "Determination of the Performance of a
        Dual Fuel Diesel Engine with the Addition of Various Liquid Fuels
        to the Intake Charge", presented at the S.A.E. Int. Congress and
        Exposition, Detroit.  S.A.E. paper No. 830265, (1983).
12. M. M. Metwally, "Analytical and Experimental Studies of Homogeneous and
        Diffusion Combustion Involving Gaseous Fuels", Ph.D. thesis in Mech.
        Engg., University of Calgary, (1978).
13. G. A. Karim and I. Wierzba, "Comparative Studies of Methane and Propane
        as Fuels for Spark Ignition and Compression Ignition Engines".
        Transactions of the S.A.E., Vol. 92, p. 3.676, (1984).

14. G. A. Karim and S. R. Klat, "The Knock and Autoignition Character-
    istics of some Gaseous Fuels and their Mixtures", J. of the Inst.
    of Fuel, Vol. 39, p. 109, (1966).
15. G. A. Karim, S. R. Klat and N. P. W. Moore, "Knock in Dual-Fuel Engines",
    Proc. of Inst. of Mech. Engrs, Vol. 181, p. 453, (1967).
16. G. A. Karim, N. P. W. Moore and M. Khan, "Gross Chemical Kinetics from
    Motored Piston Engines", Transactions of the S.A.E., Vol. 79,
    p. 164 (1970).
17. D. Downs, A. D. Walsh and R. W. Wheeler, "A Study of the Reactions that
    Lead to Knock in the Spark Ignition Engine", Phil. Trans. R. Soc.,
    p. 517 (1951).
18. A. E. Felt and W. A. Steele, "Combustion Control in Dual Fuel Engines",
    S.A.E. Transactions, Vol. 70, p. 644, (1962).
19. G. A. Karim, "An Analytical Approach to the Uncontrolled Combustion
    Phenomena in Dual Fuel Engines", J. of the Inst. of Fuel, Vol. 37,
    p. 530, (1964).
20. G. A. Karim, "The Ignition of a Premixed Fuel and Air Charge by Pilot
    Fuel Spray Injection with Reference to Dual-Fuel Combustion",
    Transactions of the S.A.E., Vol. 77, p. 3017, (1968).


BIBLIOGRAPHY

Anderson, A.C., Chen, T.N. and Hutchins, W.T., "The Development and Appli-
    cation of Design Criteria for Precombustion Chambers on Natural Gas
    Fueled Engines", A.S.M.E., Paper No. 84-DGP-1.
Antonucci, G. and Zandona, L., "Heavy-Duty Dual-Fuel Diesel Engines for
    Smoke Reduction in City Bus Service", Automotive Engineering Congress,
    SAE Paper No. 740121, Feb. 1974.
Bonvecchiato, Gustavo and Magistris, Pietro, "The Italian Experience",
    published in METHANE Fuel for the Future, Plenum Press, New York,
    pp. 223-244, 1982.
Brinson, L., "High Performance Gas Burning Engines", Paper B1, 7th Int.
    Congress on Combustion Engines, 603, (CIMAC), London 1965.
Bro, K. and Pederson, P.S., "Alternative Diesel Engine Fuels: An Experi-
    mental Investigation of Methanol, Ethanol, Methane and Ammonia in
    D.I. Diesel Engine with Pilot Injection", SAE Paper No. 770794,
    Sept. 1977.
Conn, E.L., Beadle, R.H. and Schauer, G.A., "The Two Cycle Dual Fuel Diesel
    Engine with Automatic Fuel Conversion", Meeting of Oil and Gas Power
    Division of the ASME, 1949.
D'Amour, R.A., "Natural Gas Engines as Prime Movers for Air Conditioning",
    SAE Paper No. 876A, New York, 1964.
Dedeoglu, N., "Model Investigations on Scavenging and Mixture Formation in
    the Dual-Fuel or Gas Engine", Sulzer Tech. Rev., 3/1969.
Derry, L.D., Dodds, E.M., Evans, E.B. and Royle, D., "Effect of Auxiliary
    Fuels on the Smoke-Limited Power Output of Diesel Engines", Proc.
    Inst. of Mech. Engs., 168, p. 280, 1954.
Dienstdorf, J. and Klaunig, W., "Dual-Fuel Engine with a High Specific
    Output", Diesel and Gas Turbine Progress, Jan.-Feb., 1973.
Downs, D., Walsh, A.D. and Wheeler, R.W., "A Study of the Reactions that
    Lead to Knock in the Spark Ignition Engine", Phil. Trans. R. Soc.,
    A243, 1951.
Eke, P.W.A. et al, "Critical Factors in the Application of Dual Fuel
    Engines", Inst. Mech. Eng. Proc., Vol. 184, Pt. 3, No. 1, 1969-70.
Fleming, R.D. and Bechtold, R.L., "Natural Gas (Methane) Synthetic Natural
    Gas and Liquified Petroleum Gases as Fuels for Transportation",
    SAE Paper No. 820959, West Coast Meeting, San Francisco, 1982.
Goninan, W.G., "Use of Natural Gas as a Primary Vehicular Fuel for a
    Public Utility Fleet", SAE Paper No. 750074, Detroit, Feb. 1975.

Graham, P.J., "The New Zealand Experience", published in METHANE Fuel for
    the Future, Plenum Press, New York, pp. 209-221, 1982.
Henderson, R.D. and Hallinan, J.C., "Development of High-Compression-Ratio
    Gas Engine", ASME Paper No. 59-OGP-5, New York, 1959.
Joseph, J., "Natural Gas Fuels Bus Diesel", Diesel and Gas Turbine Progress,
    Aug. 1971.
Karim, G.A., "A Review of Combustion Processes in the Dual Fuel Engine -
    The Gas Diesel Engine", Prog. Energy Combustion Science, Vol. 6,
    pp. 277-285, 1980.
Karim, G.A., "On the Emission of Carbon Monoxide and Smoke from Compression
    Ignition Engines, including Natural Gas Fuelled Engines", presented
    at the Int. Air Pollution Conf. of Int. Union of Air Pollution
    Prevention Assoc., Wash., D.C., Dec. 1970.  Published by Academic
    Press, N.Y., edited by H.M. Englund and W.T. Beery, pp. 617-623, 1971.
Karim, G.A. and D'Souza, M.V., "The Combustion of Methane with Reference
    to its Utilization in Power Systems", Journal of the Institute of Fuel,
    Vol. 45, No. 376, pp. 335-339, June 1972.
Karim, G.A. and Khan, M.O., "Examination of the Effective Rates of
    Combustion Heat Release in a Dual Fuel Engine", Journal of Mech. Engg.
    Science, Vol. 10, No. 1, pp. 13-23, 1968.
Karim, G.A., "Combustion in Dual Fuel Engines", Proc. of the 8th Int.
    Congress on Combustion Engines, CIMAC, Brussels, p. 59, May 1968.
Karim, G.A., Aliyu, A.I. and Khanna, S., "Some Effects of Low Ambient Air
    Temperature on the Performance and Exhaust Emission of Engines", Proc.
    of 9th Intersociety Energy Conversion Engineering Conference, San
    Francisco, August 1974, published by ASME, pp. 965-969.
Karim, G.A. and Klat, S.F., "Experimental and Analytical Studies of Hydrogen
    as a Fuel in Compression Ignition Engines", ASME Paper No. 75-DGP-19.
Karim, G.A. and Wierzba, I., "Comparative Studies of Methane and Propane
    as Fuels for Spark Ignition and Compression Ignition Engines", proc. of
    the Int. West Coast Meeting of the SAE, Vancouver, SAE Paper No. 831196,
    Aug. 1983.  Trans. of the SAE, Vol. 92, pp. 3.676-3.688, 1984.
Karim, G.A., "Some Aspects of the Utilization of L.N.G. for the Production
    of Power in I.C. Engines", Proc. Int. Conf. on LNG, Inst. of Mech.
    Eng., London, pp. 360-61, March 1969.
Karim, G.A., "Some Considerations of the Safety of Methane (CNG), as an
    Automotive Fuel - Comparison with Gasoline, Propane and Hydrogen
    Operation", presented at the SAE Int. Congress and Exposition, Detroit,
    SAE Paper No. 830267, Feb. 1983.
Karim, G.A., "Methane and Diesel Engines", published in METHANE Fuel for
    the Future, Plenum Press, New York, pp. 113-130, 1982.
Lalk, T.R. and Blacksmith, J.R., "Dual Fueling of a Single Cylinder Diesel
    Engine with Simulated Low-Btu Lignite Gas as the Primary Fuel",
    SAE 820316, 1982.
Land, M.L. and Carameros, A., "Turbocharged Two-Stroke-Cycle Gas Engines",
    ASME Paper No. 65-OGP-6, 1965.  Also J. Engng Pwr, Trans. Am. Soc.
    Mech. Engrs, Vol. 87 (Series A), p. 421, 1965.
Lowe, W. and Rawlins, R., "Fuel Injection Development for Dual Fuel Engines",
    CIMAC, Vol. 1, p. 411, 1975.
Lowi, A., "Supplementary Fueling of Four-Stroke Cycle Automotive Diesel
    Engines By Propane Fumigation", SAE Paper No. 841398, 1984.
Lyn, W.T. and Moore, N.P.W., "Combustion of Weak Mixtures of Methane and
    Propane by Pilot Oil Ignition in a Compression-Ignition Engine",
    Fuel, Vol. 30, No. 7, pp. 158-166, 1951.
Milkins, E.E., Watson, H.C., Sprigg, G. and Considine, M., "Emission Control
    by Dual-Fuelled Operation of a Diesel Engine", Paper presented Aust.
    and N.Z. Clean Air Soc., Rotarura, N.Z., 1975.
Mitchell, R.W.S. and Whitehouse, N.D., "The Development and Performance of
    a Range of Dual Fuel Engines", The Canadian Mining and Metallurgical
    Bulletin, Vol. 48, No. 519, p. 427, 1955.
Moore, N.P.W. and Mitchell, R.W.S., "Combustion in Dual-Fuel Engines",

Proc. of the Joint Conf. on Combustion, I. Mech. E./ASME, p. 300, 1955.

Moore, N.P.W. and Lyn, W.T., "Combustion of Weak Mixtures of Methane and Propane by Pilot Oil Ignition in a Compression-Ignition Engine". Fuel, Vol. 30, p. 158, 1951.

O'Neal, G.B., "The Diesel-Gas Dual-Fuel Engine", Symposium Paper, Non-petroleum Vehicular Fuels III, Institute of Gas Technology, 1982.

Porter, J.W., "Environmental and Safety Aspects of Natural Gas-Fueled Vehicles", Symposium Paper, Nonpetroleum Vehicular Fuels II, Institute of Gas Technology, 1981.

Ritter, T.E. and Wood, C.D., "An Unthrottled Gaseous Fuel Conversion of a 2-Stroke Diesel Engine", Auto. Engg. Congress and Exposition, Detroit, SAE Paper No. 750159, Feb. 1975.

Seal, M.R., "High Performance Methane Engines", Compressed Natural Gas Conf. Proceedings P-129, SAE Paper No. 831074.

Song, S. and Hill, P.G., "Dual Fuelling of a Prechamber Diesel Engine with Natural Gas", ASME Paper No. 85-DGP-3, presented at the Energy Sources and Technology Conference and Exhibition, Dallas, Texas, Feb. 1985.

Storment, J.O. and Baker, Q.A., "Dual Fueling of a Two-Stroke Locomotive Engine With Alternate Fuels", SAE Paper No. 810252 (SP-480), 1981.

Tison, R.R. and Sprafka, R.J., "Developments in Methane Vehicle Technology", paper presented at the 9th Energy Technology Conference, Wash., D.C., Feb. 1982.

Ullman, T.L., Hare, C.T., and Baines, T.M., "Emissions From Direct Injected Heavy Duty Methanol-Fuelled Engines (One Dual Injection and One Spark Ignited) and a Comparable Diesel Engine", SAE Paper No. 820966, West Coast Meeting, San Francisco, 1982.

Van der Weide, Jouke, "The Netherlands Experience", published in METHANE Fuel for the Future, Plenum Press, New York, pp. 253-266, 1982.

Whitehouse, N.D., "Advances in British Dual Fuel and Gas Engines", D.E.U.A., Pub. No. 353, 1972.

Wong, J.K.S., Messenger, G.S., Moyes, B.W. and Chipplor, W., "Conversion of a Two-Stroke Detriot Diesel Allison Model 12V-149T Diesel Engine To Burn Natural Gas With Pilot Injection of Diesel Fuel For Ignition", SAE Paper No. 841001, 1984.

AUTOMOTIVE APPLICATIONS OF STIRLING ENGINES


G. Walker and O.R. Fauvel

Department of Mechanical Engineering
University of Calgary
Alberta, Canada


ABSTRACT

Stirling engines are heat engines receiving heat from an external
source, chemical combustion, thermal storage, nuclear or solar, converting
a fraction to work and rejecting the remainder as waste heat at low temperature.

They operate without noise, have low exhaust emissions, can operate
on any liquid or solid fuel, have low cycle torque variation, a flat part-
load characteristic and can be made as efficient and compact as internal
combustion engines.

With this array of advantageous characteristics Stirling engines appears suitable for use as vehicle engines. Over the past decade, various
vehicles have effectively demonstrated this capability.

Nevertheless Stirling engines are unlikely to pass into general automotive service so long as liquid fuels are available and internal combustion engines can be used. Stirling engines are inherently more expensive
and more complicated than internal combustion engines. At high power densities they must use light gas (hydrogen or helium) working fluids at
relatively high pressure and high speed with difficult sealing problems
and a relatively high maintenance requirement. The cooling system must
have twice the capacity of that required for a diesel engine of equivalent
power. Moreover, there is no established production and use of Stirling
engines in large numbers and little prospect of the massive capital and
operating funds necessary to accomplish this.

Future automotive applications for Stirling engines operating with
alternative fuels are foreseen including:

a)    the small commuter car with thermal storage/Stirling engine propul-
      sion. The thermal battery may be recharged overnight with low cost
      electric energy or natural gas combustion. The combination would be
      half the size and weight of the lead acid battery/electric motor car.

b)    coal and biomass fired Stirling engines in the power range 0.5–5 MW
      for locomotives, marine propulsion, stationary power and the large,

off-highway vehicles used in mining, construction, forestry and agriculture.

Many non-automotive applications for Stirling engines are established or are in prospect including miniature cryogenic refrigerators for missile guidance and night vision equipment, natural gas and propane fired drivers for refrigeration, heat pumps, cogeneration sets and irrigation pumps, trickle chargers and low capacity generators for long term unattended operation in remote locations, underwater and space power systems, solar energy conversion, marine propulsion and yacht or recreational vehicle power generation.


INTRODUCTION

A Stirling engine is a heat engine receiving heat at elevated temperature from an external source, i.e. chemical combustion, thermal storage, nuclear or solar energy, converting a fraction to work and rejecting the remainder at a lower temperature.

Stirling engines have many advantageous characteristics. These include silent operation, low exhaust emission levels, the ability to operate on any liquid fuel, gasoline, diesel oil, kerosene, propane or methane, low cyclic torque variation and a flat part-load characteristic similar to that of a diesel engine. It is conceivable they could be used in automotive service, indeed, over the past decade or more, several cars, trucks and buses with Stirling engine propulsion have effectively demonstrated this capability.

Nevertheless, despite this favourable climate it is difficult to believe the Stirling engine will ever pass into general automotive service operating on liquid or gaseous fuels. The primary reason for this pessimistic albeit realistic, view is the existence, wide availability and very highly developed production technology and service experience of the internal combustion engines presently in use. Many hold the view that, so long as the liquid fuels are available, internal combustion engines are so well entrenched as to be unassailable for general automotive service and that neither the Stirling, Rankine or Brayton engine offers serious competition.

Stirling engines can be made as compact as spark ignition engines and as efficient as diesel engines with the same flat part-load characteristic. However, to achieve this, they must be pressurized with hydrogen to high pressure (20 MPa) and must operate at relatively high speed (60 Hz). Hydrogen is difficult to contain in any circumstances and under these conditions the dynamic seals necessary are relatively complicated and short-lived so that maintenance costs are high with periodic seal replacement and hydrogen replenishment. Moreover, Stirling engines are more complicated than diesel engines, incorporating several heat exchangers, and use relatively exotic materials (stainless steel versus cast iron) for the hot parts so the capital cost is generally reckoned to be two to five times the cost of the equivalent internal combustion engine even at comparable production levels. This latter qualification is, of course, hard to justify. The Stirling engine is nowhere in series or quantity production (except for model engines and miniature cryogenic refrigerator for military use) whereas internal combustion engines are made in very large quantities (some 20 million per year). There seems to be little chance of getting ove the production 'hump' without massive government assistance and no prospect of this is presently in sight.

Limited applications for liquid fuelled Stirling engines are foreseen

in military applications where the advantages of silent operation and low
temperature (low infrared emission) exhaust are self-evident.  These include power generation, and propulsion for a variety of nocturnal surveil-
lance vehicles, coastal vessels, rubber assault boats and the like.

Potential civil applications for automotive Stirling engines exist in
underground mining where the advantage of a low temperature, low emission
exhaust is important from the health aspect.

The real potential for automotive applications of Stirling engines appears to lie in the longer term at both the small and large ends of the
power range and in combination with alternative fuels.

Small Stirling engines (10-30 kW) might be used in light commuter
cars.  The engines would be energised from a 'thermal battery' of lithium
fluoride or other thermal storage medium.  The 'battery' could be recharged
overnight using off-peak, low cost electric power or natural gas combustion.  Meijer (1970) and co-workers predict the size and weight of the
Stirling engine thermal battery combination to be about half that required
by the electric motor, lead acid battery combination.

At the other end of the power scale (0.5-5 MW) applications are foreseen (Walker et al., 1986) for coal-fired Stirling engines for marine
propulsion, locomotives and the large off-highway rubber-tyred vehicles
used for mining, forestry, construction and agriculture.

Many attractive applications for Stirling engines exist outside the
automotive field.  These have been recently reviewed by Walker et al.
(1986) and include miniature cryogenic refrigerators for infrared night
vision, missile guidance and medical or engineering (energy conservation) thermograph systems, natural gas-fired drivers for air conditioning
refrigerators, power generation and cogeneration units, irrigation pumps,
space and underwater power systems, trickle chargers and low power generation for unattended operation in remote locations, solar energy conversion
and boat engines or recreational vehicle power generation.


DEFINITION

A Stirling engine is a form of heat engine operating on a closed,
regenerative, thermodynamic cycle in which the same working fluid is suc-
cessively compressed and expanded at different temperature levels to effect the conversion of heat to work.  The heat is supplied at high
temperature from an external source and the system is, of course, subject
to the Second Law of Thermodynamics so the conversion is not total.  Some
of the heat must be rejected at low temperature.  The basic technology of
Stirling engines has been summarized by Walker (1980), Reader et al. (1983)
and West (1986).  An elementary review of fundamental aspects, and
the advantages and disadvantages of Stirling engines, is included here
in Appendix I.


PRESENT USE

Stirling engines may also be used as refrigerators and heat pumps as
well as power systems.  Use as a cryocooler, a refrigerator capable of
achieving cryogenic temperatures, i.e. less than 110 K, is in fact the
principal commercial application for Stirling engines at present.  In
miniature form they are used in missile guidance and night vision equipment (Walker, 1986).

Stirling engines were invented in the early 19th Century and were widely used along with steam engines until both were displaced about the turn of the century by the internal combustion engine and electric motor.

Renewed interest in Stirling engines dates from the late 1930's when workers at the Philips Research Laboratories, Eindhoven, seeking a small, thermally powered, electric generator for the large, high power consumption, radios of that era, investigated the possibilities of the Stirling engine. They were completely successful increasing the speed and power levels of Stirling engines an order of magnitude compared with historical machines. They established the foundation for a research and development programme on both Stirling power systems and cryocoolers that continued 40 years until Philips abandoned the field in 1979. At its peak, as many as 100 professionals with their attendent support staff and facilities were involved. This is by far the most extensive and comprehensive research and development programme on Stirling engines in history.

The small power generator never went beyond a pre-production batch (in the late 1940's) of several hundred machines. About that time the invention of the transistor and dramatic improvements in storage batteries eliminated the old style, valved, radio sets and with it the perceived need for small power generators. Little effort appears to have been made to find alternative applications for the small generator and it remains probably the best air engine ever taken to an advanced stage of development. However, by that time interest at Philips had moved on to larger Stirling engines of higher power and also to Stirling cryocoolers. Prototype engines up to 90 h.p. per cylinder were made including several four cylinder versions of 360 horsepower (Walker, 1980). The cryocooler development programme began in 1948, the first units were sold in 1954 and went on to become an outstanding commercial success. After many years of competitive development and testing the miniature Stirling cryocooler has emerged as the preferred system for use in infrared sensor and electronic systems.

## Stirling Automotive Engines

Throughout the 1950's and early 1960's interest at Philips and their licensee, General Motors, focussed principally on the marine applications of Stirling engines. No serious consideration was given to their use for automotive propulsion although General Motors did build demonstration cars equipped with Stirling engines, one using a high temperature thermal storage 'battery' of aluminum hydroxide and another with a small liquid fuelled Stirling engine operated intermittently driving an electric power generator to recharge the batteries of the 'hybrid' electric vehicle (Walker, 1980).

Subsequently, about 1965, new licensees, MAN-MWM in West Germany and United Stirling in Sweden, studied the potential of Stirling engines, as alternatives to diesel engines, for use in buses, highway trucks and underground mining equipment. The Stirling engine was attractive for these applications because of the characteristically low noise level and clean exhaust. United Stirling held that, with a conventional bus engine costing approximately 10 percent of the overall cost, bus fleet owners would be prepared to pay an additional 10 percent to gain the advantage of low noise and environmental pollution. This additional 10 percent would, of course, permit the cost of the Stirling engine to be double the cost of the diesel it replaced.

To this end, a 200 horsepower Stirling engine, the Boxer engine,

intended for marine use, at an advanced stage of prototype development at
Philips, Eindhoven, was adapted to a four cylinder in-line configuration.
Several versions of this engine were constructed and demonstrated in buses
and trucks and it underwent extensive test-bed development.  The best
known of buses, a DAF vehicle equipped with the Philips engine, made several promotional tours to the United States.

This engine was four separate single-acting Stirling engines mounted
on a common crankcase.  Meticulous cost estimating and careful redesign of
this engine by experienced and highly competent engineers at United Stirling
resulted in an estimated 'best price' for the engine that was not twice,
but three times, the cost of the equivalent diesel.  This disappointing
result put the engine beyond further serious consideration as a bus engine.

## Siemens-Stirling Automotive Engines

Development work on the four cylinder, in-line engine continued, at
United Stirling, for use as an underwater (submarine) power system but subsequent automotive engine interest was concentrated on a different type of
Stirling, the double-acting Siemens-Stirling engine (see Appendix 1).  The
simplifications of the double acting engine compared with multiple cylinder versions of single cylinder engines justified the redirection of effort.

Various Siemens-Stirling engines were made.  One with a swash plate
drive started life as a torpedo motor at General Motors.  It was later
(mid-1970's) adapted by the Philips Research Laboratories at Eindhoven for
the successor licensee, the Ford Motor Co. in an automotive development
programme.  This development was eventually sponsored by the U.S. Energy
Research and Development Administration (ERDA) forerunner of the present
U.S. Department of Energy (DOE).  Several vehicles were equipped with
variants of this engine and it received extensive test bed development.
Ford abandoned the Philips licence, and the DOE programme, in 1979 but
an improved version of the four cylinder swash plate engine remains an
active development project of Stirling Thermal Motors, Ann Arbor,
Michigan (Meijer, 1986).

In the late 1960's General Motors concentrated their development work
on a four cylinder in-line Siemens-Stirling bus engine that reached the
advanced stage of dynamometer test prior to installation in a test vehicle
before the dramatic, and unexpected decision, in 1971, by General Motors
management not to renew the Philips licence and to withdraw entirely
from the field (Walker, 1980).

Several Siemens-Stirling engines with the potential for heavy duty
automotive use were made by the German licensees MAN-MWM although the
major thrust of their effort appears to have been in marine, principally
underwater power systems (Walker, 1980).

The Swedish licensees, United Stirling, a consortium of the Swedish
defence company F.F.V. and the shipbuilders Kockums, concentrated their
development effort on four cylinder Siemens-Stirling engines of about
40 kW capacity.  Using crank/connecting rod drive mechanisms rather than
a swash plate or wobble plate drive they advanced the technology of
Stirling engines to a very high level and carried out many design and
conceptual studies, experimental developments and vehicle conversions concerned with Stirling engine use for stationary power, marine use, automotive applications and solar power generation.

In 1978, a proposal prepared by the U.S. company Mechanical Technology
Inc. to transfer the United Stirling technology to the United States in a

cooperative programme in concert with a vehicle maker, the American Motors
Corporation.  The proposal was accepted by the U.S. Department of Energy
principally to meet the well-known American government penchant for a
'second-source' supplier.  Later on, when Ford abandoned their DOE sponsored program with Philips the MTI/United Stirling/AMC program became
the only automotive Stirling engine programme in North America and, so far
as is known, the world.  Work performed under this programme has been
extensively reported including the review presented at the conference
(Nightingale, 1986).  Beremand (1986) has reviewed associated research
carried out in support of the programme.

The present D.O.E. automotive Stirling program is scheduled to con-
clude in 1987.  Few expect the programme to be renewed although some
form of Stirling engine technology support programme will probably be
maintained.  There presently appears no prospect for a) any significant
production of Stirling engines using distillate (gasoline or diesel oil)
fuels for automotive use, or b) any substantical research/development programs for distillate fuelled automotive Stirling engines.

Interest in Stirling engines for other applications continues at a
high level evidenced by the 60 papers presented at the Third International
Stirling Engine Conference (Naso, 1986) and the 50 or so papers presented
annually in North America at the Intersociety Energy Conversion Engineering
Conference (A Chem Soc, 1986).

A comprehensive account of the Stirling engine research and development activities of Philips and their licencees with many illustrations
and photographs of the engines and components was given by Walker (1980).


SOME IMPORTANT QUESTIONS

Consideration of the recent history, summarised above, of automotive
Stirling engines raises important questions.  Why did Philips, General
Motors and Ford all abandon Stirling engine research and development.  It
turns out, as with other important technology advances, personalities
are important.  It is significant that Philips abandoned the Stirling
engine field within a year of the retirement of Fritz Philips, a top
corporate executive of the Philips Company and a devotee of Stirling engine technology.  Following his departure Philips' new management quickly
decided that power system development was outside the corporate game
plan and disbanded the research and development team although continuing
production of the highly successful Stirling cryocooler systems.  Fritz
Philips is said to be an influential member of the Stirling Thermal
Motors (Europe) Ltd., formed to exploit the swash-plate Siemens-Stirling
engine (Meijer, 1986).

In a similar way, Arthur Underwood, head of the General Motors
Technical Center, Warren, Michigan, and another long-term Stirling en-
thusiast, retired in 1970.  Within the year General Motors failed to
renew the Philips license and departed abruptly from the Stirling engine
field.  General Motors research on Stirling engines was the most extensive
and comprehensive outside of the Philips company.  It involved the
Electromotive Division at La Grange, Illinois (on large 1000 horsepower
Stirling marine engines) and the Allison Division (on aerospace Stirling
power systems) funded by the US Air Force.  However, the main work was
concentrated at the G.M. Tech. Center, Warren, Michigan (see Walker,
1980).  The final project, a four cylinder bus engine, constituting a
synthesis of much exploratory and developmental research, was on dynamometer test prior to installation in a motor coach.  Out of the blue, the
team were instructed one day by G.M. senior management to 'stop work on

Stirling, start tomorrow on emission controls'. Hearsay has it that G.M.
management were at that time entirely preoccupied with an apprehended
disaster involving failed school bus brakes, a disaster that, in fact,
never materialised. It was easier not to renew the Philips Stirling
licence than to negotiate renewal. With Underwood gone, the mainspring
of the G.M. effort was broken and tragically they departed the field.

The Ford Motor Company quickly (about 1972) became the successor
licensee to General Motors. Soon they received substantial financial support from the U.S. Department of Energy for the 100 kW four cylinder
swash-plate Siemens-Stirling engine for automotive use conceived earlier
at General Motors for use as a torpedo motor.

The research/development effort at Ford was nothing like the scale
of the General Motors programme. The actual engine development was carried out at Philips, Eindhoven with Ford primarily responsible for
vehicle installation and testing. However, the combined team was making
good progress with excellent prospects in sight when Henry Ford, Jun. took
the decision, in the throes of the automotive recession, to stop further
work on Stirling engines (and also gas turbines) so as to concentrate
corporate effort on stratified charge internal combustion engines.

The vestigal remnant of this unfortunate sequence of events is the
small company Stirling Thermal Motors, Ann Arbor, Michigan, led by Dr.
Rolf Meijer, leader of the Philips team for many years and very closely
involved with the General Motors and Ford programmes. Presently Stirling
Thermal Motors is working in concert with a European consortium Stirling
Motors Europe to produce a very advanced 40 kW four-cylinder Siemens-
Stirling swash plate engine in quantity for use as a driver for heat
pumps and stationary power generation.

The United Stirling effort at Malmo, Sweden has continued over 20
years embracing a wide variety of Stirling applications, marine propulsion,
stationary power, automotive, solar energy, mining and underwater power.
In recent years effort has concentrated on solar energy and underwater
power. Recently the solar energy interest was sold to McDonnel-Douglas
and a new company Sub Power AB formed to exploit the underwater power
work. Some 40 people remain engaged full-time in the United Stirling
programme.

So far as is known, the German licensees MAN-MWM never pursued automotive applications for Stirling engines but concentrated their work almost entirely in underwater power systems. The programme is said to be
continuing in close confidence sponsored by the German Ministry of
Defence and to involve large engines up to 1000 kW.

Another long term contributor to Stirling engine developments, the
Ohio based company Sunpower, continues work on Stirling engines on a broad
front. They have never engaged in automotive Stirling development.

The biggest concerted research and development effort currently in
progress is the 'Moonlight' programme of the Japanese Ministry of International Trade and Industry (MITI) for 3 kW and 30 kW Stirling drivers
for residential and commercial heat pumps. Details of this programme
and other worldwide efforts will be found in the Proceedings of the Third
International Stirling Engine Conference (NASO, 1986).


CONCLUSION

Despite the several advantageous characteristics of Stirling engines

it is unlikely that, while liquid fuels are available, they will displace
internal combustion engines for general automotive use.

Future automotive applications of Stirling engines using alternative
fuels are anticipated in small commuter cars with a thermal storage battery
and in large coal or biomass fired locomotives, and off-highway vehicles
used in mining, construction, forestry and agriculture.


ACKNOWLEDGEMENT

Research on Stirling engines at the University of Calgary is supported
by the Natural Sciences and Engineering Research Council of Canada.


REFERENCES

American Chemical Society, 1986, Proc. 21st Intersociety Energy Conversion
      Engineering Conference, Am. Chem. Soc., Washington, D.C.
Beremand, D., 1986, Overview of the Automotive Stirling Engine Program,
      Proceedings 21st Intersociety Stirling Engine Conference, San Diego,
      August, Am. Chem. Soc., Washington, D.C.
Meijer, R., 1970, Prospects of the Stirling Engine for Vehicular Propul-
      sion, Philips Tech. Rev., Vol. 31, pp. 168-185, see also The Philips
      Stirling Engine as a Propulsion Engine, Proc. 5th IECEC, Sept. 20-25.
Meijer, R., 1986, Design Philosophy, Design and Preliminary Test of a New
      Stirling Energy Conversion Unit, Paper 54, pp. 819-836, Proc. Third
      Intl. Stirling Engine Conf., Rome, Italy, June.
Naso, V., 1986, Editor, Proc. Third Intl. Stirling Engine Conf., University
      of Rome, La Sapienza, Via Eudossiana, Rome, Italy.
Nightingale, N., 1986, Development Status of an Automobile Stirling En-
      gine, Proc. Intl. Symp. on Alternative and Advanced Auto. Engineering,
      Vancouver, B.C., Aug. (Plenum Publishing Co., New York, N.Y.).
Reader, G. and Hooper, C., 1983, Stirling Engines, E.F. Spon and Co.,
      London, England.
Walker, G., 1980, Stirling Engines, Oxford Univ. Press, Oxford, England.
Walker, G., 1986, Development Trends in Miniature Stirling Cryocoolers,
      Paper No. 50, pp. 409-423, Proc. Third Intl. Stirling Engine Conf.,
      Rome, June.
Walker, G. and Fauvel, R., 1986, Applications for Stirling Engines:  A
      Personal View, Paper No. 51, pp. 171-196, Proc. Third Intl. Stirling
      Engine Conf., Rome, June.
West, C., 1986, Principles and Applications of Stirling Engines, Van
      Nostrand, New York.


APPENDIX I

IDEAL STIRLING CYCLE

The ideal Stirling cycle is illustrated in Figure A.1.  Consider a
cylinder containing two opposed pistons, with a regenerator between the
pistons.  The regenerator may be thought of as a thermodynamic sponge, alternately releasing and absorbing heat.  It is a matrix of finely-divided
metal in the form of wires or strips.  One of the two volumes between
the regenerator and the pistons is called the expansion space, and is
maintained at a high temperature $T_{max}$.  The other volume is called the
compression space, and is maintained at a low temperature $T_{min}$.  There
is, therefore, a temperature gradient $(T_{max}-T_{min})$ between
the ends of the regenerator, and it is                     assumed that there is
no thermal conduction in the longitudinal direction.  It is assumed the

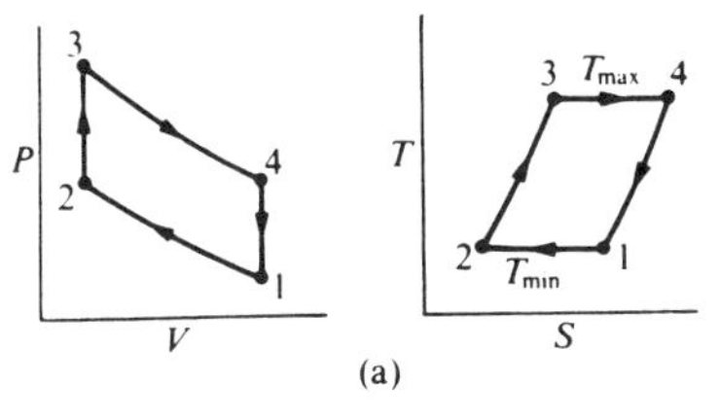

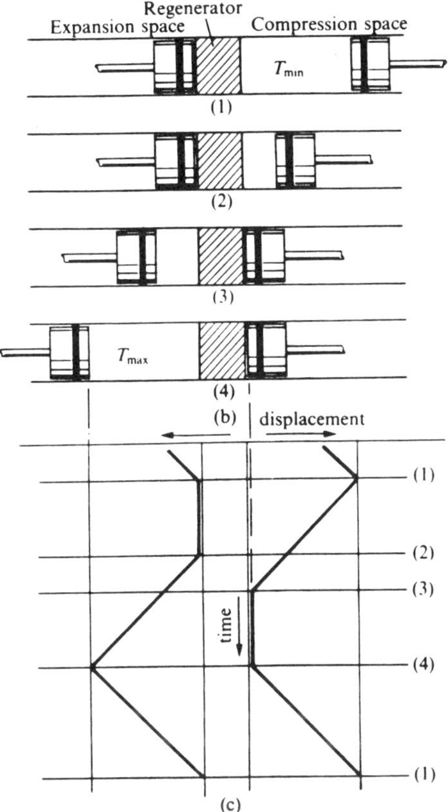



Fig. A.1  The Stirling Cycle
   (a) P-V and T-S diagrams
   (b) Piston arrangement at the terminal points of the cycle
   (c) Time-displacement diagram


pistons move without friction or leakage loss of the working fluid enclosed between them.

To start the cycle, assume the compression-space piston is at the outer dead point, and the expansion-space piston is at the inner dead point, close to the face of the regenerator.  All the working fluid is then in the cold compression space.  The volume is a maximum, so that the pressure and temperature are at their minimum values, represented by 1 on the P-V and T-S diagrams, shown in Fig. A.1.  During compression (process 1-2), the compression piston moves towards the inner dead point, and the expansion-space piston remains stationary.  The working fluid is compressed in the compression space, and the pressure increases.  The temperature is maintained constant because heat $Q_c$ is abstracted from the compression-space cylinder to the surrounds.

In the transfer process 2-3, both pistons move simultaneously, the compression piston towards (and the expansion piston away from) the regenerator, so that the volume between them remains constant. Therefore, the working fluid is transferred, through the porous metallic matrix of the regenerator, from the compression space to the expansion space. In passage through the regenerator, the working fluid is heated from $T_{min}$ to $T_{max}$, by heat transfer from the matrix, and emerges from the regenerator into the expansion space at temperature $T_{max}$. The gradual increase in temperature in passage through the matrix, at constant volume, causes an increase in pressure.

In the expansion process 3-4, the expansion piston continues to move away from the regenerator towards the outer dead point; the compression piston remains stationary at the inner dead point, adjacent to the regenerator. As the expansion proceeds, the pressure decreases as the volume increases. The temperature remains constant because heat $Q_E$ is added to the system from an external source.

The final process in the cycle is the transfer process 4-1, during which both pistons move simultaneously to transfer the working fluid (at constant volume) back, through the regenerative matrix form and the expansion space, to the compression space. In passage through the matrix, heat is transferred from the working fluid to the matrix, so that the working fluid decreases in temperature, and emerges at $T_{min}$ into the compression space. Heat transferred in the process is contained in the matrix, for transfer to the gas in process 2-3 of the subsequent cycle.

The cycle is composed, therefore, of four heat-transfer processes.

Process 1-2:   isothermal compression; heat transfer from the working fluid at $T_{min}$ to the external dump.

Process 2-3:   constant volume; heat transfer to the working fluid from the regenerative matrix.

Process 3-4:   isothermal expansion; heat transfer to the working fluid at $T_{max}$ from an external source.

Process 4-1:   constant volume; heat transfer from the working fluid to the regenerative matrix.

If the heat transferred in process 2-3 has the same magnitude as in process 4-1, the only heat transfers between the engine and its surroundings are (a) heat supply at $T_{max}$ and (b) heat rejection at $T_{min}$. This heat supply and heat rejection at constant temperature satisfies the requirement of the second Law of Thermodynamics for maximum thermal efficiency, so that the efficiency of the Stirling cycle is the same as the Carnot cycle, i.e. $\eta = (T_{max}-T_{min})/T_{max}$. The principal advantage of the Stirling cycle over the Carnot cycle lies in the replacement of two isentropic processes by two constant volume processes, which greatly increases the area of the P-V diagram. Therefore, to obtain a reasonable amount of work from the Stirling cycle, it is not necessary to resort to very high pressures and swept-volumes, as in the Carnot cycle.


PRACTICAL REGENERATIVE CYCLE

For the ideal cycle it was assumed that all the processes were thermodynamically reversible, i.e. that the fluid was everywhere at the same instantaneous equilibrium condition. It was assumed also that the processes of compression and expansion were isothermal which requires infinite rates of heat transfer between the cylinder walls and the working fluid.

114

It was further assumed that the whole mass of working fluid in the
cycle was, at any particular time, all in the compression space or the
expansion space.  The effects were neglected of any voids in the regenerative matrix, the cylinder clearance space, and any pockets in the cylinder
or connecting ducts.

The two pistons were caused to move in some idealized discontinuous way
to achieve the prescribed working fluid distribution.  All aerodynamic and
mechanical friction effects were ignored.  Finally regeneration was
assumed to be perfect.  This implies infinite rates of heat transfer between
the working fluid and the regenerative matrix and that the heat capacity of
the regenerative matrix is infinite.  Under such conditions the temperatures of the gas and the matrix are, at any point, the same and remain constant regardless of the direction of fluid flow.

In practical Stirling engines all these factors combine to reduce the
thermal efficiency to well below the Carnot value of the ideal cycle.  The
actual thermal efficiency may be quoted as a fraction of the theoretical
Carnot efficiency; this ratio is called the relative efficiency:

$$\eta_{rel} = \text{actual thermal efficiency/Carnot thermal efficiency}$$

A value in excess of 0.4 for the relative efficiency is evidence of a well-
designed machine.  The maximum achievable value is about 0.7.

To illustrate the discussion of the ideal cycle, a mechanical arrange-
ment was assumed of two opposed pistons, with an interposed regenerator.
The two-piston machine is one of several different mechanical arrangements.
A practical version of a two-piston machine is shown in Fig. A.2.  It

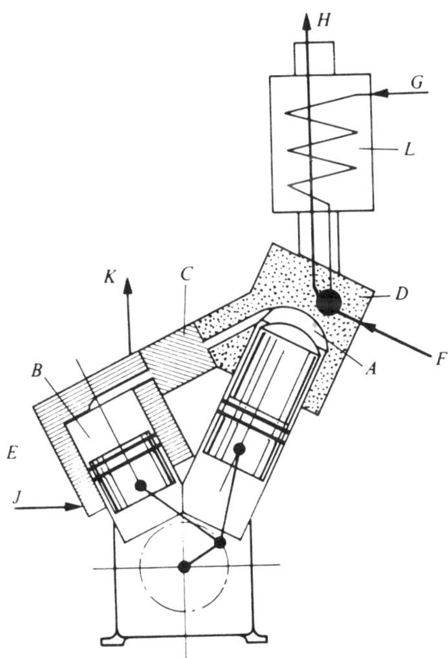


Fig. A.2  Diagram of Practical Two-Piston Stirling Engine
A - expansion space, B - compression space, C - regenerator,
D - heater, E - cooler, F - fuel inlet, G - air inlet, H - ex-
haust products of combustion, J - water inlet, K - water outlet,
L - exhaust gas inlet-air preheater

consists of a Vee engine, with both pistons coupled to a common crankshaft.  The spaces above the pistons constitute the compression and expansion volumes; they are coupled by a duct, containing the regenerator and
additional heat exchangers.

In the operation of this engine, a significant departure from ideality
arises as a consequence of the continuous, rather than discontinuous,
motion of the pistons.  This results in a P-V diagram which is a smooth
continuous envelope as shown in Fig. A.3.  The four processes of the ideal
cycle are not sharply defined.

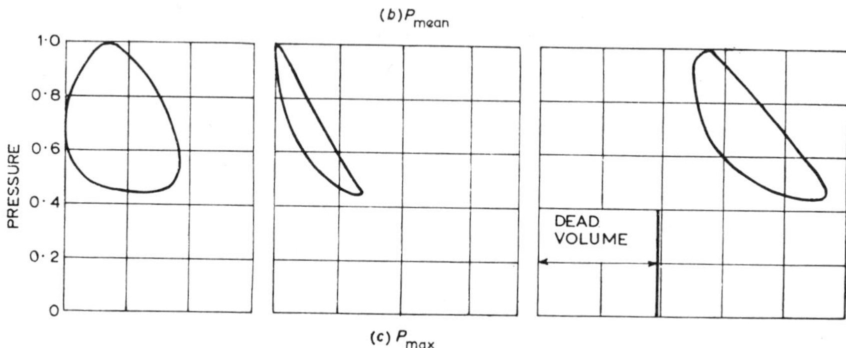


Fig. A.3   Pressure Volume Diagrams for Practical Engines

The processes of compression and expansion do not take place wholly
in one or other of the two spaces, so that three P-V diagrams may be
drawn, one for the compression space, one for the expansion space, and one
for the total enclosed volume, which includes the 'dead' space.  The 'dead'
space is defined as that part of the working space not swept by one of the
pistons, and includes cylinder clearance spaces, void volumes of the
regenerator and other heat exchangers, and the internal volume of associated
ducts and ports.  The P-V diagram for the expansion space represents the
total positive work of the cycle, whereas the diagram for the compression
space represents the compression (or negative) work of the cycle.  The difference in the areas of these diagrams is the net cycle output, the 'indicated' work available for overcoming mechanical-friction losses and for
providing useful power to the engine crankshaft.

In a cycle where the processes of compression and expansion are isothermal and there are no friction losses, the difference in the area of
the expansion- and compression-space diagrams will be found to be exactly
equal to the area of the P-V diagram for the total working space.  In a
practical engine, of course, this equality does not obtain, because
aerodynamic-flow losses in the regenerator and other heat-exchangers cause
differences in the pressure of the working fluid in the compression and
expansion spaces.  Flow losses are important, because they cause a decrease
in the area of the expansion-space P-V diagram, resulting in a decrease
in the net cycle output and efficiency.

The sinusoidal piston motion results in the working fluid being distributed in a cyclically time-variant manner throughout various temperature ranges, and it is not possible to draw a meaningful T-S diagram for
the total mass of the working fluid.  It is possible to draw T-S diagrams
for particular particles of the working fluid, as they move from one
temperature range to another, but no convenient way has been found to
combine these multiple diagrams.

116

The processes of compression and expansion are not isothermal, another
major departure from ideality. In an engine, running at a reasonable speed
(say, 1000 rev/min), it is likely that the processes are nearer adiabatic
(no heat-transfer) than isothermal (infinite heat-transfer). In order to
improve the situation special heat exchangers are often incorporated, including (a) a heater, adjacent to the expansion space, imparting heat to
the working fluid, and (b) a cooler, adjacent to the compression space,
abstracting heat from the working fluid. Despite the advantages of improved
heat transfer, the provision of such heat exchangers imposes some penalties. Additional aerodynamic-flow losses are likely, with the consequent
deleterious effect on performance, as discussed above. The dead volume
will be increased by the void volume of the heater and cooler, and this has
a critical effect on the performance of regenerative engines. Furthermore,
the working fluid is heated, not only when flowing from the regenerator
to the expansion space, but also when flowing from the expansion space to
the regenerator. Similarly, the working fluid is cooled when flowing from,
as well as to, the compression space. The provision of one-way flow
systems is possible, but adds much complication to the machine.

Considerations of increased flow loss and void volume (along with
considerations of cost, size and weight) combine to produce a compromise
heat exchanger design. Consequently, substantial differences may exist
between the temperatures at which heating (combustion products) and cooling
(water or air) is available and the temperatures experienced by the working fluid. In a fossil-fuelled, water-cooled regenerative engine the
temperatures of the combustion products and cooling water might be 2800 K
(5040°R) and 280 K (504°R), respectively. The metallurgical limit of the
materials used for the expansion cylinder and heater may be 1000 K (1800°R).
This provides for a steep temperature gradient of 2800 to 1000 K (5040 to
1800°R) between the combustion products and cylinder wall, with the potential for high rates of heat transfer. Further temperature gradients of
(say), 100 K (180°R) between the working fluid and the expansion space, and
50 K, between the working fluid and the compression space, might exist, so
that the cyclic temperature excursion of the working fluid varies from
(280+50) = 330 K (594°R) to (1000-100) = 900 K (1620°R). Whereas the
Carnot- (or Stirling-) cycle efficiency for the system might be calculated
as

$$\eta_c = (2800-280)/2800 = 2520/2800 = 90 \text{ percent}$$

to give a more realistic picture it should be calculated as

$$\eta_c = (900-330)/900 = 570/900 = 63 \text{ percent}$$

This example demonstrates one of the major difficulties in the commercial application of Stirling engines - one shared by the gas turbine and
the steam engine - the question of materials. Some parts of the machine
(the heater and expansion space), are exposed, continuously, to a high
temperature, and are subject, therefore, to the metallurgical limit of
the heater and expansion cylinder materials.

The allowable temperature excursion of the working fluid in a
Stirling engine is limited to a fraction of that permissible in an internal-
combustion engine using an Otto or Diesel cycle, where the maximum cycle
temperatures are attained only momentarily. Thus, although regenerative
cycles between given temperature limits are thermodynamically more efficient than Otto or Diesel cycles, in practice regenerative engines are
compared with gas (or oil) engines operating with radically different
temperature limits.

Not all the heat available from combustion of the fuel and air can be

transferred to the working fluid, since this would require a very large
heater. The heat passing to exhaust in the combustion products of a
Stirling engine represents a direct loss, because it must be paid for in
terms of gallons of oil (or cubic feet of gas burned), but has served no
useful purpose in the engine. An important engine accessory, therefore,
is another heat exchanger (the exhaust/air preheater), used to warm the
incoming air by heat transferred from the exhaust gas. This heat exchanger
can be of the recuperative type or the regenerative type. In the recuperative type, the two fluids, exhaust gas, and incoming air are separated
by walls into separate ducts. In the regenerative type, the fluids flow
alternately, and usually in contraflow, through the same porous matrix.
It is important to distinguish carefully between the regenerative heat/
exchanger, incorporated as an integral part of the engine, and the recuperative (or regenerative) heat exchanger, used as an accessory of the
engine for exhaust/air preheating.

The continuous motion of the reciprocating elements, the nonisothermal
compression and expansion processes, the limited heat transfer in cooling
and heating devices, the exhaust-stack loss, the increased dead space,
and aerodynamic-flow loss together constitute the principal reasons for
the failure of most practical Stirling engines to fulfil their designer's
hopes and ambitions. Other causes of disappointment include deficiencies
in regenerator operation- high mechanical-friction losses, temperature
equalization as a result of relatively massive conduction paths, and fluid
leakage owing to imperfectly designed (or imperfectly operating) seals.


## ADVANTAGES OF STIRLING ENGINES

As a power system (converting heat to work) Stirling engines have
unique advantages.

### Multifuel Capability

They can use any kind of heat supplied from an external source. It
can be combustion heat from gas or liquid fuels, coal, biomass, etc. or
solar heat when combined with a solar concentrator, stored heat, i.e. a
thermal battery of lithium fluoride, and nuclear heat, a radioisotope or
fission source.

### Quiet Operation

In a Stirling engine there are no valves or periodic explosions so
the mechanisms operate at very low noise levels.

### Good Combustion Characteristics

In combustion heated Stirling engines the combustion process takes
place external to the engine and occurs continuously at atmospheric pressure in a chamber with hot walls. The combustion process therefore tends
to proceed to completion (no unburnt hydrocarbons) with high temperatures.
The propensity to develop nitrous oxides ($NO_x$), as a consequence of the
high temperatures, can be offset by recirculating a fraction of the exhaust products.

### Low Temperature Exhaust

In combustion heated Stirling engines exhaust gas/inlet air pre-
heaters are necessary to maintain reasonable efficiencies. Any heat passing to the exhaust has not 'entered' the engine and so must be recuperated
from the exhaust to the inlet air. This absolute necessity for an

118

exhaust gas inlet air preheater has the advantage that exhaust temperatures are reduced to low values so the infrared emissions from the exhaust
are reduced, an advantageous characteristic for military purposes.

The low exhaust temperatures enhance the attraction of the engine
for underground operations. The disadvantages of the exhaust heat exchanger
are, of course, increase in the size, mass, cost and complexity of the
unit.

## Low Torque Variation

The cyclic pressure variation of the working fluid in a Stirling
engine is near sinusoidal with a maximum/minimum pressure ratio usually in
the range 1.3 to 2. The engine also has a 'two-stroke cycle' so that it
experiences a full range of operation every revolution. Compared with a
four-stroke spark ignition engine the cyclic torque variation is very
modest indeed. A two cylinder Stirling engine has approximately the same
cycle torque variation as a six cylinder four-stroke internal combustion
engine. This not only simplifies engine design but also facilitates engine
balancing.

## Wide Range of Engine Arrangements

The essential elements of a Stirling engine consist of two spaces,
a hot expansion space and a cold compression space whose volumes can be
varied cyclically but out of phase, and which are mutually coupled
through three heat exchangers, a heater, regenerator and cooler. These
apparently simple elements can be arranged in an incredible variety of
mechanical configurations (see Walker, 1980). The range of variation
is simply too great to consider in detail here but does provide for great
flexibility in design.

Basically engines may be divided into single-acting or double-acting
types. The double-acting square-four Siemens-Stirling arrangement is
much favoured for the multi-cylinder engines.

Single-acting machines can be generally classified as:

(a)   two-piston machines,

(b)   piston-displacer machines with i) piston and displacer in the same
      cylinder, or ii) piston and displacer in separate cylinders.

Sometimes the piston and cylinder are coupled by a kinematic mechanism, to each other, and to the crankshaft; these are disciplined piston
engines. In other cases, free-piston Stirling engines, (see Walker et al.,
1985), there is no mechanism and no crankshaft. The piston and displacer
move entirely under the interaction of fluid forces and any restraint
(or load) placed upon them. In other cases, Ringbom-Stirling engines,
there is a free displacer and a crank-controlled piston. In yet others,
Martini-Stirling engines, the displacer is driven separately and independently from the piston which is coupled to the crankshaft and delivers
work to it.

## Flat 'Part-Load' Characteristics

Stirling engines share with diesel engines the advantage of a 'flat
part-load characteristic', that is, the efficiency remains more or less
constant over very wide load changes. The efficiency of a Stirling engine
in fact increases with decrease in speed to a maximum value at approximately
one-third the speed at maximum power.

## Low Internal Wear and Lubricant Consumption

The combustion products in a Stirling engine are not in contact with
the moving parts.  Therefore, there is no contamination of the lubricating
oil as in a diesel engine.  Moreover, the working fluid experiences a near-
sinuosidal harmonic variation so the rates of pressure change are low
compared to a diesel engine.  Consequently, there are no hammer-like blows
on the pistons and the running gear that characterize the diesel engine.
The rates of wear and the demand imposed on bearings and lubricant in
Stirling engines are therefore relaxed.

## No 'Top-Ring Lubrication' Problem

The most difficult lubrication problem in the internal combustion engine is the lubrication of the upper piston ring separating the hot
combustion gases from the crankcase.  Effective sealing of the hot high
pressure gases is required.  Maintenance of an oil film on the cylinder
walls is necessary to sustain a hydrodynamic film preventing metal to metal
contact of the piston ring and cylinder liner.  This problem simply does
not exist in Stirling engines.  The piston rings work at near ambient
temperatures on water cooled walls sealing gases that are uncontaminated
and at near ambient temperatures.

## Use of Water as an Engine Lubricant

The relaxed lubricating requirements permit the use of water as an
engine lubricant rather than the oil customary in engines.  The use of
water eliminates the hazard of internal explosion with air as the working
fluid and the blockage of the regenerator matrix.  It also permits effective resolution of the seal problem in Stirling engines (see below).


## DISADVANTAGES OF STIRLING ENGINES

Stirling engines have few disadvantages compared with internal combustion engines but these few are sufficiently important to make the Stirling
engine virtually non-competitive for many conventional engine applications.

## Cost

The primary disadvantage is undoubtedly cost.  In the automotive
size range there appears little prospect for a Stirling engine to cost
less than twice the price of a diesel engine of equivalent power, even
with the major presumption of similar rates of production; a presumption
that in itself is quite unrealistic at present and in the foreseeable
future.  Recent studies suggesting equal or even lower costs for automotive Stirling engines compared with diesels simply lack credibility.

The high cost of Stirling engines compared with diesel or gasoline
engines is intrinsic and inherent.  It arises from the fact that combustion in Stirling engines is external and so heat must be transferred
to the working fluid from the combustion products through a special heat
exchanger, the heater.  Moreover, the hot parts of the engine are exposed
continuously to the maximum cycle temperature.  Thus the maximum temperature is limited by the high temperature strength properties of the hot
parts.  The engine thermal efficiency is highly dependent on the maximum
cycle temperature so to sustain a reasonable level (30%) of engine efficiency it is necessary to use relatively expensive materials (stainless
steel or high alloy) for the hot parts.  These materials are not trivial
to fabricate or join.

Combustion in internal combustion engines takes place inside the engine cylinder.  The fuel, air and combustion products are the actual working fluid and are periodically discharged from the cylinder and replaced by fresh charge.  The maximum working fluid temperature is attained intermittently and so briefly that very high temperatures may be used even though low cost cast iron is used routinely for the hot parts and no complicated shapes are necessary.

Thus, although the thermodynamic cycle of operation of the Stirling engine is inherently more efficient that the internal combustion engine, between the same temperature limits, in practice the engines do not operate over the same temperature limits.  The momentary, but very much higher, temperatures permissible in the internal combustion engine more than offset the thermodynamic advantage of the Stirling engine cycle.  Thus, even the best Stirling engines can only achieve thermal efficiency similar to those routinely achieved in diesel engines (35%).

Efforts to use high temperature ceramic materials for Stirling engine hot parts to gain efficiencies of 50% or more can be countered by their use in advanced 'adiabatic' diesel engines where the necessary shapes and fabrication requirements are much less complicated.

Another contribution to increased cost is that Stirling engines require a cooling system having a thermal capacity approximately twice that of the internal combustion engine of comparable power output.  The heat supplied to a diesel engine, by combustion of the fuel in the cylinder, is converted to work (about 1/3), passed to the engine cooling system (about 1/3) or blown to exhaust (about 1/3).  In a Stirling engine of comparable efficiency, the same one-third of the heat supplied will be converted to work but very little of the heat must be allowed to escape in the exhaust for in doing so it will not have 'passed through' the engine and so will be a direct loss.  Usually an exhaust gas inlet air preheater is provided to reduce the 'stack losses' to about five percent of the heat supplied. Thus the remainder (100-33-5 = 62 percent) of the heat supplied must be rejected from the engine through the cooling system, approximately twice as much as in the diesel engine.

In automotive applications, trucks, railway locomotives, etc. the size of the 'radiator', for air-cooling the engine cooling water, is already uncomfortably large.  To suggest the need for one twice the size is very unattractive as well as uneconomic.

There is a further important consideration.  Whereas the efficiency of a diesel engine increases as the coolant temperature increases the Stirling engine operates less efficiently.  Now for an air-cooled radiator to function properly the temperature of the liquid engine coolant must, of course, be higher than the ambient air temperature.  Thus in automotive use this temperature difference causes a decrease in Stirling engine efficiency compared with a marine application where direct liquid cooling at the ambient temperature is available.  Efforts to economise on the size, weight and cost of the increased 'radiator' will simply result in an increase in the temperature of liquid engine coolant with the inevitable loss of Stirling engine efficiency.

Stirling engines have other heat exchangers.  One is the regenerator, a finely divided matrix of metal wires, screen or balls.  The regenerator is located between the heater and the cooler in the duct connecting the expansion and compression spaces.  Another heat exchanger is the exhaust gas/inlet air preheater referred to above.  This can be either regenerative (thermal wheel) or recuperative (compact plate-fin brazed assembly). Regenerative types require to be rotated or, if static, have flows switched

periodically.  Recuperative types are so compact as to be very hard (near
impossible) to clean and rapidly become 'fouled' with many combustible
fuels.

It is likely that the cost penalty for Stirling engines compared with
diesel engines would be less for large engines (0.5 MW) than for automotive
size engines.  Large and heavy duty diesel engines are not manufactured in
the same volumes as automotive engines and cost considerably more per unit
power.  They are made in the way that large Stirling engines would likely
be made, one at a time, or at most in batches of a dozen or more.  It is
perhaps conceivable that Stirling engines could come close to diesel cost
in these circumstances.

Savings on the cost of fuel (coal or natural gas compared to diesel
oil) in lubricant oil consumption, and in reduced maintenance costs would
justify a substantial premium in the capital cost of the Stirling engine.

<u>Seals</u>

Another major problem of Stirling engines are the seals used to 1) contain the working fluid, and 2) prevent the ingress of lubricant from the
crankcase.

To increase the power density of Stirling engines is is necessary to
raise the pressure of the working fluid, sometimes to a high level (100 to
200 atmospheres).  Air can be used and minor leakage is then not a big
problem for the working fluid can be replenished by a small engine driven
air compressor.  Unfortunately, air is a relatively 'heavy' gas (molecular
weight 28.8) and is not suitable for the high power density Stirling engines (60 kW/litre displacement) typical for automotive use.

For such engines the light gases helium and hydrogen (molecular
weight 4 and 2, respectively) are necessary.  These allow operation of
highly pressurised engines (200 atmospheres) at high speed (60 Hz)
achieving power densities three or four times that possible with air in
lower speed, lower power engines and at higher efficiencies.  Parentheti-
cally, it is important to note the use of light gases has no advantage
in terms of power density or efficiency at lower speeds and in low power
density engines.

Use of the light gases in high speed, highly pressured Stirling
engines poses a fluid sealing problem that has not been effectively solved
despite the investment of a king's ransom over the past 40 years by highly
competent teams of research scientists.  Seals are required that will
effectively contain the highly pressurised, low molecular weight (and
therefore, highly mobile) gases, seals that are non-lubricated, have low
power losses (i.e. low friction) when operating at relatively high speed
and which have a long life with a low maintenance requirement and may be
readily replaced by relatively unskilled personnel.

Such seals do not presently exist.  Extraordinary ingenuity and
resourcefulness have been invested in seal development for Stirling
engines over many years and significant improvements in seal performance
have been made.  Nevertheless, completely adequate seals for light gas,
high power density Stirling engines are not available.  This probably is
the single most important reason why Stirling engines using light gases
as the working fluid appear to have a relatively limited future.  Engine
users are now so familiar with long-lived, low maintenance internal combustion engines that for conventional automotive engine usage, they simply
would not accept the cost and inconvenience of seal replacement, and
working fluid replenishment aspects of light gas Stirling engines even if

they could be induced to pay the initial two or three times capital cost
premium.

In addition to preventing the egress of working fluid, another very
important function of seals in Stirling engines is to prevent the ingress
of lubricant from the crankcase.  Any oil that enters will eventually end
up in the fine interstices of the regenerator and carbonise to block the
flow of working fluid through the matrix causing a decrease in engine output.

Little of the above discussion of sealing problems relates to
Stirling engines using air as the working fluid.  Air is readily available
worldwide and costs nothing.  Leakage of the working fluid may be readily
replenished by a small engine driven air compressor.  Moreover, it is a
relatively heavy gas and therefore easier to seal.

The possibility of oil ingress to the working space with compressed
air at high temperatures in the heater poses a hazard, graphically termed
'dieseling', of catastrophic internal explosion.  This may be avoided by
the use of water as the engine lubricant.  Water in the working space
poses no threat to the integrity of the engine.  It naturally accumulates
in the liquid state in the cold regions of the engine and may there be
expunged from the working space by 'pumping seals'.

The presence of water in the cold space provides the opportunity for
the seals to be lubricated, thus reducing the friction losses, and in
fact, converted to hydraulic rather than pneumatic seals thus reducing the
sealing problem to relatively trivial proportions.  It is anticipated that
future water lubricated, air engines will have positively lubricated seals
included in the lubrication system.

Another inducement to use water as the engine lubricant is the
opportunity for enhancement of the engine power density with water injection to the hot space, and, consequently, use of a two-phase, two-component
air/water working fluid.  Theoretical studies have indicated the possibility of increasing the engine power density by several times with this
system.  It appears particularly suitable for locomotive or large truck
application where occasional bursts of power are required for accelerating
or climbing with longer periods of idling or low power operation.

## Control Systems

Control systems for Stirling engines take various forms.  Some of these
are complicated, some are relatively simple.  The customary form of engine
control, regulation of the fuel supply, is routinely used for most Stirling
engines but the engine response is very slow because of thermal inertia.
Therefore, a supplementary control system is often necessary to provide
rapid response of the engine to sudden load changes.

One popular control system varies the mean pressure level of the
working fluid.  This requires a storage reservoir of highly pressurised
gas for admission to the engine when an increase in power is required.  In
addition, a small engine-driven compressor is necessary to draw working
fluid from the engine to pump back into the reservoir when a reduction in
load is required.  This pump-up procedure is normally fairly slow-acting
but instantaneous reduction in the power level can be achieved by putting
adjacent (but out of phase) engine cylinders into communication, thereby
'spoiling' the pressure characteristic.

The same effect can be achieved using 'dead volume control' wherein
a series of reservoirs are progressively incorporated or removed from the

working space by the action of solenoid valves.  Increase in the 'dead volume' of the working space reduces the range of pressure variation and the power output of the engine.  Other control systems vary the phase angle between the pistons and displacers and yet others vary the stroke of the reciprocating elements.  Further details of all these systems was given by Walker (1980).

THE DEVELOPMENT STATUS
OF AN
AUTOMOTIVE STIRLING ENGINE

Noel P. Nightingale

Assistant General Manager
Stirling Engine Systems Division
Mechanical Technology Incorporated
Latham, New York

ABSTRACT

The Stirling engine represents an alternative power plant with
superior efficiency and multifuel capability over existing
engines, but success in applying the Stirling to automotive use
has remained elusive. Its potential has been realized through the
development of an automotive Stirling engine, designated the
Mod II. The Mod II engine represents the culmination of the Automotive Stirling Engine Development program begun in 1978. The
program was managed by the Department of Energy and administered
by the NASA-Lewis Research Center in Cleveland, Ohio. The goals
of the program were to develop an automotive Stirling engine and
to transfer Stirling technology to the United States. The Mod II
Stirling engine demonstrates the achievement of these two goals.
Installed in a General Motors 1985 Chevrolet Celebrity car, this
engine has a predicted combined fuel economy on unleaded gasoline
of 17.5 km/ℓ (41 mi/gal) -- a value 50% above the fleet average.
The Mod II Stirling engine is a four-cylinder V-drive design with
a single crankshaft. The engine is also equipped with all the
controls and auxiliaries necessary for automotive operation.

AN ALTERNATIVE POWER PLANT: THE CHALLENGE

This paper presents the culmination of years of work by many dedicated
individuals. It describes an engine that places the United States at the
forefront of a new, dynamic technology. The need for this engine was recognized by Congress in the mid-1970s when it sought to protect our nation from
the vulnerability of a dependency on a sole type of fuel. An alternative
power plant -- one with superior efficiency and multifuel capability over
existing engines -- was envisoned.

The Stirling engine is this alternative.

Invented in the early nineteenth century, the Stirling engine was
regarded as a laboratory curiosity and was not taken seriously by the engineering community. What hampered its development? Two reasons are evident.
As a heat engine, the Stirling must operate at high temperature, e.g., 700°C

(1292°F), and the long-life, high-temperature materials necessary were not available.  Second, early Stirling engines were slow-running machines that produced low power and therefore could not compete with the more versatile spark ignition and diesel engines.  These reasons are no longer valid, as evidenced by the work described in this paper.

In fact, Stirling engine technology now contains advancements as rapid and significant as those in microchip technology and this leap forward will invalidate any existing misconceptions of Stirling in the general technical community.  Although designed for an automotive application, the basic concept of this engine can be used across a broad range of applications.  It represents, therefore, not a subtle change in the technology but a watershed achievement.

There are still those who feel that the Stirling engine will never be a practical power plant and that the need for such an engine has disappeared forever.  In reply to those readers, the following quote from the U.S. Gas Turbine Committee of the National Academy of Science in 1940 is offered for their consideration:  "Even considering the improvement possible ... the gas turbine could hardly be considered a feasible application to airplanes because of the difficulty in complying with the stringent weight requirements."

STIRLING ENGINE DEVELOPMENT: A HISTORY

As established in 1978, Title III of Public Law 95-238, the Automotive Propulsion Research and Development Act directed the Secretary of Energy to create new programs and to accelerate existing ones within the Department of Energy (DOE) to ensure the development of advanced automotive engines.  The act was based on congressional findings that existing automotive engines failed to meet the nation's long-term goals for energy conservation and environmental protection.  Similar congressional findings established that advanced, alternative automotive engines could, given sufficient research and development, meet these goals and offer potential for mass production at a reasonable cost.

To this end, Congress authorized an expanded research and development effort to advance automotive engine technologies such as the Stirling cycle. The intent was to complement and stimulate corresponding efforts in the private sector and, in turn, encourage automotive manufacturers to serious-ly consider incorporating such technology into their products.  The Automo-tive Stirling Engine (ASE) Development Program evolved from this legislation.  The program began at Mechanical Technology Incorporated (MTI) in Latham, New York, in March 1978.  Funding was provided by DOE and admin-istration by the National Aeronautics and Space Administration Lewis Research Center (NASA-LeRC), Cleveland, Ohio, under Contract DEN3-32. United Stirling of Sweden (USAB) was a major subcontractor to MTI due to their extensive background in Stirling engine technology.

The ASE program set out to meet a substantial challenge -- the success-ful integration of a Stirling engine into an automobile with acceptable drivability.  At the outset of the program, the main objectives were to develop an automotive Stirling engine and to transfer European Stirling engine technology to the United States.

The detailed program objectives addressed various facets of engine development.  First, the automotive Stirling engine must demonstrate at least a 30% improvement in EPA combined urban/highway fuel economy over a comparable spark ignition engine.  Second, the engine must be installed in an American-manufactured car representative of a reasonable portion of the

U.S. automotive market. Further, to ensure the most meaningful fuel economy comparison, the acceleration rate of the Stirling-powered vehicle must match that of the spark ignition-powered vehicle, as must the Stirling's drivability in terms of braking, smooth acceleration with no noticeable peaks or lows, and quick accelerator response. Other engine development efforts addressed emission levels, power train reliability and life, competitive initial and life-cycle costs, and noise and safety character-istics to meet 1984 federal standards.

Stirling engine technology has been significantly advanced by the ASE program. These advances are best measured through a summary of the achieve-ments of the four engine classes developed throughout the program: the Reference engine, the P-40, the Mod I, and the Mod II. While interrelated to some extent, each engine had distinct objectives and achievements as summarized below.

* Reference Engine System Design. A "paper" engine, the Reference engine was maintained and modified throughout the program. This engine repres-ented that Stirling engine providing the best fuel economy, while also meeting or exceeding the ASE program objectives. In 1983, the Reference engine design was modified significantly to improve engine cost and manu-facturability. The most dramatic change was from a two-shaft, U-drive system to a single-shaft, V-drive system.

* P-40 Engine. As the program's baseline engine, the P-40 was active from 1978 to 1981. Technology transfer in the program began with the USAB P-40 engine. An ideal baseline for Stirling engine technology familiariza-tion, the P-40 provided engine test experience, vehicle integration expe-rience, and data on vehicle performance.

* Mod I. As the first generation automotive Stirling engine, the Mod I was designed early in the program and began testing in 1981. The overall aim of this engine was to achieve maximum engine performance consistent with the current proven technology. The Mod I represented a composite of early program efforts in component development and baseline engine technology. Although the family of engines is referred to as Mod I, there were actual-ly two major divisions within this configuration: the Mod I and the upgraded Mod I. Ten Mod I or upgraded Mod I engines were built and tested throughout the program. Highlights of their contributions include total accumulated test time in excess of 15,000 hours, extensive seal life and endurance testing, and successful engine/vehicle integration for evalu-ation of control, emissions, and fuel economy performance.

* Mod II. The culmination of the ASE program objectives, the Mod II is a substantial departure from the other engine configurations and incorpo-rates technologies developed throughout the program. The first Mod II engine was tested in January 1986, and it will be installed in a 1985 Chevrolet Celebrity to formally demonstrate the program objectives by September 1987.

Throughout the program, regardless of what engine system was being designed or developed, a comprehensive component development program was active. The component development activities at any given time were dictated by the Reference engine and successful advances were reflected in the hardware of the various engine designs.

THE AUTOMOTIVE STIRLING ENGINE: AN ACHIEVEMENT

The design and demonstration of the Mod II Stirling engine represent the realization of the ASE program goals. The Mod II reflects the advance-ments made in Stirling technology and specifically addresses those problems that heretofore had prevented the Stirling from achieving widespread acceptance (see Table 1). The Mod II not only fulfills the promise of superior fuel economy but nullifies arguments that Stirling engines are heavy, expensive, unreliable, and demonstrate poor performance.

Table 1. Automotive Stirling Engine Technology Progression

|  | 1978 | 1986 |
|---|---|---|
| **Weight/Power, kg/kW (lb/hp)** | 8.52 (14) | 3.35 (5.5) |
| **Manufacturing Cost*** | $5000+ | $1200 |
| **Acceleration, s** <br> **0-60 mi/h** | 36 | 12.4** |
| **Combined Fuel Economy,** <br> **mi/gal** | 19 | 41 |
| **Rare Metals** | Cobalt | None |
| **Seal Life, h** | 100 | 2000+ |

*Based on 300,000 units per year
**3125-lb car

The 1985 Chevrolet Celebrity was chosen as the baseline vehicle. This General Motors A-body car has a manual four-speed transmission, a 2.66 drive axle gear ratio, and an EPA inertia test weight of 1361 kg (3000 lb). A front-wheel-drive car, the Celebrity is representative of the majority of cars sold in the United States; the A-body line accounted for 20% of all GM sales in 1984* (see Table 2).

The Mod II-powered Celebrity has a predicted combined fuel economy on unleaded gasoline of 17.5 km/ℓ (41 mi/gal) versus 13.2 km/ℓ (31 mi/gal) for the spark ignition-powered Celebrity.

The comparison for the highway and urban mileages is equally impressive. Highway mileage is predicted to be 24.7 km/ℓ (58 mi/gal) for the Mod II versus 17.1 km/ℓ (40 mi/gal) for the spark ignition engine; urban mileage is 14.1 km/ℓ (33 mi/gal) versus 11.1 km/ℓ (26 mi/gal). Confidence for these Mod II predictions is based on experience to date with the earlier generation Stirling engines used in the program.

The Celebrity is a highly efficient vehicle, with a fuel economy well above the fleet average for U.S. automobiles. Thus, the Mod II predicted fuel economy is not only 32% above that of the Celebrity but also 50% above the fleet average (see Figure 1).

Comparisons of fuel economy should not be made, however, without a simultaneous comparison of vehicle performance, a factor that, until recently, has impeded the Stirling engine from competing with spark ignition engines (see Table 3 and Figure 2). Earlier Stirling engines were heavy and exhibited poor transient characteristics as compared with their maximum power levels. One standard used by the automotive industry to compare performance is the time required to accelerate from stop to 97 km/h (60 mi/h). Currently, the Mod II provides very competitive performance for the

---

*Selection of the manual transmission purposely deviated from the popular configuration. The stock gear ratios are better suited to the Stirling application, and the shift schedule could be changed easily and optimized for Stirling engine operation.

Table 2.  Chevrolet Celebrity Specification

**GENERAL VEHICLE DESCRIPTION**

| | |
|---|---|
| Model Year: | 1985 4-Door Notchback, Chevrolet Celebrity |
| EPA Car Class: | Midsize |
| Maximum Passengers: | 2 Front/3 Rear |

| | Spark Ignition | Mod II |
|---|---|---|
| **ENGINE SPECIFICATIONS** | | |
| Type | Inline 4 | V4 |
| Displacement, in.$^3$ | 151 | 30.3 |
| Power, kW (hp) | 69 (92) at 4800 r/min | 62.3 (83.5) at 4000 r/min |
| Torque, ft-lb | 134 at 2800 r/min | 156.5 at 1000 r/min |
| **CELEBRITY FUEL ECONOMY** | | |
| Urban without Cold-Start Penalty, mi/gal | 27.2 | 41 |
| Cold-Start Penalty, g | 37 | 125 |
| Urban, mi/gal | 26 | 33 |
| Highway, mi/gal | 40 | 58 |
| Combined Mileage, mi/gal | 31 | 41 |
| **CELEBRITY ACCELERATION** | | |
| 0-60 mi/h, s | 13.0 | 12.4 |
| 50-70 mi/h, s | — | 10.5 |
| 0-100 ft, s | — | 4.1 |
| Gradeability, % | 30 | 30 |

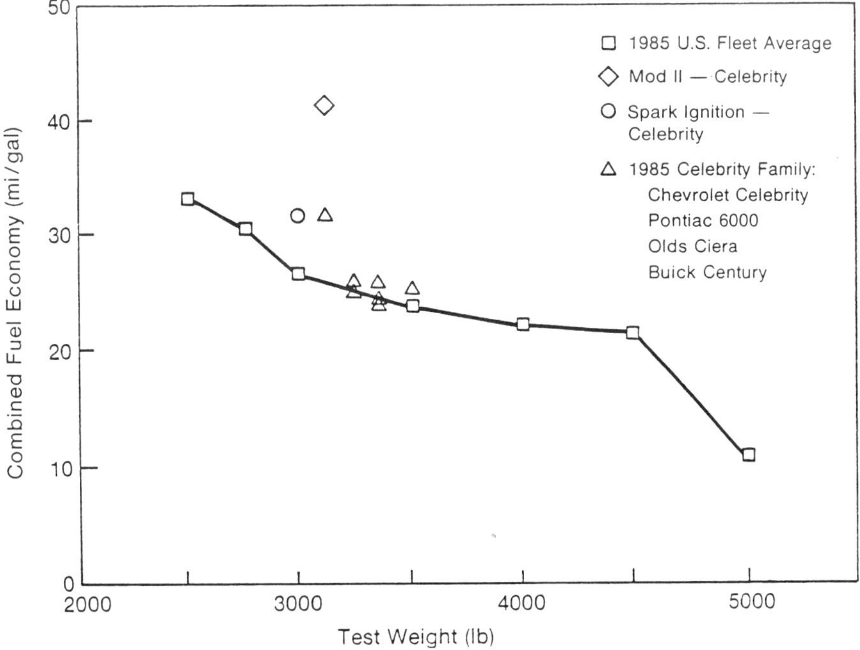


Fig. 1. Comparison of spark ignition and Mod II fuel economies.

Table 3. Mod II Engine Specification

| PART LOAD POINT (p = 12 kW, n = 2000 r/min) | | IDLE POINT (p = 2 MPa, n = 400 r/min) | |
| --- | --- | --- | --- |
| Indicated Power, kW | 15.7 | Net Power, kW | 0.5 |
| Friction, kW | 2.0 | Net Efficiency, % | 8.0 |
| Auxiliaries, kW | 1.6 | Fuel Flow, g/s | 0.159 |
| Net Power, kW | 12.0 | | |
| External Heat System Efficiency, % | 90.4 | | |
| Net Efficiency, % | 33.2 | | |

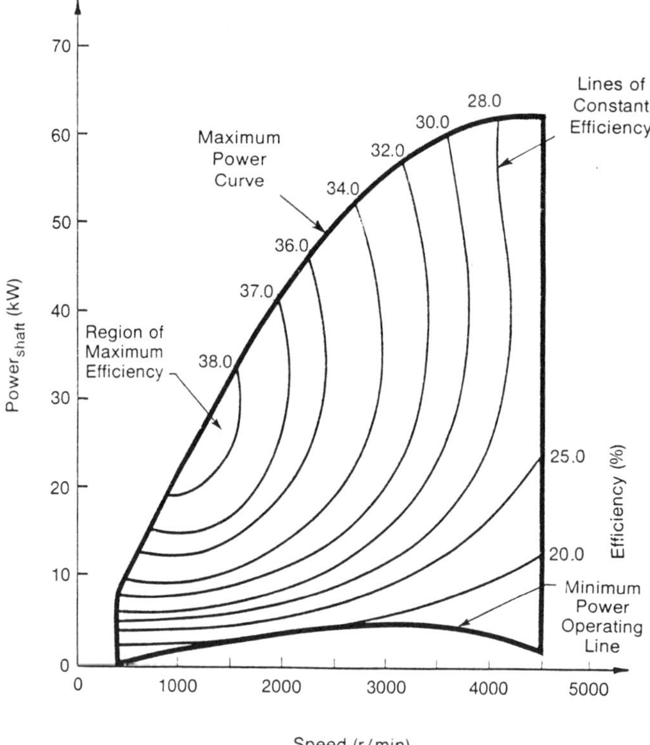


Fig. 2. Mod II engine performance map.

Celebrity, with a projected acceleration time of 12.4 seconds. This is as compared with 13.0 seconds for the spark ignition-powered Celebrity and 15.0 seconds for the generally accepted industry standard.

The Mod II engine is closely matched to the spark ignition engine, having a maximum design speed of 4000 r/min versus 4800 r/min for the spark ignition engine and maximum power of 62.3 kW versus 69 kW. The Mod II offers superior low-speed torque performance, with a peak torque rating of 156.5 ft-lb at 1000 r/min versus 134 ft-lb at 2800 r/min for the spark ignition engine as shown in Figure 3.

The acceleration of a vehicle is a function of engine torque over the acceleration period. Relative to the spark ignition engine, the Mod II provides quicker acceleration at low engine speeds (due to higher torque)

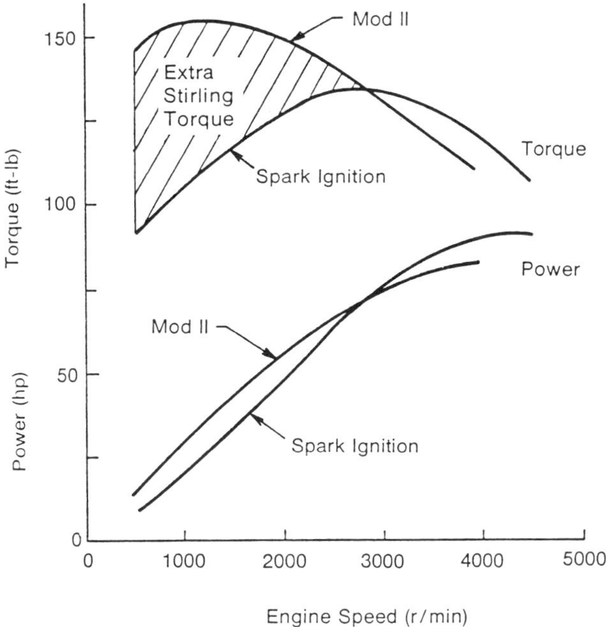



Fig. 3. Comparison of Mod II and spark ignition engine
torque and power.

and slower acceleration at high engine speeds (due to lower maximum power
and lower torque at maximum power). Integrated over an acceleration period
of 0-97 km/h (0-60 mi/h), the total acceleration times of the two engine
types are approximately the same.

The Mod II engine is optimized to provide maximum fuel economy in an
EPA combined urban/highway driving cycle. The engine average operating
condition for this application occurs at a fraction of the maximum power
point (1000 r/min, 10 kW as opposed to 4000 r/min, 62 kW). Technology
development during the course of the ASE program has identified the means to
tailor the highly efficient engine operating regime to match that of the
installation (or application) requirements.

The Mod II is designed to fit in the same engine compartment as the
spark ignition engine, using a stock transaxle (see Figure 4). Very few
peripheral changes to the frame and compartment are required. This is especially important for a front-wheel-drive car because of the sensitivity of
its handling characteristics to suspension and power train mounting geometry.

Along with demonstrating a Stirling engine that delivers superior fuel
economy and performance, the ASE program accumulated more than 15,000 test
hours on Stirling engines, as stated earlier. Such testing has yielded
valuable information on the life and reliability of various engine components, as well as data on the formation of emissions and soot during transient operations such as engine start-up and vehicle acceleration. Further,
the specific weight of the Stirling engine has been reduced from 8.52 kg/kW
(14 lb/hp) to 3.35 kg/kW (5.5 lb/hp), while the cost of manufacturing has
been reduced to be competitive with spark ignition and diesel engines.

Fig. 4. Mod II installation in Celebrity engine compartment.


ENGINE DESIGN SUMMARY: THE MOD II

The Mod II Stirling engine utilizes a four-cylinder V-block design with a single crankshaft and an annular heater head (see Figure 5). There are three basic engine systems. First, the external heat system converts energy in the fuel to heat flux. Next, the hot engine system contains the hot hydrogen in a closed volume to convert this heat flux to a pressure wave that acts on the pistons. Finally, the cold engine/drive system transfers piston motion to connecting rods and the reciprocating rod motion is converted to rotary motion through a crankshaft. The engine is also equipped with all the controls and auxiliaries necessary for automotive operation.

### External Heat System

The external heat system converts the energy in the fuel to heat flux into the closed working cycle. It consists of a preheater, inlet air and exhaust gas manifolds, insulation cover, combustor assembly, fuel nozzle, and flamestone (see Figure 6). To maximize fuel economy, this system requires high efficiency combined with a low hot mass to reduce cold-start penalty.

The air needed for combustion is delivered to the engine combustion chamber from the combustion air blower through two opposing inlet tubes. The inlet air flows through the preheater, where its temperature is increased by heat transferred from the combustion exhaust gas, into a plenum between the insulation cover and the combustor. From there it flows at high speed through ejectors into the combustor mixing tubes, carrying with it part of the combustion gas. In effect, this combustion gas recirculation utilizes air flowing through multiple ejectors to entrain exhaust gas flow, which then flows via a mixing section into the combustion zone. By recirculating exhaust gases through the combustor, flame temperature can be reduced, which, in turn, reduces the amount of nitrogen oxide emissions produced in the combustor.

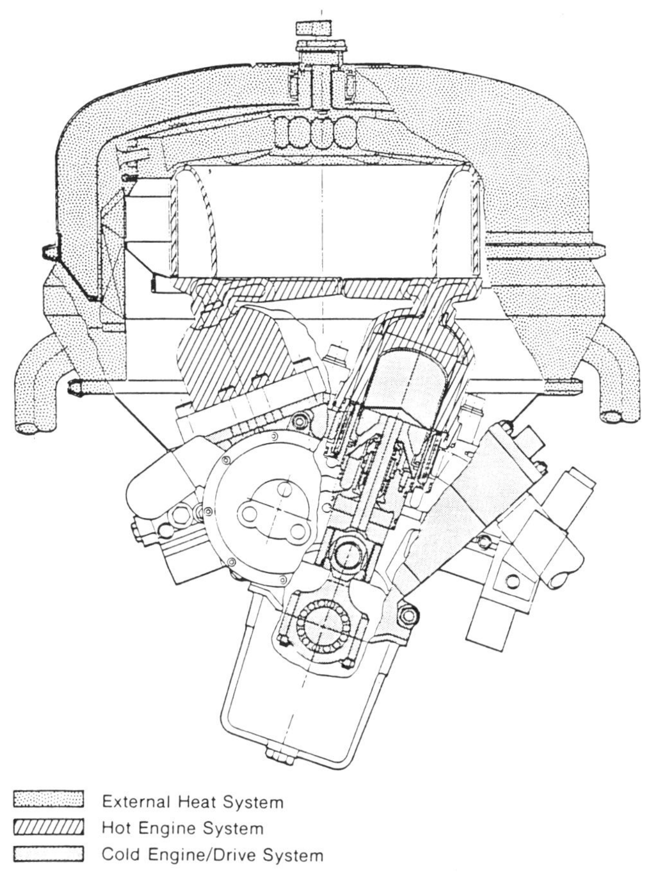



Fig. 5. Mod II Stirling engine.

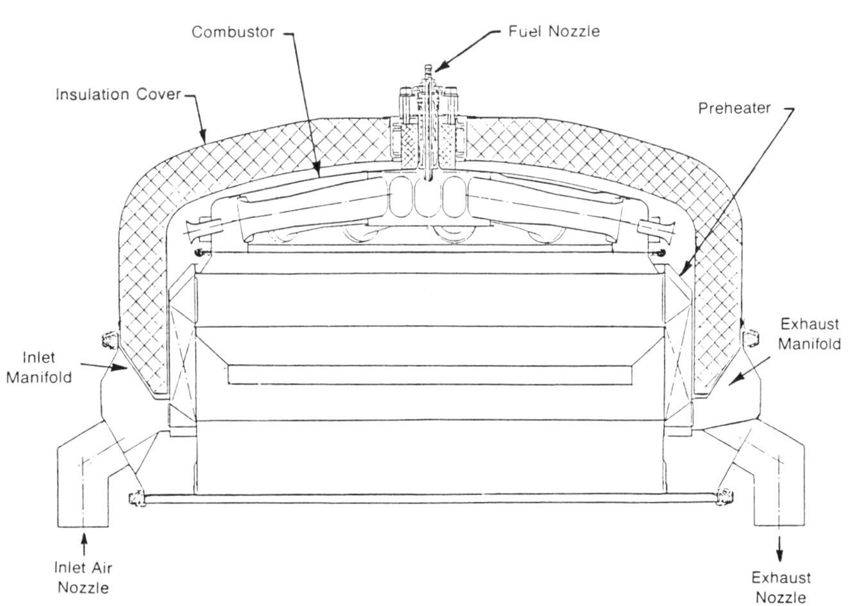



Fig. 6. External heat system.

The air/combustion gas mixture then enters the swirler region of the combustor where air-atomized fuel is injected through the fuel nozzle and ignited. The burning fuel releases heat in the combustor, increasing the gas temperature to a maximum level. This causes the combustion gas to accelerate toward the heater head.

At the heater head, the combustion gas passes through gaps between tubes and fins, transferring heat from the combustion gas to the Stirling cycle through the thin walls of the heater head tubes. After passing the heater, part of the gas mass is recirculated through the combustor mixing tubes, but the majority is forced through the preheater into the exhaust manifold. The exhaust gas temperature is reduced in the preheater as heat is transferred through its walls to the inlet air. The exhaust gas leaves the engine through two opposed outlet tubes that extend into tail pipes. Because the continuous combustion system of a Stirling engine produces such low emissions and is so quiet and clean, the tail pipes do not require any catalytic converter or muffler.

General specifications and conditions for the external heat system are:

| | |
|---|---|
| Fuel massflow: | 0.15 to 5.2 g/s (0.02 to 0.69 lb/min) |
| Excess air factor: | 1.15 to 1.25 |
| Airflow: | 2.9 to 86.5 g/s (0.38 to 11.44 lb/min) |
| Atomizing airflow to fuel nozzle: | 0.36 to 0.8 g/s (0.05 to 0.11 lb/min). |

Hot Engine System

The hot engine system consists of two heat exchangers that are directly involved in the operation of the Stirling cycle: the heater head and the regenerator. Both contain hydrogen and impart heat to the hydrogen that, in turn, provides the force to drive the pistons and thus powers the engine crankshaft.

The heater head transfers the heat contained in the hot combustion gas provided by the external heat system to the hydrogen. The heater head is constructed of many fine tubes. Hot combustion gas passes over the external surfaces of these tubes while the hydrogen passes through the internal surfaces of the tubes. It is the metal temperature of the tubes that sets the metallurgical limit of the heater head design.

A matrix of fine wire mesh, the regenerator is also a heat exchanger by virtue of its construction. As the hydrogen flows from the hot heater head to the cold cooler, it passes through the regenerator where a transfer of heat occurs. This transfer is accomplished by the wire mesh absorbing the energy. When the pistons push the hydrogen in the opposite direction, from the cooler to the heater head, it passes through the regenerator and absorbs heat from the wire mesh.

It is this transfer-absorption phenomenon that enables the Stirling cycle to operate efficiently. After passing through the regenerator on its way to the heater head, the hydrogen is already hot and therefore requires less heat to raise it to operating temperature. Additional heat to raise the hydrogen to operating temperature comes from heat transferred through the heater head tubes from combustion gas as explained above. From this point, the Stirling cycle repeats itself. The hydrogen is heated and expanded, which provides the force to drive the piston.

Heater Head. A heater head has three functions. First, it delivers the hydrogen to the top of the piston to convert high pressure forces into work through the downward motion of the piston. Second, it passes the

134

hydrogen through a finite length of tube so that heat can be transferred to it from the combustion gas. This is normally accomplished by passing hydrogen through many separate tubes, whose external surface is heated by the combustion gas. The internal surface of the tubes is cooled by the hydrogen which picks up heat from the tube wall and carries this energy to do the work of the Stirling cycle. Third, the heater head delivers the hydrogen to the top of the regenerator through which it must flow on its way to the cooler.

There are many variations in heater head configurations to address these functions. The Mod II hot engine system is an annular configuration (see Figure 7).

As shown, the regenerator and cooler are concentric with the piston and separated from it by a partition wall that also separates the regenerator from the expansion space above the piston. The hydrogen travels from the expansion space to the volume above the regenerator through the heater head (see Figure 8).

As hydrogen flows out of the expansion space, it is collected in a manifold. Thus, the first function of the heater head is met. An array of tubes is placed in the front manifold through which hydrogen will pass and pick up heat from combustion gas. The tubes extend upward and then back downward to bring the hydrogen to the volume above the regenerator.

Since the Mod II has four cylinders, it has four heater head assemblies that are mounted on the top of the engine block. Because of the V-drive configuration, each assembly is at an angle to the other. The front and rear row tubes of each heater head assembly combine to form a circle. The resulting configuration is symmetric to the combustion volume and preheater of the external heat system. This helps achieve an even flow through the heater head tubes and preheater and results in no abnormal variations in temperatures from one area of the engine to another.

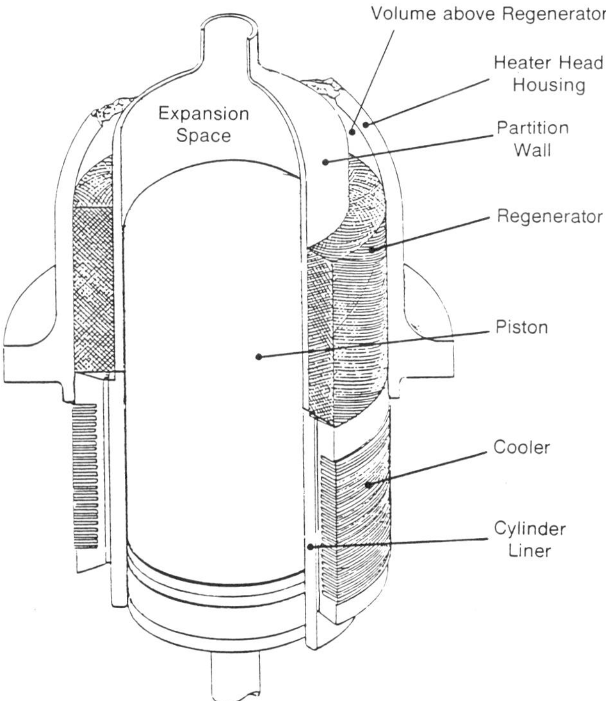


Fig. 7. Typical annular hot engine system.

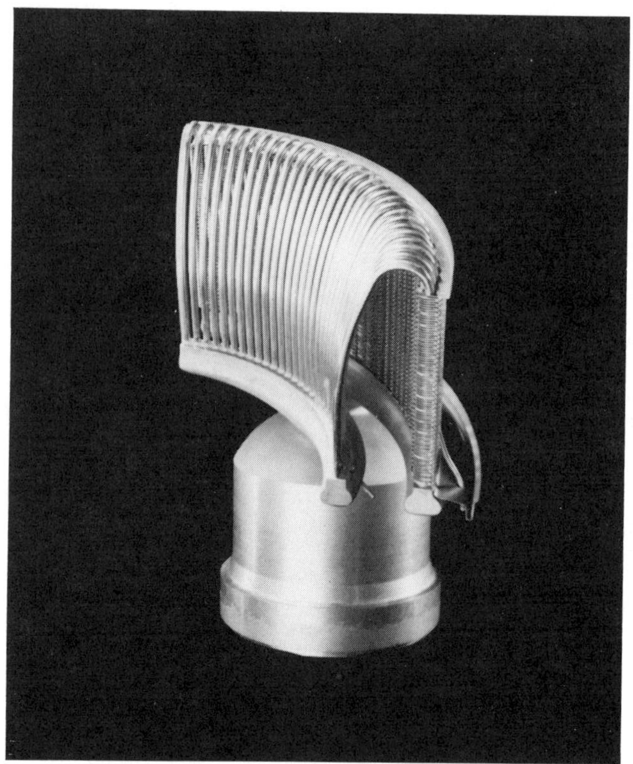

Fig. 8. Typical annular heater head.

The heater tubes attach horizontally on the inner face of the manifold. On the regenerator side, bosses are provided on the top face of the housing where the tubes attach.

Regenerator Assembly. The regenerator assembly consists of the regenerator matrix and partition wall. Individual wire-mesh screens are stacked, pressed, and vacuum sintered into a single annular biscuit. This ring is turned on its inner diameter and slipped over the thin metallic partition wall. The partition wall separates the cylinder/expansion space from the regenerator flow channel, and its cylindrical shell is flanged outward at the cooler end to act as a spacer between the regenerator and cooler. This assembly is vacuum brazed and then final machined to ensure inside-to-outside concentricity (see Figure 9).

## Cold Engine/Drive System

The cold engine/drive system transfers piston motion to connecting rods and then converts the reciprocating rod motion to rotary motion. It consists of the engine block, gas cooler/cylinder liners, seal housing assemblies, piston and connecting rod assemblies, crankshaft, bearings, and lubrication and cooling systems (see Figure 10). Note that the gas cooler is the third and final heat exchanger in the closed Stirling cycle. The hydrogen transfers its heat into the cooler through an array of tubes that are cooled on their outside surface by water. This water is provided by the vehicle cooling system and cycles through the radiator.

Fig. 9. Typical regenerator assembly.

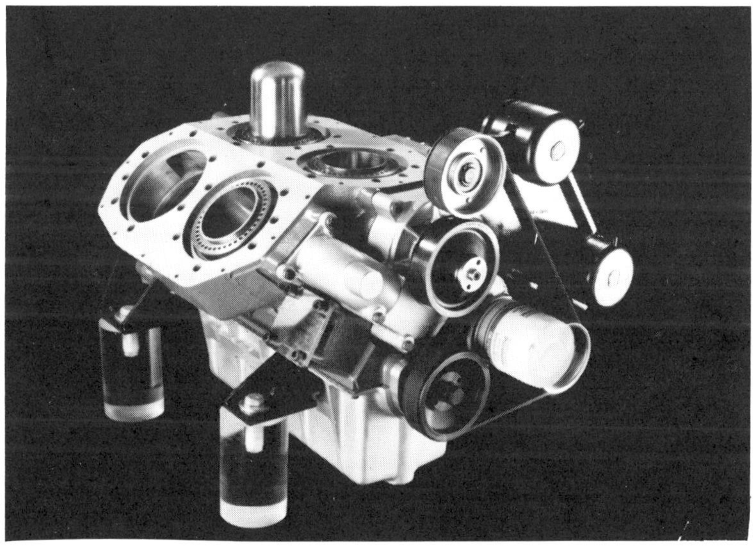

Fig. 10. Cold engine/drive system.

The cold connecting duct, which is cast in the engine block, affects Stirling cycle performance. Dead volume and pumping losses are intensified in this cold region because of the increased hydrogen density. Minimizing both requires minimizing cylinder-to-cylinder spacing to provide short ducts of adequate flow area. Adequate provision for cooling water flow must also be made. To achieve maximum engine performance, all cylinders should run as close as possible to the same low temperature; therefore, coolant flow balance is very important. Minimum cooling water pumping power is

obtained by arranging the coolers in parallel flow, but because the total
pressure drop is much lower in that pattern, balancing becomes more difficult. Attention must be paid to the symmetry of water passages and concentration of pressure drop at the coolers in order to ensure the best
performance. These demands shape the engine block design since the space
claims of cold ducts and water passages are the dominant features.

Engine Block. The basis for engine construction is the unified cast
iron engine block. This single structural element establishes the basic
geometry of the engine and incorporates a water jacket, cold duct plates,
crosshead liners, and a crankcase. It also provides alignment to critical
components and an attachment base for the assembly of all other parts.
Control lines are directly embodied in the casting, greatly reducing
external plumbing complexity.

The cast block is a four-cylinder unit with the cylinders arranged two
each on two banks separated by an angle of 40° (around the crankshaft axis)
(see Figure 11).

Fig. 11. V-block casting.

Gas Cooler/Cylinder Liner. The cylinder liner on which the piston
rings slide contains the cycle pressure (see Figure 12). The wear surface
must be hard; thick and strong enough to contain the pressure and thin and
conductive enough to allow its water-cooled exterior to carry away the heat
of ring friction.

Seal Housing Assembly. The seal housing assembly consists of a sliding
seal, loading spring, cap seal, supply bushing, seal seat, and other small
parts (such as O-rings) in a housing-and-cap container (see Figure 13). The
sliding seal acts to seal hydrogen up to a pressure of 10 MPa (1450 psi)
against an ambient-pressure crankcase with lubricating oil.

Piston and Connecting Rod Assembly. A lightweight piston design inte-
grates the piston base, piston dome, and piston rod into a single,
shrink-fit welded component that is assembled to the connecting rod and
crosshead unit. The crosshead is a separate part, acting in concert with
the wrist pin to form a joint between the connecting rod and piston rod.

The piston configuration includes two sets of rings per piston. The
sets are adequately separated to allow any hydrogen leakage to vent through
the piston rod and relieve any pressure buildup between the rings. Venting
the gap between the rings minimizes leakage of the hydrogen past the rings.

Fig. 12. Cooler assembly.

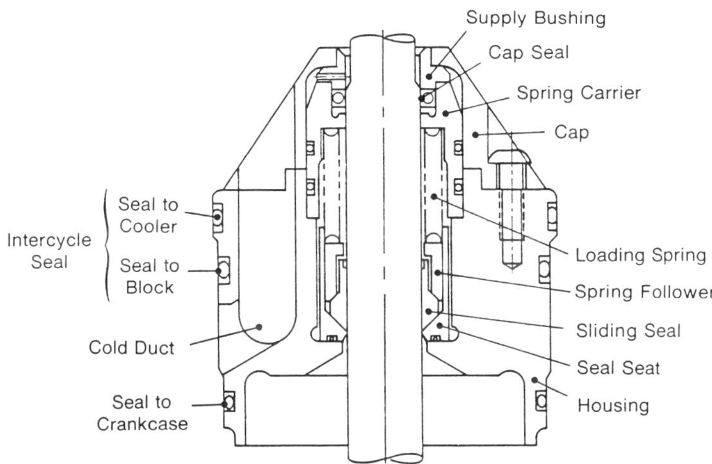


Fig. 13. Seal housing assembly.

Each set is composed of two rings; one ring is solid and the other is split. This configuration is termed a split-solid piston ring. The solid ring minimizes hydrogen leakage between it and the cylinder wall during engine operation. The split ring ensures cylinder wall contact during a cold start when a solid ring would not normally seal against the cylinder wall.

Crankshaft. One crankshaft carries the crankpins for all four pistons and the hydrogen compressor crankpin on three main bearings. Two power-piston crankpins are supported between each main bearing, while the compressor is overhung. The extremely low stroke of the Mod II (30 mm) compared to spark ignition engines of similar power and speed (60 to 100 mm) allows greater overlap between adjacent journal sections and provides an extremely stiff crankshaft, especially in torsion. Simultaneously, the reciprocating elements are joined by a short, forked connecting rod in order to minimize the overall height of the engine. The result is a very compact crankcase, with a shaft that is stiff enough to allow balancing forces and torques to be carried through it. That stiffness allows the use of a unit balance.

Cooling System. The majority of the cooling system is consistent with automotive practice, but the higher heat rejection and higher pressure required by the restriction of tubular coolers in the coolant flow path requires a special pump (see Figure 14). The water pump is constructed from standard-size Gerotor elements in a specially designed aluminum cast housing. The central element is stainless steel, and the outer rotor is plastic to give lubricity against the stainless steel and aluminum parts in the water environment. The aluminum parts subject to sliding are hard anodized to prevent corrosion and galling.

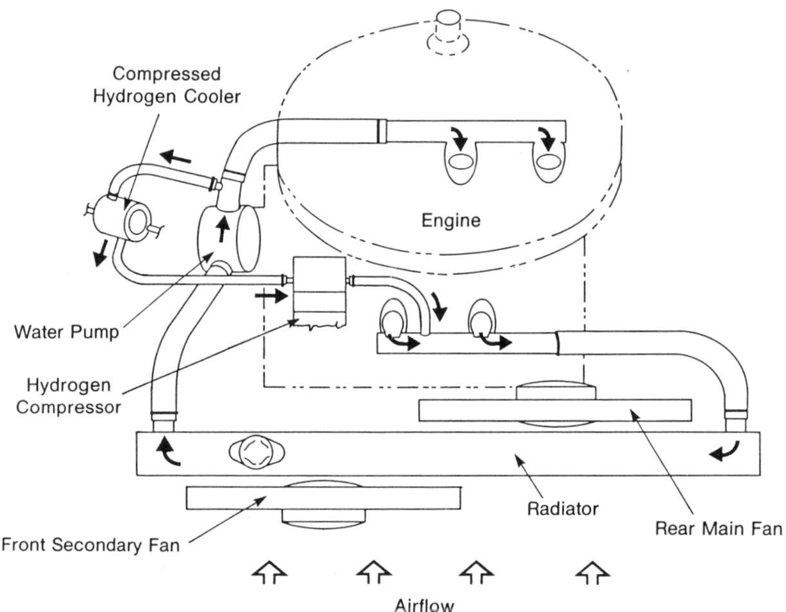


Fig. 14. Cooling system.

## Control Systems and Auxiliaries

The control systems and auxiliaries are not part of the basic Stirling engine but contribute significantly to the total cost of the engine system and dominate its reliability. Furthermore, analyses of the Mod II system have highlighted the importance of these components to the transient response of the engine and their impact on fuel economy. The controls and auxiliaries of the Mod II incorporate the most recent advances in this technology. The advanced design of these components has, to a large extent, enabled the Mod II to achieve projected fuel economy superior to that of the spark ignition engine.

The engine control systems consist of the digital engine control system, the combustion control system, and the mean pressure control system (see Figure 15). Auxiliaries include the combustion air blower, alternator, fuel atomizing air compressor, fuel pump, and starter.

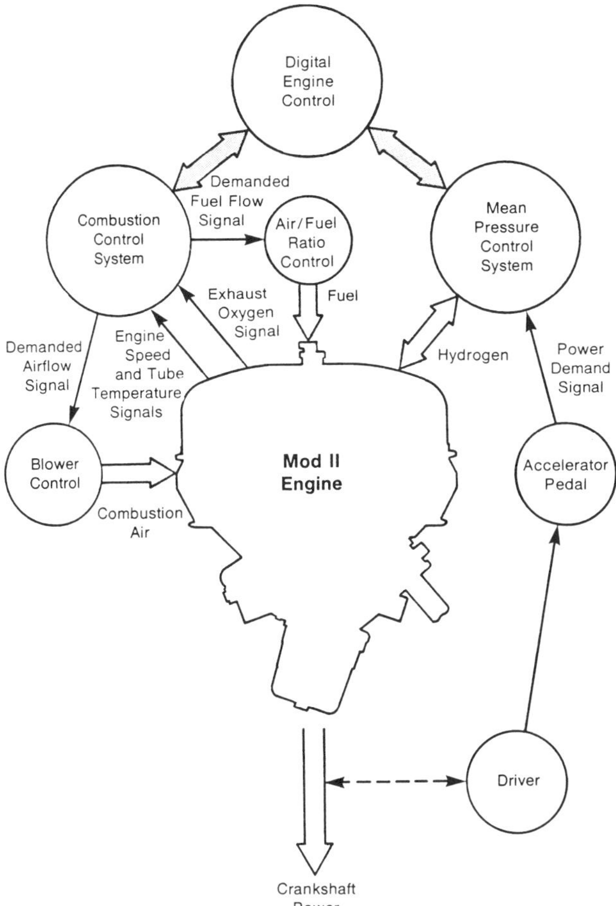


Fig. 15. Engine control systems.

The logic followed in controlling engine power can best be understood by the following example. When the accelerator pedal is depressed to call for more power, the digital engine control senses this change and commands an increase in hydrogen pressure within the closed Stirling cycle. The mean pressure control system then admits pressurized hydrogen into the engine. The resulting stronger pressure wave produces higher engine torque and an acceleration to a higher crankshaft speed. Speed and pressure sensors inform the digital engine control of these increases. Since more heat is being extracted from the engine, the heater head temperature drops. The digital engine control then increases the combustion air blower speed and thus airflow to return heater head tube temperature to its set point. An airflow meter continuously informs the digital engine control of the new airflow rate. The digital air/fuel control converts the airflow rate into a desired fuel flow to be proportioned into the combustor.

A reduction in engine power starts with a transfer of hydrogen from the engine cycle to the hydrogen storage tanks. To reach idle condition, an engine-driven hydrogen compressor is required to pump from the low engine pressure to the higher tank pressure. During this pump-down period, engine power in excess of that desired is dissipated by short-circuiting hydrogen flow from the maximum pressure point to the minimum pressure point of the Stirling cycle.

THE STIRLING ENGINE: A FORECAST

The Mod II represents an engine option ready for prototype or preprod-
uction engineering -- a Stirling engine that is a viable alternative power
plant.  Through the excellence of a government and industry team, the ASE
program goals were achieved.  With the technology transfer complete, the
United States emerges as an internationally recognized leader in Stirling
engine technology suitable for a range of applications.  Examples include
light-duty and heavy-duty vehicles, generator sets, irrigation pumps, solar
electric units, heat pumps, industrial prime movers, submarines and other
marines uses, and farm equipment. Whereas the windfall of Stirling technology to other uses may be a somewhat subjective assessment, it serves to
underscore a key national benefit -- that the ASE program provided the
United States with a core technology having a host of spin-off products.


ACKNOWLEDGMENT

The work reported in this document was performed by Mechanical Technology Incorporated (MTI), 968 Albany-Shaker Road, Latham, New York 12110, as
prime contractor to the National Aeronautics and Space Administration's
Lewis Research Center, Cleveland, Ohio 44135, under Prime Contract No.
DEN3-32, Automotive Stirling Engine Development Program.  The program is
sponsored by the U.S. Department of Energy, Conservation and Renewable Energy, Office of Vehicle and Engine R&D.


BIBLIOGRAPHY

Assessment of the State of Technology of Automotive Stirling Engines.  NASA
        CR-159631, 79ASE77RE2, September 1979.
Mod II Basic Stirling Engine (BSE), Volume I - Design Review Report.
        85ASE444DR4, MTI 85TR24, 2 April 1985.
Mod II Stirling Engine System (SES) Design Review Report.  85ASE465DR5, MTI
        85TR47, August 1985.
Reference Engine System Design (RESD) Summary Report.  NASA CR-174674,
        84ASE356ER59, MTI 84TR11, 1 June 1984.
The Automotive Stirling Engine - Mod II Design Report.  NASA CR-175106,
        86ASE518SR1, MTI 86TR14.

# THE ADIABATIC ENGINE FOR ADVANCED AUTOMOTIVE APPLICATIONS


Roy Kamo

Adiabatics, Inc.
Columbus, Indiana, USA


## ABSTRACT


The high temperature adiabatic diesel emerges as a
possible contender for future automotive powerplants.
The powerplant under consideration is the insulated
diesel type engine without a cooling system.  A waste
heat recovery system is considered for further improvement.  Considerable progress has been made in this
technology by many organizations. Currently the thermodynamics, high temperature tribology, materials, design
emissions and performance aspects of the adiabatic
engine are being investigated worldwide.  The possible
role of the adiabatic engine in future automotive type
engines, and the rotary Wankel type engine are also
presented.  Its potential performance improvement is
discussed.  The advantages of the adiabatic engine
concept for multi-fuel capability in direct-fired coal
combustion and other degraded fuels are covered.
Finally, the technical problem areas holding back its
commercialization are presented.


## INTRODUCTION

The feasibility of the adiabatic turbocompound truck diesel
engine was recently demonstrated by Cummins Engine Company and
the U.S. Army Tank Automotive Command of Warren, Michigan.  A
five-ton U.S. Army truck installed with one of these first
generation adiabatic engines was tested for performance, reliability and cold start.  The engine without a cooling system
demonstrated its capability quite ably.

Since the above demonstration, many organizations and
institutions have embarked on similar research endeavors.
Approaches to adiabatic engines are many and results are also
diverse.  Nonetheless, most of the major engine manufacturers
worldwide are involved in adiabatic engine technology.

The reciprocating heterogeneous combustion machine, the
rotary Wankel engine, gas turbines, and the Stirling engine can
all benefit with advanced high performance ceramics.  Higher
temperature operation of adiabatic engines with ceramics has
shown excellent fuel economy and the potential for cost reduction by eliminating cooling water, radiators, fan and water
pumps.  Reductions in $NO_x$, unburned hydrocarbons, and carbon
monoxide and particulates can also be expected.  The hot engine
with short ignition delay offers a smooth, quiet combustion diagram.  Multi-fuel capability of the hot engine is another important feature.  The density of ceramics is usually lower than
its metal counterpart and provides lightweight features.  Uncov-
ering the potential of each powerplant with advanced materials
could represent a new era for ceramic and engine industries.


ADIABATIC DIESEL ENGINE

An adiabatic engine is one in which no heat is added or
subtracted during a thermodynamic process.  Obviously, a true
adiabatic engine with zero heat loss (100% degree of adiabacity) is not possible.  However, 50% to 60% degree of adiabacity can be achieved with the use of advanced ceramic
materials[1].  In many circles, the adiabatic engine is called
the low heat rejection engine (LHRE) which more accurately
describes today's adiabatic technology.

An advanced adiabatic engine with no cooling water jacket
is shown in Figure 1.  A through bolt anchored in the crankcase
block holds the cylinder head and the cylinder liner in compression. Practically all major engine components and acces-
sories could use high performance ceramics somewhere in their
design.

The adiabatic diesel engine with waste heat utilization is
a rewarding concept with extremely challenging design problems.
It offers reduced brake specific fuel consumption in future
diesel engines as a result of the following basic revisions:

- Insulating the combustion chamber (cylinder
  liner, piston crown, and cylinder head), exhaust
  and intake ports, and the exhaust manifolds.

- Elimination of the cooling system and its
  associated parasitic losses.

- Waste exhaust heat utilization by turbocompounding.

A schematic of the basic adiabatic turbocompound diesel
engine is shown in Figure 2.

There are a number of potential advantages in operating an
adiabatic turbocompound diesel engine.  They are:

- Reduced fuel consumption
- Reduced emissions and white smoke
- Multi-fuel capability
- Reduced noise level
- Improved reliability and reduced maintenance
- Longer life

- Smaller installed volume
- Lighter weight.

Most of the above advantages have been demonstrated in the
laboratory and in the 5-ton U.S. Army vehicle.

The key to energy recovery is an efficient turbocompounding
system. Overall turbocharger efficiency of 0.64 is desired.
The free power turbine geared to the engine crankshaft should
have an efficiency of greater than 84% and mechanical gearing
efficiency of greater than 95%.

Other reasons for the use of a turbocompound system are for
improving the $BSNO_x$ and BSFC trade-off, as well as reducing
the in-cylinder peak pressures. These improvements without
sacrifice in fuel economy are achieved by retarding the engine
timing. Another reason why retarded timing could be used is
the reduced ignition delay in the adiabatic engine.

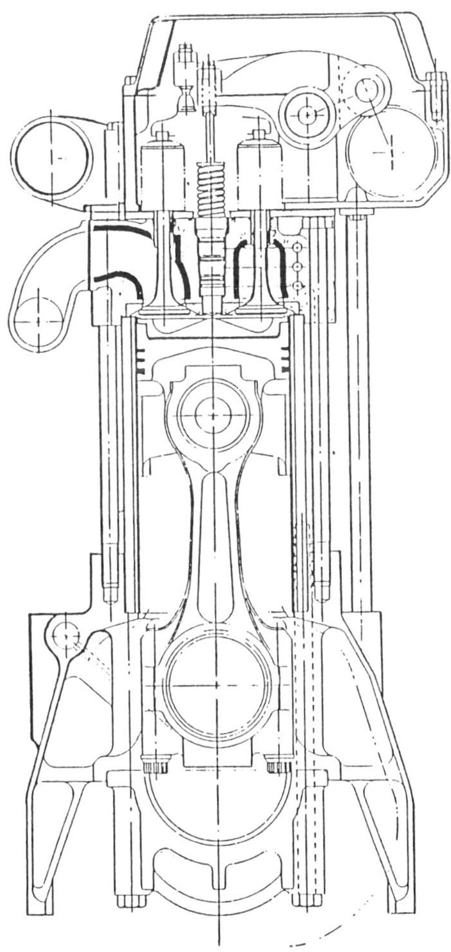

Figure 1.  Cross-section of an advanced adiabatic engine
without a cooling water jacket.

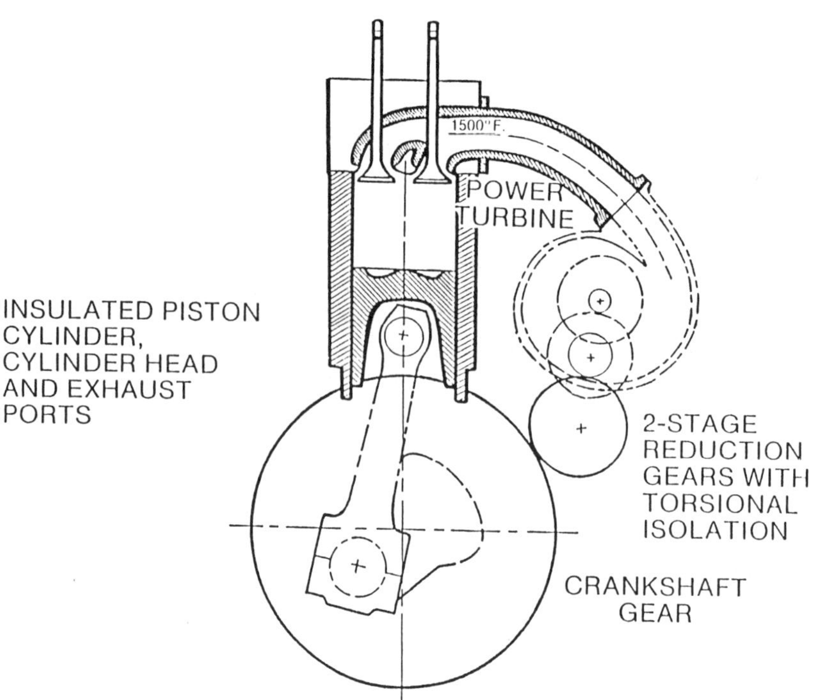


Figure 2.  Total energy recovery via Cummins adiabatic turbocompound engine.


## Engine Operating Environment

The environment within the insulated combustion chamber of the adiabatic engine depends on the load, air-fuel ratio, intake air temperature, injection timing, etc.  Figure 3 shows a least square curve fit of the thermocouple probe temperature within the combustion chamber for the peak torque condition of the engine.  The measured temperature was used in all piston, liner and head analyses.  A piston-cylinder liner temperature prediction shown in Figure 4 was calculated from measured data.

Figure 5 shows the pressure-volume (P-V) diagram of an adiabatic and a water cooled engine.  The peak cylinder pressure is about 2000 psi (pounds per square inch) for design purposes.  Because of the insulating feature and the lack of a cooling system, the adiabatic engine tends to operate at higher pressures and temperatures.

Figure 6 shows the calculated cyclic cylinder wall temperature for various degrees of insulation.  No cooling water is present in the Figure 6 calculation.  With increasing degree of insulation, the surface temperature increases and the temperature swing remains approximately constant.  The importance of cyclic analysis becomes obvious with insulation.  In the case of the cylinder liner, additional functions which cannot be overlooked for optimized engine operation are friction and wear properties.  Corrosion and erosion also cannot be overlooked.

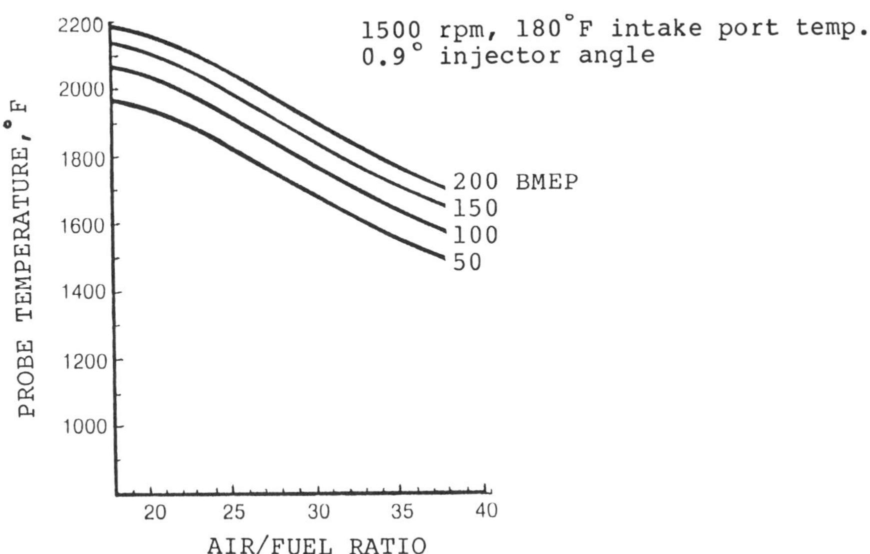



Figure 3.    Least-square fit of the thermocouple probe
temperature within the combustion chamber for
for the peak torque condition of the engine.

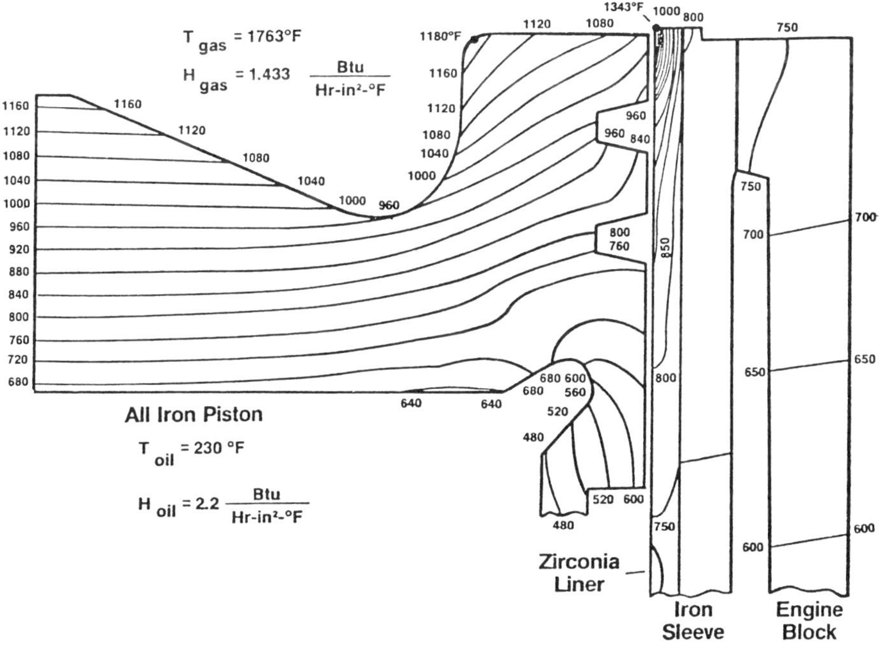



Figure 4.    Predicted piston and zirconia/cast iron
liner temperature.

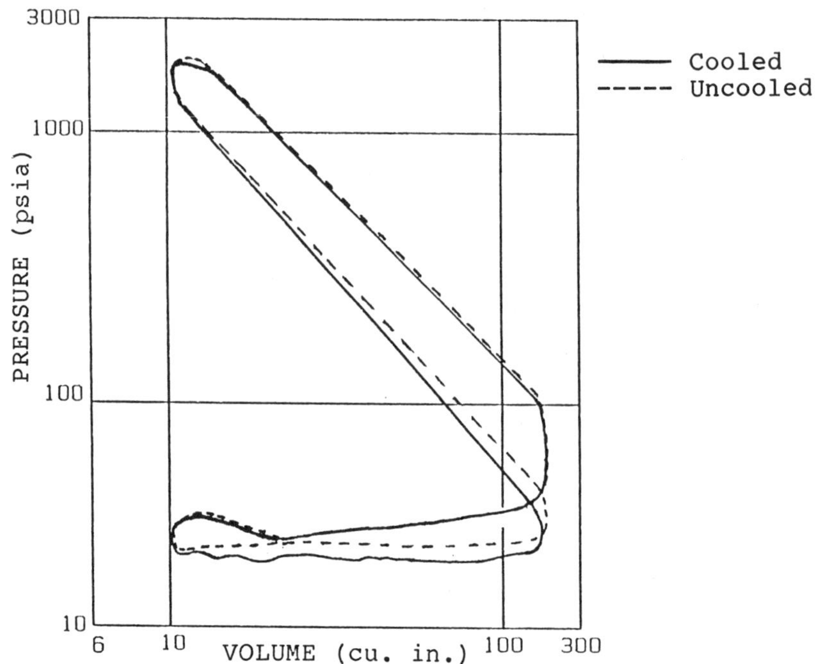

Figure 5.  Pressure-volume diagram of an
NH 450 engine; cooled versus
uncooled, at 1900 rpm and 1100 ft-lb.

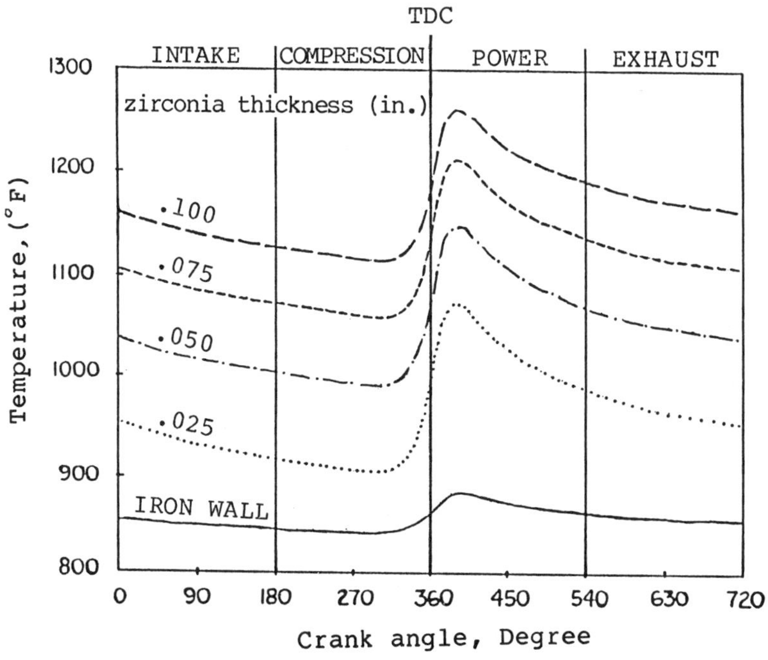

Figure 6.  Calculated cylinder wall temperatures
for various degrees of insulation.

<u>Materials</u>

The combustion chamber environment was presented above. With
wall temperatures exceeding 1100°C at torque peak engine operation, ceramics seem to offer the best compromise.  Super-alloy
metallic design requires strategic materials, i.e., molybdenum,
chromium, nickel, titanium, etc.  Thus, availability of ceramic
materials greatly influenced the selection of materials for the
adiabatic engine.  Furthermore, the alloy strength deteriorates
rapidly when compared to ceramics as shown in Figure 7.

Table 1 illustrates the three basic materials considered for
the adiabatic engine.  Glass ceramic was included solely for its
insulation effectiveness.  Unfortunately, the high performance
ceramics, i.e., silicon nitride ($Si_3N_4$) and silicon carbide
(SiC), possess relatively high thermal conductivities and complex insulation design must be provided.

The common high performance ceramics such as $Si_3N_4$ and SiC
lack insulation properties.  The zirconia and glass ceramics
lack high temperature strength.  Today, there is no satisfactory
ceramic material to meet all the requirements of an adiabatic
engine.

When the right material becomes available, it should be sub-
jected to long term durability concerns for ceramics.  Aging
properties of materials become highly critical.  The important
long term properties of materials that should be determined in
any of these promising materials are: 1) phase change, 2) high
temperature creep, 3) oxidation and wear, 4) corrosion and
deposits, and 5) others.

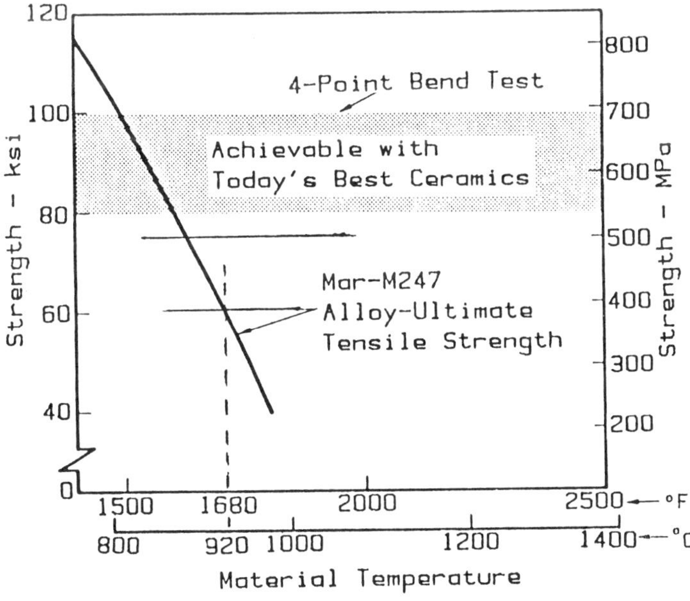


Figure 7.   Strength versus temperature data for
for the metals and ceramics.[2]

Table 1

TYPICAL PROPERTIES OF THREE SELECTED MATERIAL CATEGORIES

| MATERIAL | THERMAL CONDUCTIVITY (cal-cm $cm^2$-sec$°$F) | THERMAL EXPANSION COEFFICIENT ($10^{-6}$/$°$C) | TENSILE STRENGTH AT 1000$°$C ($kg/mm^2$) | MAX DESIGN TEMPERATURE ($°$C) |
|---|---|---|---|---|
| Metallic | .054 | 14.4 | 30 | 1000 |
| High Perf. Ceramics | .043 | 3.2 | 40 * | 1350 |
| Glass Ceramics | .004 | .7 | 9 * | 1000 |

    * Tensile strength assumed to be one half
    the bending strength.

## Ceramic Coatings

The laboratory engine testing with ceramic coated adiabatic engine components gave very good results.[3] Figure 8 shows a single cylinder test on a shrunk-fit PSZ cylinder liner which failed at 172 BMEP** load after 16 hours of testing. Figure 9 shows the same cylinder liner coated with plasma spray $ZrO_2$ and coated with $Cr_2O_3$. No failure has been experienced at targeted 195 BMEP loading. The thermal conductivity of ceramic coatings is generally lower than its monolithic version because of its porosity. Table 2 shows the thermal conductivity of monolithic zirconia, zirconia coating, and $Cr_2O_3$/$ZrO_2$ coating, respectively at two temperature conditions.[2]

## Multi-fuel Capability

In the petroleum short world, there are still many other sources of energy. These other important alternate energy sources are fossil fuels, biomass, shale, etc. The diesel engine currently is one of the most efficient known power-plants, thus, it is highly desirable to burn these alternate fuels in a compression ignition engine.

Unfortunately, many of these alternate fuels are of low cetane number and are unsuitable for most diesel engines. For example, a high compression prechamber engine is quite tolerant of low grade diesel fuels but is totally ineffective in burning coal-derived fuel oil SRC-II or powdered coal (beneficiated).[4]

In a multi-fuel engine, a shortening of the ignition lag is highly desirable in order to eliminate knock and control the rate of pressure rise in the combustion chamber. One or both of the following means are currently used to achieve shorter

---

** BMEP - Brake mean effective pressure in pounds per square inch.

Figure 8.  Photograph of a shrunk-fit PSZ cylinder liner.
Failure occurred at 172 BMEP load after
16 hours of engine testing.

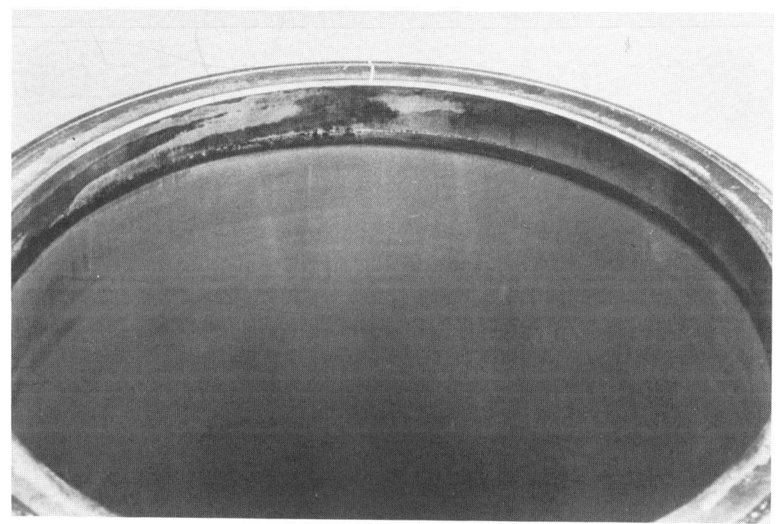

Figure 9.  Same liner as above, but coated with $Cr_2O_3$.
No failure has been experienced at targeted
195 BMEP loading.

ignition delay in diesel engines: Preconditioning the fuel;
Modification in engine design and operating conditions.

With the advent of the hot wall adiabatic engine (Figure 1)
without cooling water, the combustion chamber temperature rises
significantly beyond that needed for good multi-fuel condition.
Figure 6 showed the increase in combustion chamber wall temperature with increasing insulation of zirconia.  Even the temperature swing during the combustion period experiences higher
temperature fluctuation.

TABLE 2.  THERMAL CONDUCTIVITY OF MATERIALS
AT TWO TEMPERATURES

| MATERIAL | ROOM TEMPERATURE | 1470 °F |
|---|---|---|
| | (Btu/hr ft (°F/in) | |
| Iron | 270 | 259 |
| $M_gO$ Stabilized PSZ | 22.7 | 17.01 (800 F) |
| $Y_2O_3$ Stabilized PSZ | 16.6 | 15.4 |
| Plasma Sprayed $ZrO_2$ | 5.91/9.01 | 5.67/6.53 |
| Plasma Sprayed $ZrO_2$ Coated with $Cr_2O_3$ | 9.80 | 9.69 |

The maximum temperatures encountered by the piston top and cylinder walls were depicted in Figure 7.  These high in-cylinder
temperatures in the adiabatic engine can shorten the ignition
delay of the fuel as shown in Figure 10 and provide the necessary condition for an alternate fuel engine.

Figure 11 illustrates the brake specific fuel consumption
of an adiabatic engine burning diesel fuel No. 2 (House fuel),
shale fuel and SRC-II fuel.  SRC-II has 10% less heating value,
thus, the fuel consumption is proportionately higher.  The
theoretical 10% curve is the diesel No. 2 fuel oil curve with
10% less heating value.

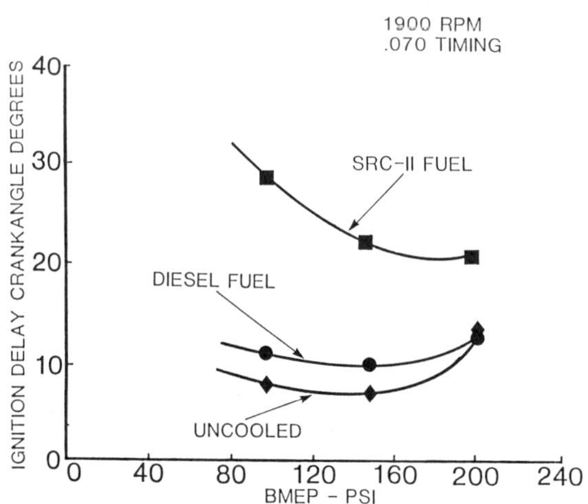


Figure 10.   Ignition delays with varying fuels versus BMEP
(cooled and uncooled engine data).

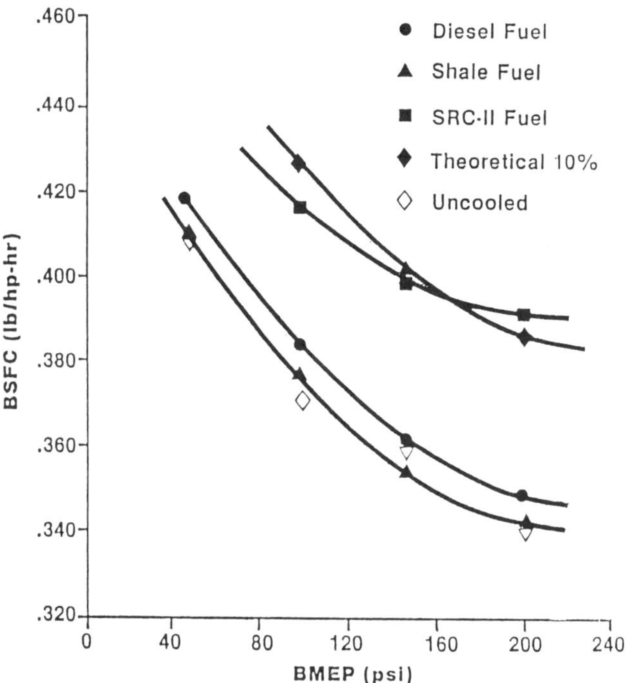

Figure 11. Efficiency comparison - house fuel versus synthetics (1900 rpm .085 static timing).

## Adiabatic Coal Diesel

Combustion of direct-fired coal has always presented mechanical problems as well as combustion efficiency problems in the past. It was reasoned that an adiabatic engine fabricated with hard ceramic surfaces should overcome many of the mechanical problems encountered in the early investigations. Furthermore, the hot adiabatic combustion chamber could potentially decrease the ignition delay period and increase the heat release rates; thereby achieving more complete combustion and higher thermal efficiency.[5] Figure 12 shows the cross section of the adiabatic coal engine and Figure 13 shows the ceramic engine components.

The adiabatic coal engine prepared the coal fuel for combustion and burned it efficiently. Figure 14 shows the rapid heat release of the coal-fired engine.

## Performance

In terms of fuel economy improvement, the adiabatic engine can offer the following improvements:

|  |  |
|---|---|
| In-cylinder | 4-5% |
| Turbocompounding | 8-9% |
| Installed basis | 5-7% |
| Total | 17-21% |

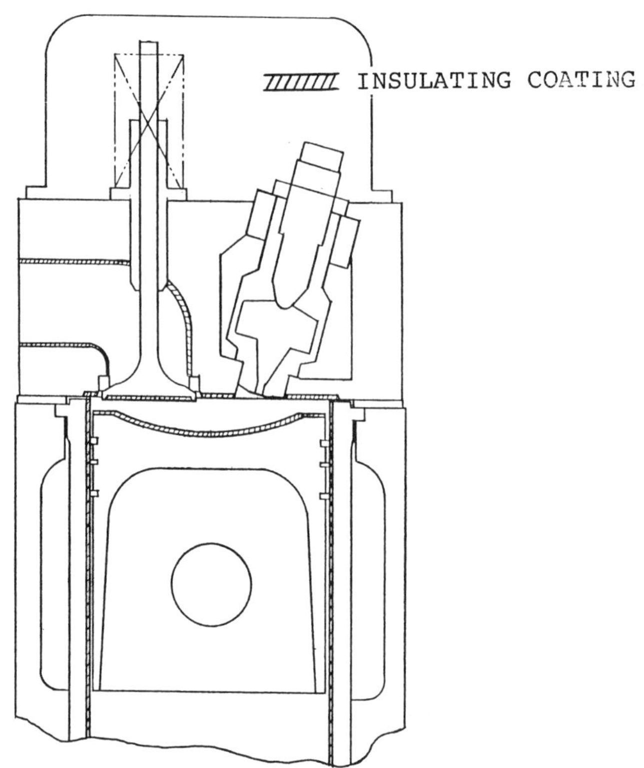



Figure 12.  Cross section of the adiabatic coal engine.

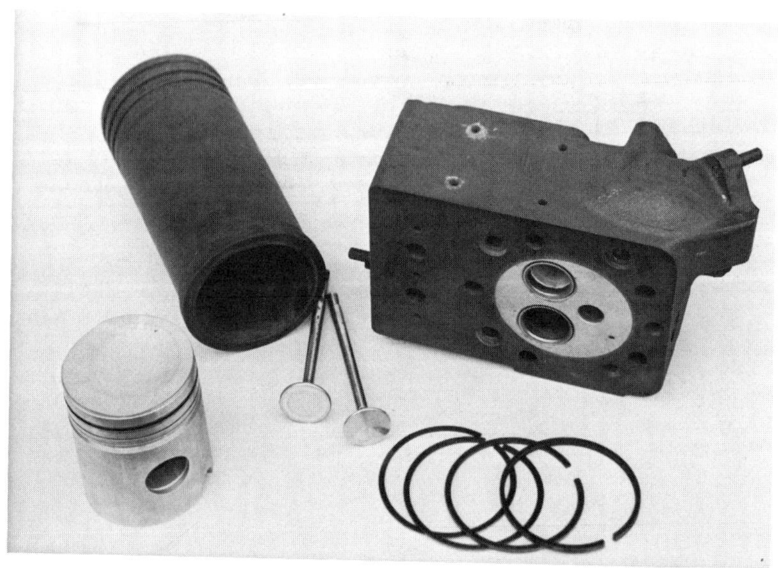

Figure 13.  Photograph of the ceramic engine components.

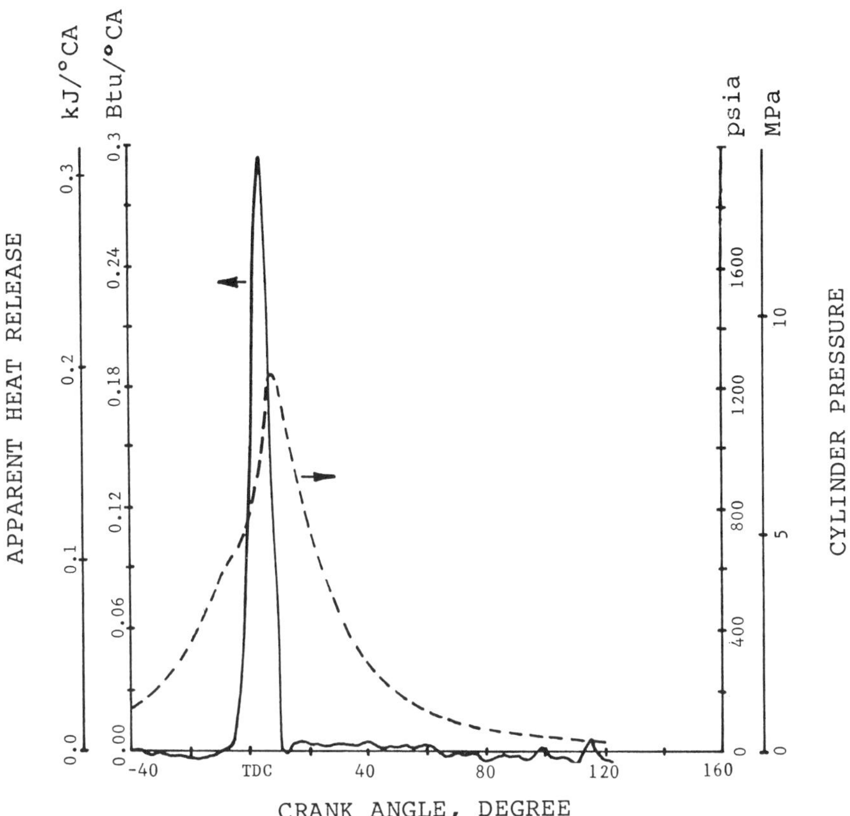


Figure 14.  Heat release and cylinder pressure diagram
for the coal fueled engine at 1000 rpm and
56.9 N.m (42 ft-lb) torque.

It should be noted above that turbocompounding can also improve
the performance of the conventional water cooled engine, thus
the real improvement in waste energy availability to turbocompound is 3 to 4%.

Figure 15 shows the comparison of brake specific fuel consumption of a conventional water-cooled turbocompound engine
and a waterless turbocompound engine (uncooled).  To compare
the two engines on a thermodynamic basis, the water pump losses
are subtracted from the cooled engine as shown in Figure 15.

Emissions

The hot walls of the adiabatic engine coupled with the
higher charge air temperature and oxidizing atmosphere, the
carbon monoxide, and unburned hydrocarbons were low as
expected.  The nitrogen oxide ($NO_x$ was expected to be higher
than the conventional engine; but, surprisingly, it showed
slightly higher $NO_x$ emissions in the exhaust gas).

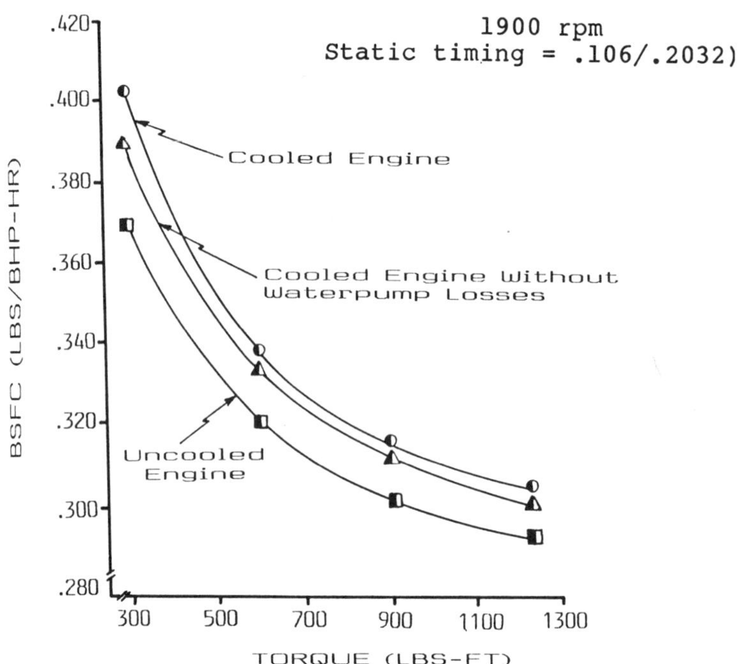


Figure 15.  Brake specific fuel consumption of a conventional cooled engine and a 20% adiabatic engine.

However, when the $NO_x$ data are plotted against BSFC data,
the $NO_x$ vs. BSFC trade-off is better for the adiabatic engine.
This point is illustrated in Figure 16.  The numbers on Figure
16 represent injection timing.  However, the important point to
remember is the start of combustion or beginning of heat
release.  Thus, for equal injection timing, the beginning heat
release or start of combustion is earlier for the adiabatic
case due to the shorter injection delay as shown in Figure 10.

The other important pollutant in diesel engine combustion
is the particulate.  Figure 17 shows the drastic reduction in
particulate levels of the adiabatic engine over the conventional water cooled engine.  The high combustion gas temperature in the adiabatic case could be responsible for this
reduction.[6]

During this investigation, the smoke level continued to
drop with each increase in insulation or adiabacity of the
engine.  Neither black smoke nor white smoke presented a
problem.


<u>Problem Areas</u>

The use of advanced ceramics in adiabatic engines with all
its associated benefits will not come easy.  The problems that
need to be resolved are: high temperature tribology; high
temperature, high strength, insulating ceramics; low cost
fabrication; low cost finishing and machining; and simple
quality control methods.  These problems are being worked on
currently and none appears insurmountable.

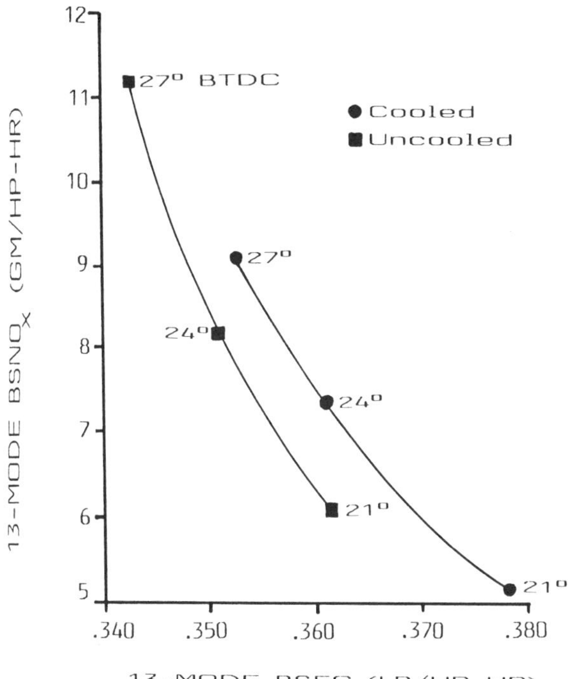



Figure 16.   BSNO$_x$ - BSFC trade-off for a cooled and uncooled engine using No. 2 diesel baseline fuel.

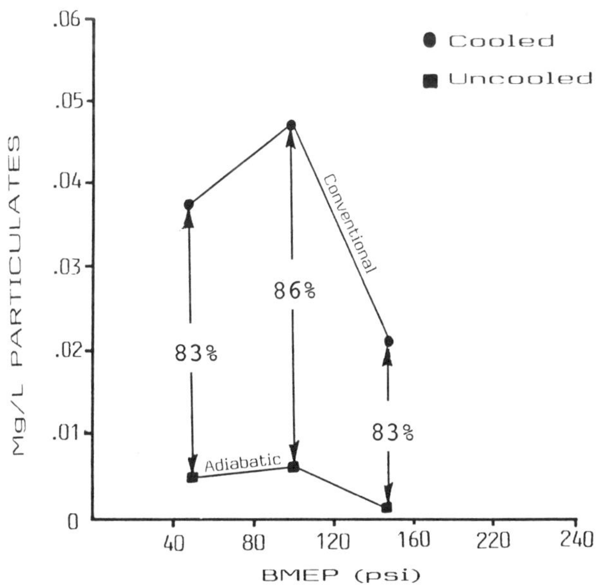



Figure 17.   Relative carbon particulate levels for an adiabatic versus a conventional watercooled engine.

ADIABATIC PASSENGER CAR ENGINE

The engine requirements for the future passenger car engine were evaluated and presented in the following order of importance:

1. Fuel economy
2. Emissions characteristics
3. Cost
4. Driveability
5. Size/Weight
6. Noise
7. Reliability/Durability
8. Multi-fuel capability.

An adiabatic engine with minimum friction features was studied for its appropriateness as an engine of the future.[7]

## Features of the Selected Engine

The engine of adiabatic design, Advanced Adiabatic Diesel, (AAD) features insulated pistons, cylinder liners, cylinder head, intake and exhaust ports, exhaust manifolds, and valves. No water or other liquid cooling system is needed, resulting in complete elimination of the conventional cooling system components such as radiator, water pump, etc. In order to insulate the hot gas stream components, extensive use of ceramic coatings and monolithic components is necessary. To reduce frictional and parasitic losses, the engine is designed to operate without liquid lubricants, also. Anti-friction ceramic rollers are used for wrist pin, crankpin, and main bearings. Solid lubricants are provided for other surfaces, such as cams and valve guides, which are normally lubricated with oil. To improve multi-fuel capability and to achieve "fast" heat release, a spark-assisted direct injection, high swirl diesel combustion system is used. Figures 18 and 19 show a cross section of AAD side view and combustion chamber, respectively.

The charge air system includes a high efficiency helical screw compressor and a ceramic expander. These two units are connected by belts to the crankshaft. Thus, a simple positive displacement compounding system is achieved. These helical screw machines have higher efficiencies compared to conventional turbomachines and have excellent part load efficiencies. Table 3 gives the major specifications of the AAD engine.

TABLE 3 - Major Specifications of the AAD Engine

| | |
|---|---|
| Rating | 70 BHP at 3000 rpm |
| Engine Cycle | 4 Stroke, DI, Diesel |
| Bore x Stroke | 77 mm x 77 mm (3.03 in. x 3.03 in.) |
| Displacement | 1.4 liters (87.4 cu. in.) |
| Number of Cylinders | 4 in-line |
| Compression Ratio | 14.0 |
| Maximum Cylinder Pressure | 200 Bars (3000 psi) |
| Length, Width, Height | 621 mm x 589 mm x 479 mm (24.4 in. x 23.2 in. x 18.9 in.) |
| Approximate Weight | 136 kg (300 lb.) |
| Maximum BMEP | 16 Bars (240 psi) |
| Firing Order | 1-3-4-2 |

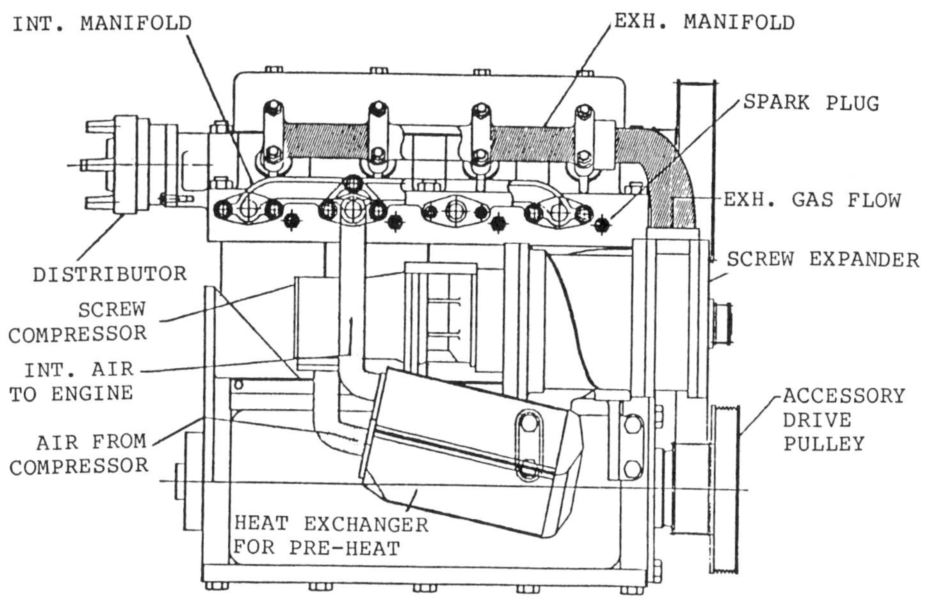


Figure 18.  AAD Engine layout - side view.

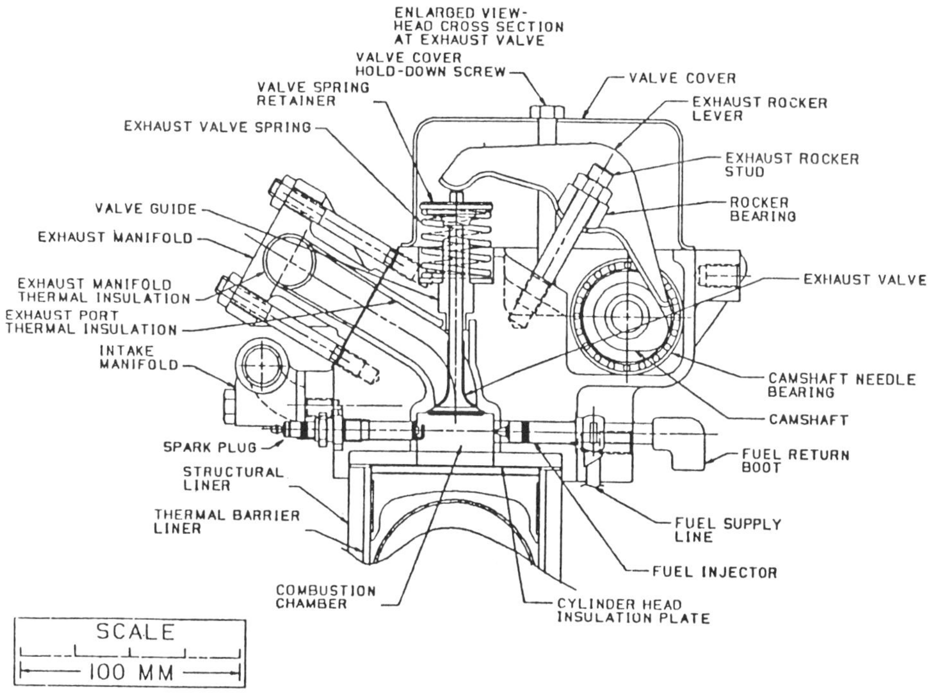


Figure 19.  AAD Engine cross section of combustion chamber.

<u>Projected Fuel Economy</u>

The engine performance analysis was conducted using the Cummins Diesel Cycle Simulator (DCS) computer program.  This is a thermodynamic model of the diesel engine cycle that has been validated by experimental data on several heavy duty engines. Some modifications were made to analyze the various advanced technologies included in the AAD engine.  Figure 20 gives the AAD engine's performance curves.  The predicted fuel map of the engine is presented in Figure 21.  These performance levels are considered to be practically achievable.  The performance levels shown in Figures 20 and 21 are less than "ideal" but significantly better than state-of-the-art diesel engines.

The impact of the AAD engine performance on vehicle fuel economy and acceleration characteristics was analyzed by the Ford Motor Company research staff.  The 1984 model Ford Tempo was chosen as the reference 3,000-pound vehicle.  The computer program to analyze the vehicle performance in a Federal Driving Cycle was utilized extensively.  The fuel map shown in Figure 21 was utilized in the vehicle analysis program, along with other vehicle related characteristics such as frontal area, drag coefficient, etc.  In addition, a continuously variable transmission (CVT) was assumed to be part of the vehicle.  The final results along with the expected performance using a state of the art IDI diesel are presented in Table 4.

Additional fuel economy, about 3%, is possible from the vehicle aerodynamic drag coefficient reduction.  A potential for this drag reduction exists due to the elimination of the radiator and the consequent lowering of the hood profile.

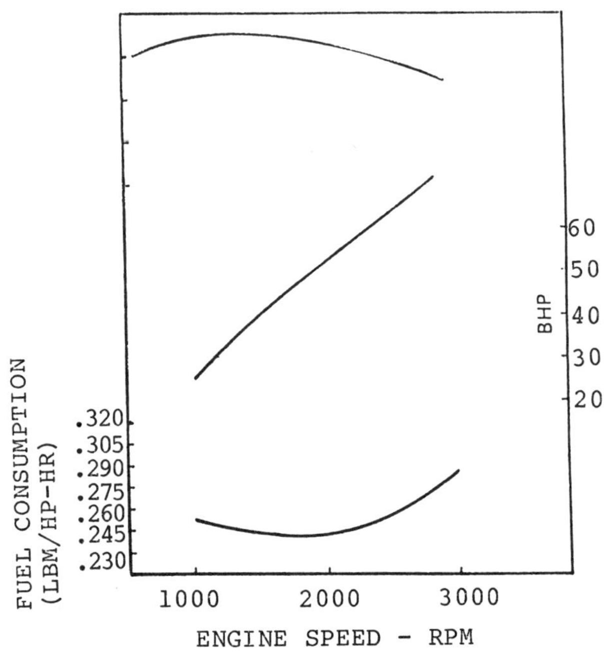


Figure 20.  AAD Engine performance curves.

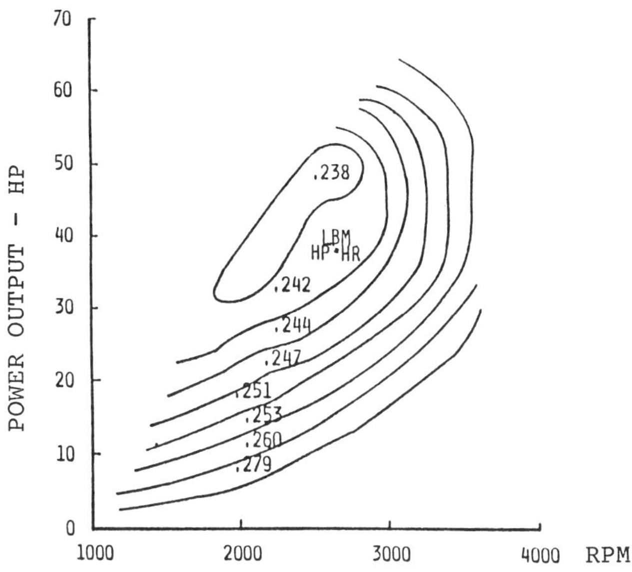


Figure 21.  Predicted fuel map for the AAD Engine.

## Table 4 - Vehicle Performance Comparisons
### (3,000 pound passenger car)

| 0-60 MPH Accel. Power Train | City | Fuel Economy, MPG Highway | Combined | Time Sec. |
|---|---|---|---|---|
| 1. Baseline IDI Diesel | 35.0 | 41.7 | 37.7 | 15.2 |
| 2. AAD Engine System | 72.2 | 96.3 | 78.8 | 13.9 |

ADIABATIC WANKEL ENGINE

Figure 22 shows the cross section of a rotary Wankel engine
The triangular shaped rotor goes through the typical four-cycle
reciprocating piston engine events of 1) intake 2) compression
3) combustion, and 4) exhaust.  For each revolution of the
rotor, three complete cycles are made, which is equivalent to a
three cylinder piston engine, thus, the high power density of
the rotary Wankel engine.

The Wankel engine simulation model predicts the heat loss
through the rotor, rotor housing and side housing.  The range
of percent heat loss through these components is as follows:

| | % Heat Loss |
|---|---|
| Rotor | 25-40 |
| Rotor housing | 50-72 |
| Side housing | 3-10 |

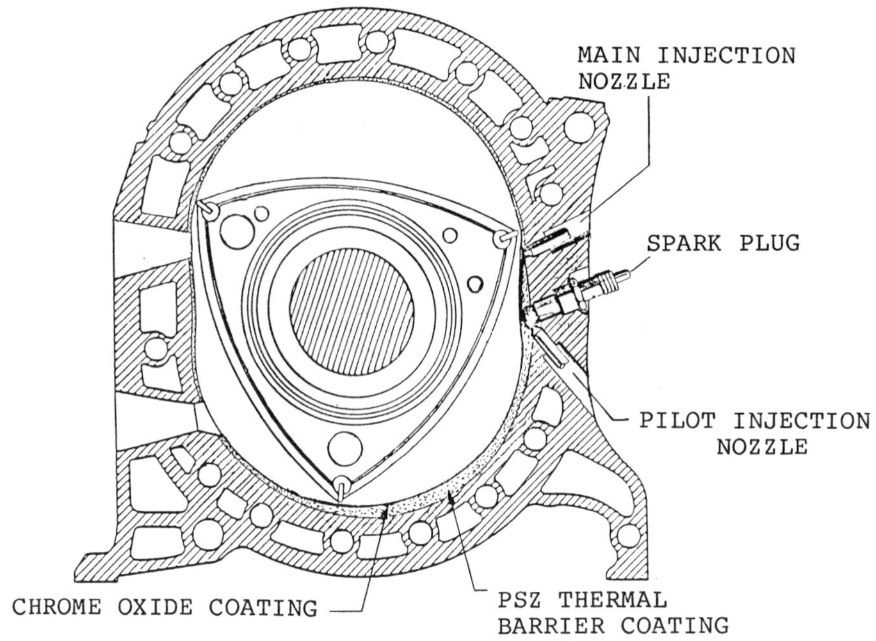


Figure 22.   Cross section of the rotary Wankel engine.


As expected, the maximum heat loss is through the rotor housing
to the cooling water.  The rotor is cooled by air in the intake
stroke and heated by burned gas in the combustion stroke.

The adiabatic Wankel rotary engine design should include
the insulated components - rotor, rotor housing and side
housing.  As noted by the percent heat loss through these components, a 50% reduction in heat loss (or 50% degree of adiabacity) may be achieved by the ceramic thermal barrier coating.

## Performance Improvement

The simulation model predictions for an adiabatic engine
are: an increase in thermal efficiency or decrease in indicated
specific fuel consumption (ISFC), an increase in the exhaust
gas temperature, a reduction in unburned fuel, no significant
change in volumetric efficiency, and a small increase in peak
pressures.  Table 5 presents the effect of the 1.6 turbocharger pressure ratio.  The model predicts a small increase in
the peak pressure.

## Component Design

The rotor of an adiabatic Wankel engine can be insulated by
the ceramic coatings and monolithic parts as shown in Fig. 23.
Three approaches have been proposed for rotor insulation:
1) ceramic coating, 2) snap ceramic ring, and 3) ceramic rotor
pocket.  The first approach consists of coating the rotor surface with a thermal barrier coating (up to 2.5 mm or 0.1 in) of
partially stabilized zirconia (PSZ) and sealing with a thin,
dense coating of Zircon or $Cr_2O_3$.  However, with the

Table 5.  Effect of Degree of Adiabacity on Wankel Engine
Performance as Predicted by the Simulation Model. [8]

(Turbocharged engine, 6,000 rpm,
Compression ratio = 7.5,
Burn Duration = 9° crank angle,
Leakage area = 0.005 cm$^2$ per apex,
Turbocharger pressure ratio = 1.6,
Baseline = 0% Degree of Adiabacity)

| Percent Change in: | Degree of Adiabacity | | | | |
|---|---|---|---|---|---|
| | 0% | 25% | 50% | 75% | 100% |
| ISFC | 0 | 1.6 | 3.4 | 5.0 | 6.5 |
| Power Output | 0 | 1.4 | 2.9 | 4.3 | 5.8 |
| Exhaust Temp. | 0 | 4.1 | 8.3 | 13.0 | 18.0 |
| Volumetric Efficiency | 0 | 0.1 | 0.5 | 0.9 | 1.2 |
| IMEP | 0 | 1.5 | 2.8 | 4.1 | 5.7 |
| Peak Chamber Pressure | 0 | 0.4 | 0.7 | 0.9 | 1.2 |

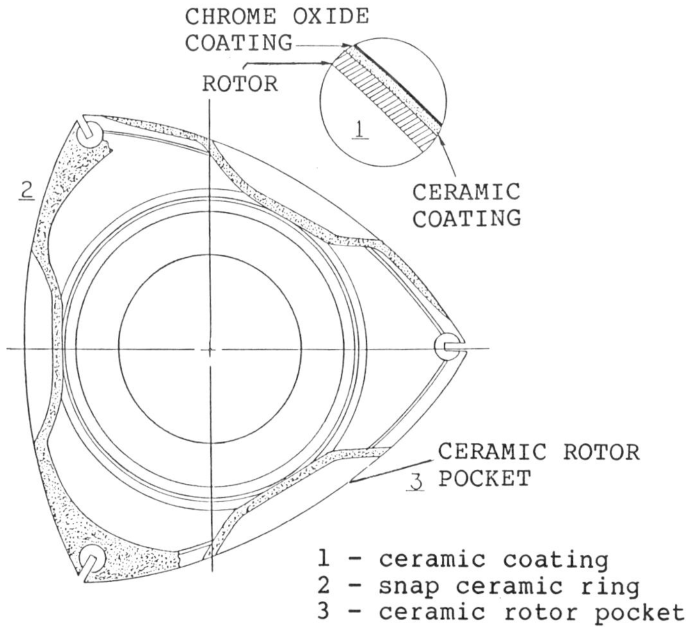


Figure 23.  Three approaches for insulating the rotors
of the adiabatic Wankel engine.

advancement in ceramic materials technology, it may be feasible
to consider monolithic ceramic parts for the adiabatic Wankel
engine in the near future at a reasonable cost.

The adiabatic Wankel rotary engine without a cooling system
will expose the apex seals to excessively high temperatures -
beyond 500°C.  Lubrication under those temperature conditions
will be difficult.  An unique approach to overcome the tribological problem of an adiabatic diesel engine was uncovered.
M2 steel with LiF+Cu in pockets showed extremely good friction
and wear data against $Cr_2O_3$.  This approach to the apex
seal may offer seals of superior design.  The side seals which
slide on the side housing would be plasma sprayed with CrC or
$Cr_2O_3$ which should give good results.


CONCLUSIONS

In conclusion:

1.    The adiabatic diesel engine will probably be first to
      demonstrate commercial reality in the new generation
      of ceramic heat engines.

2.    The ceramic coatings offer an immediate alternative
      solution to the monolithic design approach.

3.    The rotary Wankel engine can be significantly improved
      in performance and durability by the use of high
      performance ceramics.

4.    The monolithic ceramic heat engines are still a decade
      away because of:
          o   Materials availability
          o   Quality control
          o   Processing/machining
          o   Cost

5.    In the absence of adequate monolithic ceramic mate-
      rials today, sintered silicon nitride probably
      presents the best opportunity, but we must learn to
      design silicon nitride into an adiabatic product.  For
      example, where insulation is required, use air gap or
      any other insulative technique with silicon nitride.

6.    Continue developing new ceramic materials.  Consider
      changing of properties of ceramic by compromising,
      i.e., thermal conductivity for modulus of rupture.


<u>Acknowledgment</u>

The author wishes to acknowledge the help provided by Dr.
R. Kakwani and Mr. M.E. Woods of Adiabatics, Inc.  Cummins
Engine Co., where the author was employed until his time of
retirement, is also acknowledged for its support.


164

REFERENCES

1.  R. Kamo, W. Bryzik, "Ceramics for Adiabatic Turbocompound
    Diesel Engines," Presented at First International
    Conference on Fine Ceramics for Heat Engines, Hakone,
    Japan, (October 1983).

2.  J.L. Mason, A.F. McLean, "Ceramics for Gas Turbine Engines"
    in _Materials and Society_, W.R. Hibbard, Jr., ed, Pergamon
    Press, New York, vol 18, No. 2, (1984).

3.  R. Kamo, "Combustion Chamber Components for Internal
    Combustion Engine," U.S. Patent No. 4,495,907, (January 29,
    1985).

4.  R. Kamo and J. Fairbanks, "Alternate Fuels Capability of
    Adiabatic Diesel Engine Concept," Presented at CIMAC
    Conference, Oslo, Norway, (June 3-7, 1984).

5.  R. Kamo and R.M. Kakwani, "Coal Burning Adiabatic Diesel
    Engine," Presented at 13th Energy Technology Conference,
    Washington, DC, (March 1986).

6.  O.A. Uyehara, "Diesel Combustion Temperature on Soot," SAE
    Tech. Paper No. 800969, (Sept. 1980).

7.  R.R. Sekar, R. Kamo and J.C. Wood, "Advanced Adiabatic
    Diesel Engine for Passenger Cars," in _Adiabatic Engine:_
    _Worldwide Review_, SAE Publication SP-571, (Feb. 1984).

8.  R. Kamo, R.M. Kakwani, W. Hady, "Adiabatic Wankel-Type
    Rotary Engine," in _The Adiabatic Engine: Global_
    _Developments_, SAE Publications SP-650, (Feb. 1986).

# LOW HEAT REJECTION DIESEL ENGINES

R. H. Thring

Southwest Research Institute
San Antonio, Texas

## ABSTRACT

When compared with conventional diesel engines, Low
Heat Rejection (LHR) engines have the following features.
Fuel economy is improved by 5 to 10 percent in turbocharged
engines, or 9 to 15 percent with turbocompounding. Power is
reduced by up to 25 percent in naturally-aspirated engines,
but this loss can be recovered by pressure boosting. $NO_x$
emissions data vary widely, but it is concluded that they
will be increased by an average value of 15 percent. However, HC and CO emissions will be reduced by up to 50 percent. Smoke levels should be reduced and particulates will
be reduced by up to 80 percent. Noise levels should be reduced and there should be improved capability for operation
with low cetane fuels.

Problems to be solved include high temperature lubrication and high temperature materials.

## INTRODUCTION

There has been much work done and many papers published over the
past ten years on the subject of "Low Heat Rejection" or "Adiabatic"
engines. These terms tend to be used to describe the same thing,
though they imply differences. The former term is preferred by the
author, since it more accurately describes what can be achieved in
practice as opposed to a theoretical curiosity. Indeed, the term "low
heat rejection" or LHR, engines seems to be supplanting the term adiabatic engine in much of the more recent literature.

A true adiabatic engine would be one where there was zero heat
transfer between the walls of the combustion chamber and the contained
gases at all times during the cycle. This would require the walls to
be made of a material having either zero thermal capacity or zero
thermal conductivity or both—an impossible requirement. It has been
calculated that even if the combustion chamber walls could be constructed from metal only 0.006 inch thick, backed by perfect insulation, there would still be a temperature excursion during the cycle of
60°C (140°F).

The work that has been done on LHR engines has not been intended to lead to a true adiabatic engine. Rather, the objective has been to reduce the heat rejection below the levels that apply to existing engines. A substantial amount of heat is still rejected, either to the coolant or the lubricating oil, and in many cases, an oil cooler must be provided to reject this heat.

LHR engines are generally either diesel or late injection stratified charge engines. It would be undesirable to insulate the combustion in homogeneous charge engines, since the resulting high wall temperatures would cause either preignition or knock or both.

Development work on LHR diesel engines was initiated by the U.S. Army Tank-Automotive Command. The Army's goal was to eliminate the conventional engine cooling system: radiator, radiator fan, hoses, water pump. This objective is a high priority for the Army, because cooling system components are a major source of maintenance problems and are vulnerable to combat damage.

To the commercial diesel engine user, the objectives are somewhat different. While reduced maintenance is desirable, the commercial user would probably not accept reduced engine life in return for reduced maintenance. The commercial user is also deeply concerned with such aspects as fuel economy, power, smoke, noise, and exhaust emissions. How these aspects are affected by LHR operation is discussed in this paper.

Cummins Engine Company was chosen by the U.S. Army to design and demonstrate an "adiabatic" (LHR) engine. Cummins emphasized the use of ceramic materials as a key element in the successful development of the LHR engine. Many ceramics have good strength at high temperatures, many have low thermal conductivity, and the extreme hardness of many ceramic materials should be useful in reducing engine wear. As a result of this emphasis many people came to believe that LHR engines required the extensive use of ceramics. However, in the author's opinion this does not necessarily follow, since in many cases the ceramics with good mechanical properties have poor insulating properties and vice versa. The most serious problem with the use of ceramics in engines is their unreliability. Because their ductility is low, the smallest flaw causes large stresses to build up easily when the component is loaded, and when failure occurs, it is usually catastrophic, causing disintegration of the component and destruction of the entire engine.

There are many good high temperature metal alloys that have been developed, mainly for gas turbine use, that offer excellent high temperature strength and creep resistance, and do not shatter on failure like ceramics. They can be used to form the combustion chamber wall, and insulated from the surrounding metal by the use of air gaps.

However, even the avoidance of the use of ceramics does not eliminate what is, in the author's opinion, the most difficult problem to overcome in the LHR engine, namely lubrication. In an LHR engine, combustion chamber surface temperatures are very high, and the temperatures at the upper end of the liner and piston crown are also much higher than in conventional engines. For example, it is generally accepted that in an uncooled engine the liner temperature at the top ring reversal point will be in the region of 1200°F. This region is the hottest lubricated part of an engine. Research work with high temperature lubricants has so far only succeeded in extending the max-

imum tolerable temperature in this region to 600°F, substantially
lower than that required.

## EFFECTS

### Specific Fuel Consumption

Many workers in the LHR engine field claim reduced fuel consumption as a major benefit of LHR engines.  They argue that reducing heat
losses to the coolant will increase the proportion of the fuel energy
that is available to produce useful work.

It is helpful when comparing improvements in specific fuel consumption resulting from LHR operation to divide engines into three
categories:

      (1)   Naturally-aspirated (NA)
      (2)   Turbocharged (TC)
      (3)   Turbocompound (TCO)

A fourth category might be engines with thermodynamic bottoming
cycles, but these are beyond the scope of this paper.

Theoretical calculations carried out at SwRI and elsewhere[1,2,3]
indicated that in NA engines there should be an improvement in the
region of 5 to 10 percent by reducing heat rejection to zero.  However, in practice the effect of the large rise in cylinder wall temperatures is to reduce the volumetric efficiency quite severely.  Thus
the maximum power output is reduced, and hence the best mechanical
efficiency of the engine is reduced.  This effect negates any improvement in theoretical efficiency and in some cases has resulted in worse
fuel economy than in the standard engine.  An example is shown in
Figure 1, which shows some experimental results reported by Kamo, et
al.[4] from engines operated with pistons having ceramic coatings.
Figure 1 shows results from a 95 mm bore NA engine.  It can be seen
that at higher loads, fuel consumption increased by approximately 10
percent.

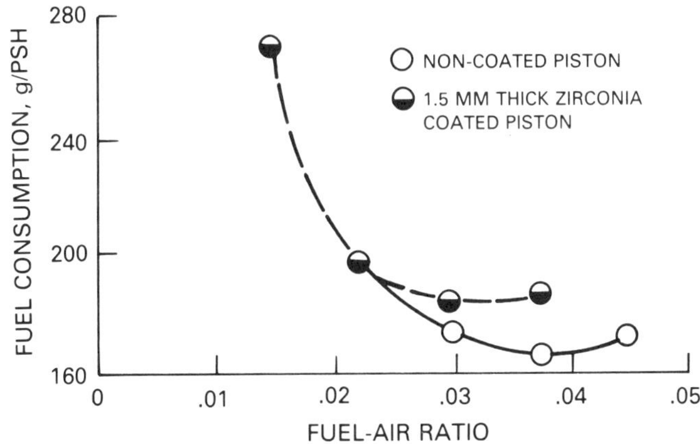



Figure 1.  Experimental Results as Functions of Air-Fuel Ratio
for a Single-Cylinder Naturally-Aspirated Engine

In turbocharged engines the situation is somewhat different, because the overall volumetric efficiency can be restored to that of the standard engine by increasing the boost pressure. Tovell[2] carried out computer simulation work which used as a baseline a turbocharged direct injection (DI) engine of 130 mm bore. Two forms of LHR engine were simulated: (1) the true adiabatic engine, where the combustion chamber walls had no thermal capacity so that their temperature was equal to the charge gas temperature at all times, and (2) the combustion chamber walls had a large thermal capacity, so that their temperature was constant throughout the cycle, but no net heat transfer occurred when integrated over the whole cycle. The resulting improvements in indicated specific fuel consumption (ISFC) were 14 and 7.5 percent, respectively. Boost pressure increases of −22 and +9 percent, respectively, were required to maintain constant power output.

Yoshimitsu[5] obtained both theoretical and experimental results with a six-cylinder, direct injection diesel engine of 105 mm bore. Figure 2 shows the theoretical results. The x-axis is degree of heat insulation given by:

$$\text{Degree of Heat Insulation} = 1 - \frac{\text{Heat rejection with insulation}}{\text{Heat rejection without insulation}}$$

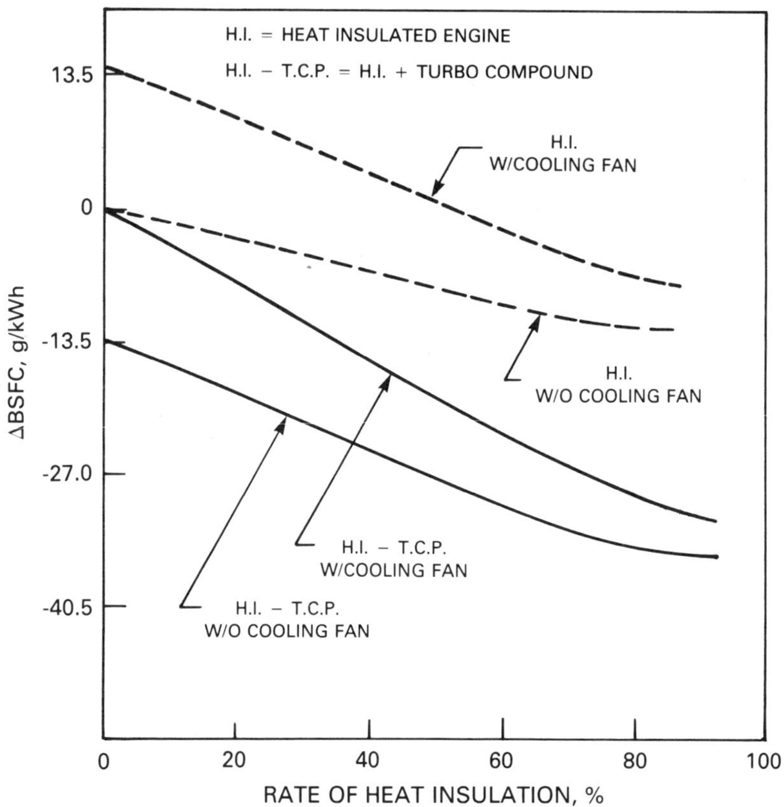


Figure 2. Simulated Engine Performance
Versus Rate of Heat Insulation

The y-axis shows the theoretical BSFC change relative to the baseline level, which was taken as the heat insulated engine without a
cooling fan.  The fuel consumption values apply to the rated maximum
power condition.  The fuel consumption improvement is greater with the
fan, since the fan power requirement is reduced as the degree of heat
insulation is increased.  Unfortunately, the value of BSFC at rated
power is not given, but assigning a value of 225 g/kWh (without the
fan) would bring about an improvement in BSFC with the fan of about 10
percent, or without it, about 5 percent.  Also shown in the figure are
the results of simulating an engine with turbocompounding and heat insulation.  Turbocompounding alone gives an improvement of 6 percent
and the heat insulation gives a further 14 percent, making a total of
20 percent.  These results are at odds with those of Savliwala et
al.[6] shown in Figure 3, which show that benefits obtained by LHR
operation are no greater with a TCO engine than with a TC engine.  One
possible reason for the discrepancy is that Savliwala's calculations
were done at constant power, while Yoshimitsu used constant swept
volume.  The author's opinion is that LHR operation results in increased exhaust energy, that can be usefully employed in a turbocompound system.  Therefore, the fuel economy benefits of LHR operation should be greater in TCO engines than TC engines.

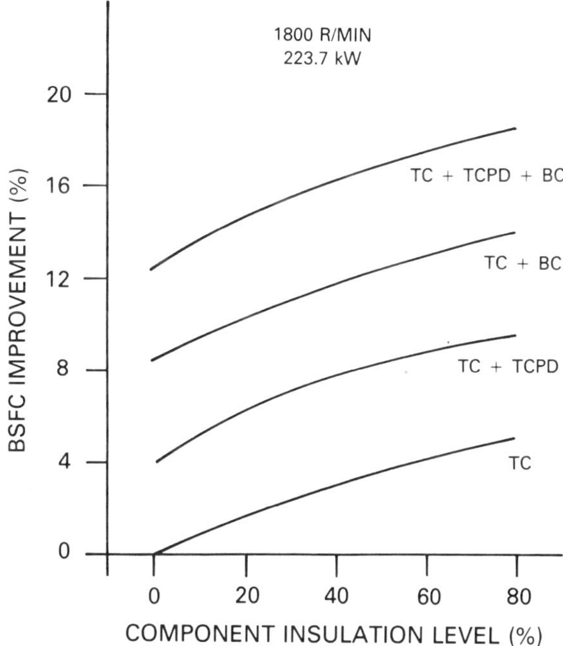


**Figure 3.  Percent Gain in Fuel Economy
as a Function of Degree of Insulation
and Exhaust Energy Recovery Scenarios**

Initial experiments reported by Yoshimitsu showed poor fuel economy due to a reduced rate of combustion, especially in the later part of the cycle. This problem was overcome by improved fuel atomization and air swirl in the combustion chamber. The results for the turbocharged engine are shown in Figure 4. The improvements in BSFC were much smaller than theory predicted, about 1 percent at best, and at light load the fuel economy decreased.

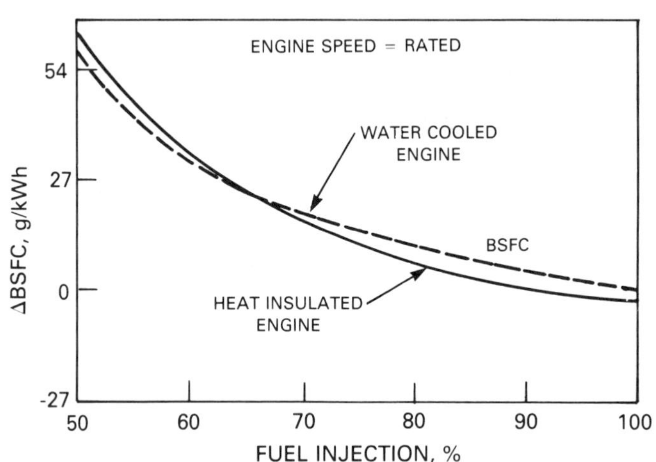


**Figure 4. Experimental Engine Performance Comparison**

Some results from a turbocompound engine were given in a later paper.[7] The engine was a six-cylinder DI engine of 125 mm bore and a swept volume of 11 liters. In baseline form it was turbocharged, aftercooled, and rated at 235 kW at 1800 rpm. The fuel consumption improvement over baseline of the LHR turbocompound engine was 13.5 percent at the rated condition. If the reduction in required fan power is taken into account, the improvement was 15 percent. The LHR operation alone contributed 11.2 percent, which was within one percent of the theoretical prediction.

Table 1 represents an attempt to summarize the economy improvement results shown in the literature. It is not possible to compare like with like in this way, since each author reports on a different engine running under different conditions. For example, the 7 to 12 percent improvement in fuel economy reported by Havstad et al. is surprisingly good for a naturally-aspirated engine. However, these results were obtained at part load, where there is plenty of excess air and the loss of volumetric efficiency is not important. In spite of the different conditions under which the results of Table 1 were obtained, the table does show some consistency in its figures.

It appears that the fuel economy improvements obtained from LHR operation can be categorized in ascending order of benefit as follows:

    (1)  Naturally-aspirated engines
    (2)  Engines with simple pressure boosting

(3)  Turbocompound engines
(4)  Engines with bottoming cycles

With naturally-aspirated engines the benefits are very small. Turbocompound engines and engines with bottoming cycles show the greatest benefits, but the extra expense and complication may be hard to justify, especially in smaller sized units (e.g., passenger cars). It seems, therefore, that the most likely candidate engine for LHR operation is the heavy-duty diesel engine with simple pressure boosting (e.g., turbocharging).

Table 1.  Comparison of Fuel Economy Improvements

| Source | Conditions | Economy Improvement (percent) |
|---|---|---|
| SwRI | Naturally-aspirated (NA), Theory | 10 |
| Tovell | True Adiabatic, Turbocharged (TC), Theory | 14 |
|  | No Net Heat Transfer, TC, Theory | 7.5 |
| Watts | No Heat Transfer, Theory | 3-20 |
| Kamo | Experimental | -10-0 |
| Yoshimitsu | Theory, TC | 5-10 |
|  | Theory, Turbocompound | 20 |
|  | Experiment, TC | 1 |
|  | Experiment, Turbocompound | 13.5-15 (11 due to insulation) |
| Bryzik | Experiment | 2-4 |
| Coers | Experiment, TC, New Cam Profile | 15 (1 due to insulation) |
| Sekar | Theory, NA | 12 |
|  | Boosted | 30 |
| Siegla | Theory, NA | 0-3 |
| Wade | Experiment | 4-7 |
| Savliwala | Turbocharged, theoretical | 5 |
|  | Turbocompound, theoretical | 10 (5 due to insulation) |
| Morel | Turbocharged, theoretical | 10 |
| Alkidas | Naturally-aspirated | 0 |
| Havstad | Naturally-aspirated | 7-12 (light load) |

Generally, the experimental results were disappointing when compared to the theoretical predictions, which range from zero to 30 percent improvement in fuel economy.  The experimental results range from

10 percent higher fuel consumption to 12 percent economy improvement, or 15 percent if turbocompounding is used. It appears that either the theoretical predictions are incorrect, or that the experimental results are not as good as they should be due to lack of optimization of the test engines.

## Smoke-Limited Power

It is generally agreed that the use of reduced cooling in naturally-aspirated diesel engines will reduce the smoke-limited power. This reduction is caused by the increase in temperature of the cylinder walls, combustion chamber and ports, which increases the temperature of air trapped in the cylinder. The increased temperature reduces the density of the trapped air, and hence its mass. Since smoke-limited power is normally reached at a fixed air/fuel ratio for any given engine, the fuel quantity has to be reduced to prevent excessive smoke, and hence the power output is reduced. Since the heat rejected to the coolant is greatly reduced, the extra energy has to be routed somewhere, i.e., into the exhaust. In a naturally-aspirated engine this extra energy is lost, but in a turbocharged engine, a turbocompound engine, or an engine with a bottoming cycle, some of this energy can be recovered. Therefore, these types of engines are ranked for improved smoke-limited power in the same order as they were for fuel economy in the previous section. While fuel economy of the naturally-aspirated engine is slightly improved, its smoke-limited power output is reduced.

Kamo et al.[8] presented the results of some theoretical calculations carried out with reduced heat loss engines. Figure 5 shows BHP versus specific heat rejection rate. Reducing the specific heat rejection rate to zero reduces BHP by about 20 percent. It was a naturally-aspirated engine, and volumetric efficiency was reduced from 84 percent to 68 percent. Simulations were also carried out for turbocharged, aftercooled engine and a turbocompound engine. Also considered was an engine with a Rankine bottoming cycle. In these latter three cases the output power was restored to or surpassed the baseline level.

Wallace et al.[9] documented experimental work with a Petter PH1W engine. The engine was operated with a heat barrier piston, having a Nimonic crown insulated by an air gap from the piston body. For a given air-fuel ratio, power output was reduced by 20 to 25 percent for the entire range of air-fuel ratios. To maintain maximum power output, it was necessary to use an air-fuel ratio of 12.76 and presumably this caused a large increase in smoke level, which would be unacceptable in practice.

Alkidas et al.[10] also reported significant losses in maximum power in a divided chamber naturally-aspirated diesel engine. They employed an air gap insulated piston, an antechamber that was partially insulated by an air gap, and high temperature coolant. Volumetric efficiency was reduced from about 82 percent to about 76 percent at 2000 rpm, resulting in a BMEP reduction from about 400 to about 350 kPa.

One of the factors that influences smoke-limited power output is the power required to overcome mechanical friction. There are some possibilities for reducing friction in LHR engines. Thring[11] reported significant reductions in friction as the liner temperature was increased in a CLR engine (Figure 6). In this case, the cylinder head was cooled but the liner was not only uncooled but had to be heated to

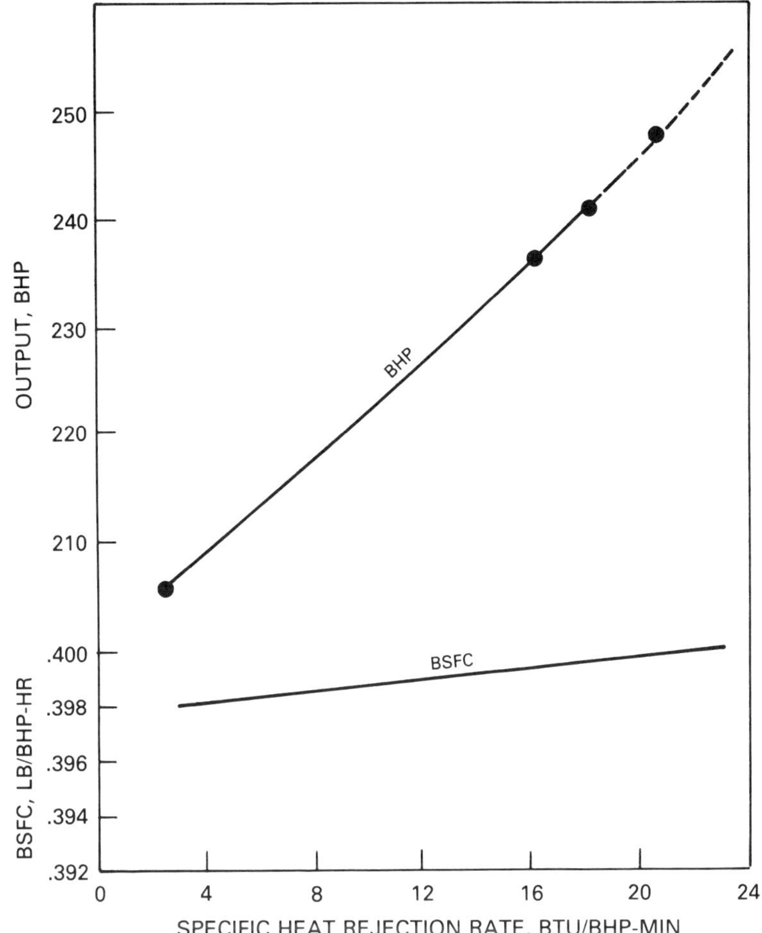


**Figure 5.  Simulated BHP
Versus Specific Heat Rejection Rate**

achieve the desired temperatures.  The same lubricant was used
throughout the test--a commercial high temperature 20W-50 oil.  The
reason for the reduction in friction was the reduction in the viscosity of the oil film on the liner, which produced a reduction in the
sliding friction of the piston and rings.

Another possibility is that if ceramic materials are used, their
low coefficients of friction could reduce engine friction levels.  For
example, Timoney et al.[12] carried out some experiments with a head-
less engine.  The friction levels measured in an unlubricated silicon
carbide engine were 30 to 50 percent lower than those measured in a
lubricated metal version of the same engine.

## Gaseous Exhaust Emissions

There is some debate about the effects that reducing heat losses will have on exhaust emissions.  For $NO_x$ emissions, there are two arguments:  One is that since wall surface temperatures will be raised, gas temperatures will be raised, and since $NO_x$ formation is a function of gas temperature, the exhaust emissions of $NO_x$ will be increased. The other argument is that since gas temperatures will be raised, combustion will occur faster and there will therefore be less time for $NO_x$ to form.  However, the rate of $NO_x$ formation rises very rapidly with temperature, much faster than it falls with decreasing time available for the reactions to take place.  Thus, it seems likely that $NO_x$ emissions will increase.

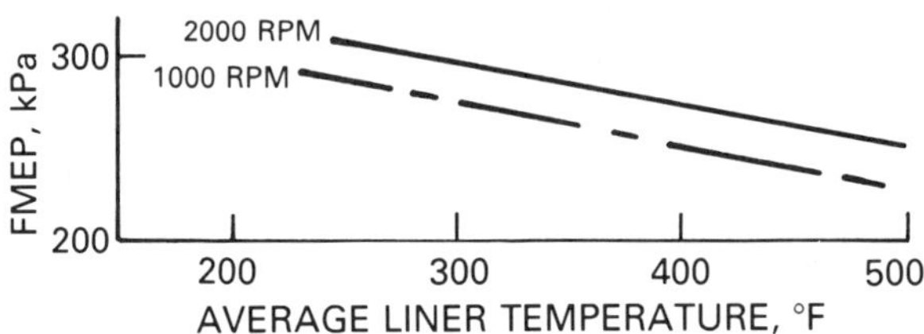


**Figure 6.  FMEP Versus Cylinder Liner Temperature**

Calculations carried out by Watts et al.[3] confirm this viewpoint.  $NO_x$ was increased about 10 percent after reducing heat transfer to 25 percent of its baseline value.

Kamo et al.[4] give some experimental results for HC, $NO_x$ and CO (Figure 7).  The tests were carried out on a Yanmar, naturally-aspirated, 12 hp, single-cylinder engine of 95 mm bore.  The piston crown had a coating of 1.5 mm thick zirconia; no other parts of the engine were insulated.  $NO_x$ emissions were increased by about 20 percent for a given fuel-air ratio, at the lower fuel-air ratios.  However, at the higher fuel-air ratios there was no effect.  There were no significant differences in HC and CO emissions.

The varied test conditions make it difficult to compare results, but some trends are evident from the comparison given in Table 2.  It can be seen that there is general agreement that the use of reduced heat rejection increases $NO_x$ emissions in the range of 5 to 100 percent, with a typical mean value of about 15 percent.  These figures do not include the effects of making other changes, such as retarding the injection timing.  HC and CO emissions are reduced up to 50 percent, though there is much less data to support this trend.

## Exhaust Smoke and Particulates

The use of insulation in engines might be expected to have some impact on the formation of smoke and particulates in the combustion chamber.  The higher gas temperatures might be expected to delay quenching of combustion reactions until later in the stroke, and therefore cause more carbon particles to be oxidized, giving rise to less smoke and particulates emissions in the exhaust.  Conversely, the higher peak temperatures might be expected to cause more pyrolysis of the fuel and therefore give rise to more smoke and particulates.

176

Tovell[2] expresses the opinion that a small reduction in smoke and particulates seems likely. However, Bryzik et al.[13] showed a substantial reduction (83 to 86 percent) in particulates (Figure 8). On the other hand, in their simulation of a low heat rejection passenger car diesel engine, Siegla et al.[14] predicted an increase in particulates. Siegla et al. were not certain that the expressions used in their model for particulate formation and oxidation accurately represented in-cylinder mechanisms. In short, the mechanisms of particulate formation in engine cylinders are incompletely understood, and so the predictions of current theoretical models may not be accurate.

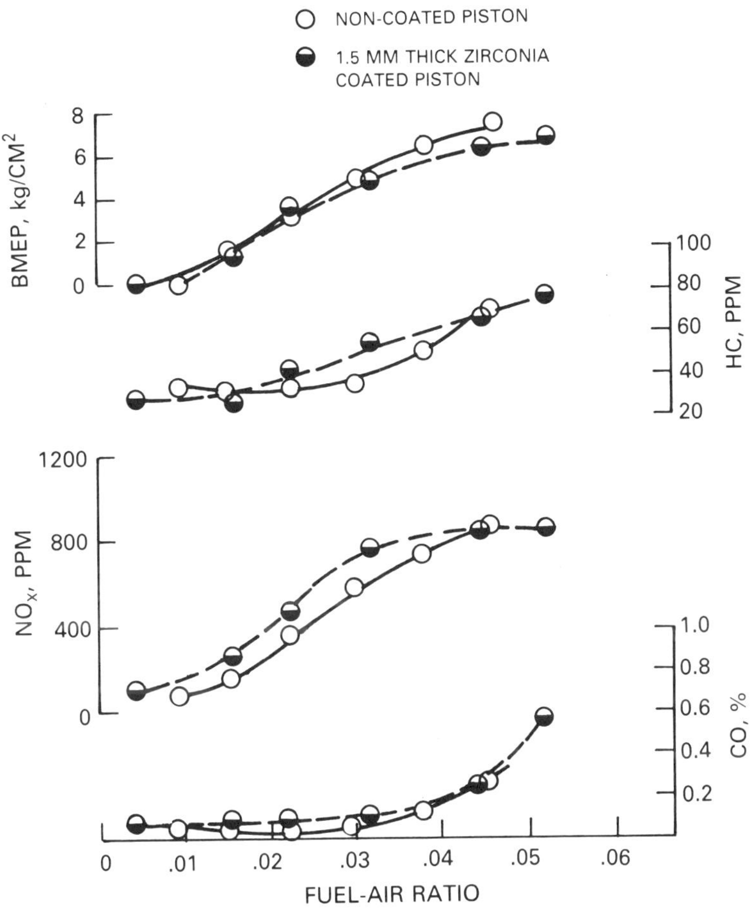


Figure 7. Experimental Emissions Results with Coated Piston

## Ignition Delay and Noise

In a conventional diesel engine, there is a delay period between the start of fuel injection and the commencement of combustion. The length of the delay period depends on many factors, such as compression ratio, fuel atomization and fuel cetane number. During the delay period, fuel continues to be injected, so that when combustion finally does commence, there is already a significant amount of fuel in the

Table 2.  Comparison of Emissions Results

| Source | Conditions | Percent Increase or Decrease | | |
|---|---|---|---|---|
| | | HC | CO | $NO_x$ |
| Watts | Theory | | | +10 |
| Kamo | Experimental, naturally-aspirated, 1.5 mm zirconia on piston | 0 | 0 | +20 |
| Bryzik | Experimental, 13-mode (fixed timing) | | | +20 |
| Tovell | Theory | | | +30-40 |
| Toyama | 13-mode, turbocharged (fixed timing) | | | +1% ($NO_x$ + HC) |
| Siegla | Theory, IDI | | | +100 |
| Cole | Experimental, IDI | -0 to -40 | -0 to -50 | +5-10 |
| Alkidas | Experimental, IDI | -40 to +30 | 0 to -60 | +5 |

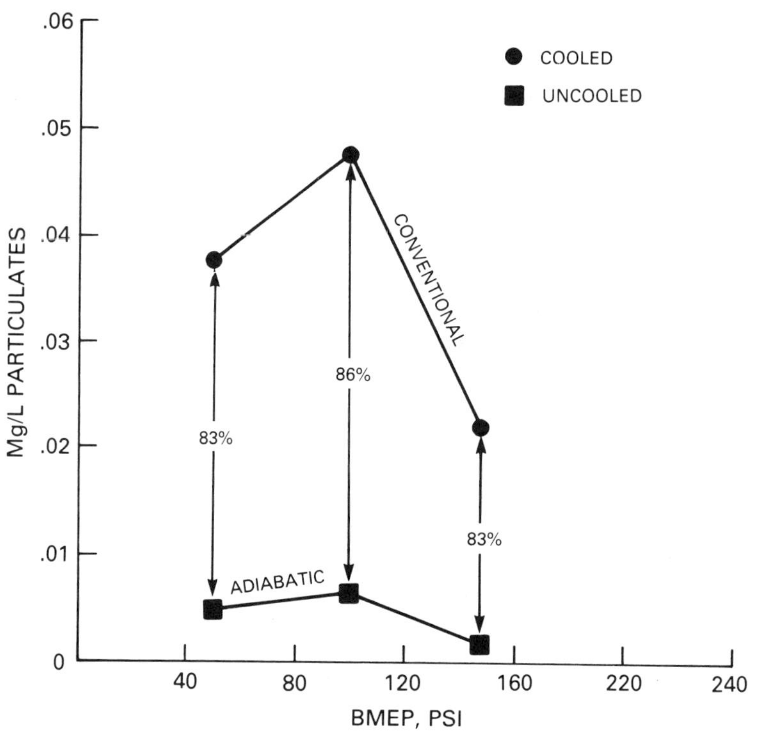


Figure 8.  Adiabatic Engine Particulate Levels

combustion chamber. This fuel burns very quickly, causing rapid pressure rise and, consequently, noise emissions from the engine. The length of the delay period can be reduced by increasing compression ratio or heating the intake air. In insulated engines where gas temperature is generally higher, it would be expected that delay times and noise emissions would be reduced. It is also generally true that decreased cetane number causes increased delay periods. This presents a problem when operating conventional diesel engines on low cetane fuels. Insulated engines should therefore be more tolerant of low cetane fuels.

Wade et al.[15] reported reduced rates of pressure rise in an uncooled single-cylinder DI engine. Figure 9 shows a comparison of the apparent rate of heat release (AROHR) for uncooled and water-cooled engines at a representative part-load condition. It appears that ignition delay was reduced by about two degrees crank angle, which represents a reduction of about 40 percent. This change resulted in a 50 percent reduction in the initial premixed combustion spike, as can be seen in the figure.

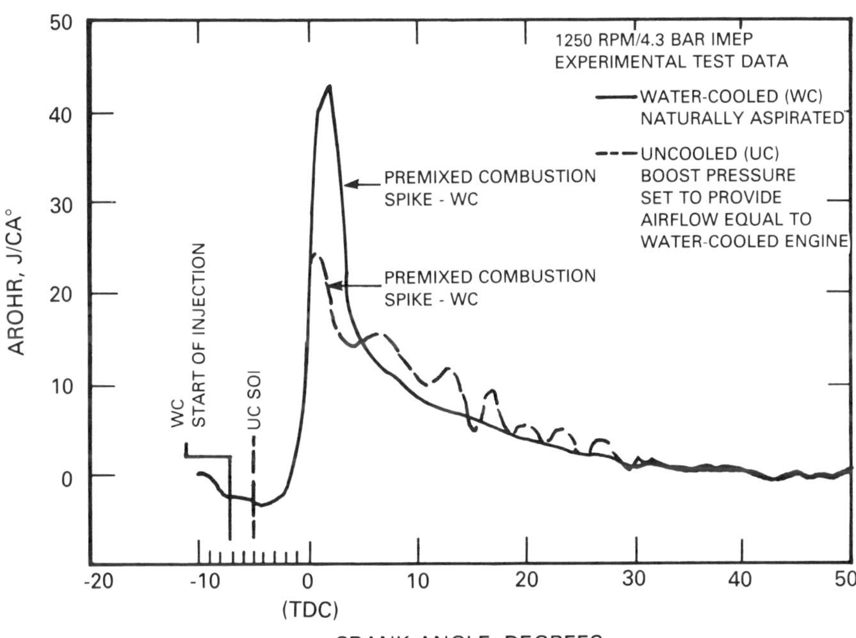


Figure 9. Comparison of Apparent Rates of Heat Release;
Single-Cylinder DI Diesel Engine, 80 mm Bore x
88 mm Stroke

Yoshimitsu et al.[5] made use of reduced delay times to run their engine on a low cetane fuel, SRC II. This fuel is coal-derived with a cetane number of less than 10. They also ran their engine with a fuel of cetane number 30. The heat insulated engine ran well on the 30 cetane fuel, and moderately well on the SRC II, except at low speed, low load, and idle, where some heavy knocking and misfiring occurred. The water-cooled engine, however, ran poorly at low speed/low load on

the 30 cetane fuel, and had heavy knocking and misfiring at all speeds
and loads on SRC II.

Thring[11] also reported significant reductions in ignition delay
as the cylinder liner temperature was increased in a CLR engine
(Figure 10).

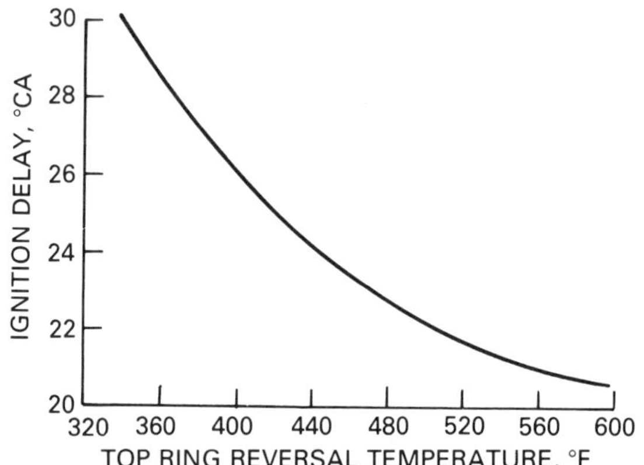

Figure 10.  Ignition Delay Versus
Top Ring Reversal Temperature

## METHODS OF ACHIEVING LHR OPERATION

The main part of this paper has considered the merits and
demerits of LHR engines.  However, there are some practical difficulties in achieving LHR operation that must be considered.  Modern
turbocharged diesel engines currently operate close to their
materials' limits in terms of high temperature stress.  Reducing heat
rejection will increase the high temperature stress on the materials
still further.  It will therefore be necessary to go to new materials.
These new materials may be metallic alloys or they may be ceramics.
The gas turbine industry has developed many new metallic alloys with
excellent high temperature strength and creep resistance.  They are,
however, expensive.  Ceramics offer a much cheaper alternative.  They
have good strength at high temperatures but they have one serious
problem: their ductility is low.  Therefore, if the material has a
small flaw (and it is difficult to make ceramics without small flaws)
when it is stressed the resulting stress concentration will cause a
crack to commence propagation.  In a metal the material at the crack
tip will undergo plastic deformation and relieve the stress, but in a
ceramic the crack will propagate right through the component, resulting in catastrophic failure of the component and sometimes destruction of the engine.  Since it is very difficult to detect such
flaws non-destructively, it is also very difficult to predict failure
in ceramic components.

Another major problem associated with LHR operation is lubrication. This problem is arguably more severe than the materials problem. Research at SwRI has revealed that special lubricants can be made to operate at 600°F top ring reversal temperature, but for insulated engine operation it will be necessary to have top ring reversal temperatures in the region of 1200°F. Several possible solutions have been proposed for this problem, but none has been successfully demonstrated outside the laboratory. One possible solution is to operate unlubricated; another is to employ solid lubricants in the top ring reversal area; and another is to use sacrificial lubricants, where particles of solid lubricant might be carried to the lubrication site by a fluid, which then was allowed to evaporate and burn.

For the future, it would seem sensible to proceed in stages. Stage 1 would be selective cooling, where some parts of the engine would be cooled while others would not. Stage 2 would be uncooled engines, and Stage 3 would be insulated engines. Each stage becomes progressively more difficult to achieve in practice, but offers greater potential benefits.

## CONCLUSIONS

The improvement in fuel economy from LHR operation is negligible in NA engines, expect possibly at light loads. TC LHR engines have 5 to 10 percent better economy than cooled TC engines. TCO LHR engines have 4 to 11 percent better economy than cooled TCO engines, which is 9 to 15 percent better than cooled TC engines.

LHR operation causes power to be reduced by up to 25 percent in NA engines. However, this loss can be recovered in pressure boosted engines by increasing the boost pressure.

Exhaust emissions of $NO_x$ are increased. The percentage in the literature varies very widely, but an average value is about 15 percent. This figure applies if no other changes are made, e.g., retarded injection timing. Hydrocarbon and CO emissions should be reduced by up to 50 percent, though there is much less information available about these emittants.

Smoke levels should be reduced, and particulates should be reduced by up to 80 percent.

Ignition delay is reduced, which is good for reduced noise and should allow operation on reduced cetane fuels.

The most likely candidate for LHR operation is the heavy-duty diesel engine with simple pressure boosting (e.g., turbocharging).

Notwithstanding the above benefits, there are many practical problems to be solved before LHR engines can be put into production. The problem of high temperature lubrication may be the most difficult to solve. The invention of a ductile ceramic would go a long way toward solving the materials problems. More research is needed.

## REFERENCES


1. R. H. Thring, "Adiabatic Diesel Engines," Fourth Quarterly Report, SwRI Project No. 03-8011, March 1985.
2. J. F. Tovell, "The Reduction of Heat Losses to the Diesel Engine Cooling System," SAE Paper 830316.

3.   P. A. Watts and J. B. Heywood, "Simulation Studies of the Effects of Turbocharging and Reduced Heat Transfer on Spark-Ignition Engine Operation," SAE Paper 800289.

4.   R. M. Kamo, M. Woods, T. Yamada and M. Mori, "Thermal Barrier Coating for Diesel Engine Piston," ASME Paper 80-DGP-14.

5.   T. Yoshimitsu, K. Toyama, F. Sata and H. Yamaguchi, "Capabilities of Heat Insulated Diesel Engine," SAE Paper 820431.

6.   M. N. Savliwala and N. S. Hakim, "Statistically Optimized Performance Predictions of Low Heat Rejection Engines with Exhaust Recovery," SAE Paper 860315.

7.   K. Toyama, T. Yoshimitsu, T. Nishiyama, T. Shamauchi and T. Nakagaki, "Heat Insulated Turbocompound Engine," SAE Paper 831345.

8.   R. Kamo and W. Bryzik, "Adiabatic Turbocompound Diesel Engine," ASME Paper 84-DGP-16.

9.   F. J. Wallace, T. K. Kao, W. D. Alexander, A. Cole and M. Tarabad, "Thermal Barrier Pistons and Their Effect on the Performance of Compound Diesel Engine Cycles," SAE Paper 830312.

10.   A. C. Alkidas, "The Influence of Partial Suppression of Heat Rejection on the Performance and Emissions of a Divided-Chamber Diesel Engine," SAE Paper 860309.

11.   R. H. Thring, "Low Heat Rejection Engines," SAE Paper 860314.

12.   S. Timoney, and G. Flynn, "A Low Friction, Unlubricated SiC Diesel Engine," SAE Paper 830313.

13.   W. Bryzik, and R. Kamo, "TACOM/Cummins Adiabatic Engine Program," SAE Paper 830314.

14.   D. C. Siegla, and C. A. Amann, "Exploratory Study of the Low-Heat-Rejection Diesel for Passenger-Car Application," SAE Paper 840435.

15.   W. R. Wade, P. H. Havstad, E. J. Ounsted, F. H. Trinkler and I. J. Garwin, "Fuel Economy Opportunities with an Uncooled DI Diesel Engine," SAE Paper 841286.

PRESENT STATUS AND FUTURE VIEW OF ROTARY ENGINES


Akihito Nagao, Hiroshi Ohzeki, and Yoshinori Niura

Mazda Motor Corp.
Hiroshima, Japan


INTRODUCTION

The fundamental mechanism of the rotary engine (RE) was invented by
Dr. Felix Wankel who had carried out thorough investigations into the
smoothness of centroid motions of moving parts.  The RE has no
reciprocating piston and valvetrain which are the sources of vibration and
the obstructions to increased engine speed.  The RE aspirates a larger
amount of air per unit volume of engine package than the reciprocating
piston engine does, because the charge flows continuously from an intake
stroke chamber through an exhaust stroke chamber.  Hence, the inherent
features of RE are less vibration, compactness, light weight and high
performance.

It has been 25 years since Mazda started to develop the RE.  Mazda
has produced more than 1,500,000 automotive REs taking advantage of the
features mentioned above.  Considering the actual production, it can be
said that the RE is playing an important role as an automotive engine.  At
present, a number of companies and institutes are carrying out research
and development on advanced and multi-fueled REs for wide use in various
fields, such as in aerospace[1], marine, and industrial use[2], in
addition to automotive use.

When we review the process of development of the RE for practical
use, the subjects may be classified into three technical areas.  The first
is the development of gas seals and a sliding surface lubrication system,
which is necessary to reduce the friction and wear of seals and housings.
The second area is the development of combustion and emission control
technologies to improve fuel economy under emission standards.  The third
area is the development of high performance technologies to intensify the
inherent features of the RE.

This paper concerns the automotive rotary engine and explains its
existing and future technologies.  The main emphasis is placed on the
techniques essential to the RE:  lubrication, gas seals, combustion, and
supercharging.

## DEVELOPMENT OF RE TECHNOLOGY

The development of automotive RE technologies which Mazda introduced for practical use is shown in Table 1 [3,4]. During the second half of the 1970's, extensive research was made on the Lean-burn RE to obtain better fuel economy under the stringent exhaust emission standards that appeared after the oil crises. In the 1980's, the development of a higher performance RE began. This work is bringing out many other potentials inherent in the RE.

### Lean Burn RE

In the early period, a thermal reactor system[5] was used as an emission control system, because it was fit for the emission characteristics of the RE, such as less NOx, more HC and comparatively higher gas temperatures in the exhaust than those of the reciprocating engine, and because it had good applicability for leaded fuel. To obtain an improved, leaner reaction, a secondary air pre-heating system and a heat insulator of exhaust port were adopted. However, it turned out that making the mixture lean tended to limit the effectiveness of the thermal reactor system.

Generally, a catalyst system is advantageous for improving fuel economy because it maintains a high conversion efficiency under low exhaust gas temperature due to a leaner mixture combustion.

Accordingly, the Lean Burn RE[5] with a catalyst system was introduced for practical use at the beginning of this decade. Two main problems had to be overcome to adopt the catalyst, the reduction of HC upstream to the catalyst to prevent heat deterioration, and the development of a high efficiency catalytic converter system. These subjects were solved as follows.

Reduction of raw HC;
1) Improved gas seals with elastic sealing material
2) High energy ignition system with 4-electrode wide gap spark plug
3) Shutter valve type deceleration control system

Development of exhaust system with higher conversion efficiency;
1) Reactive exhaust manifold for better warm-up
2) 1-converter with 2-bed type catalyst
3) Secondary air switching system for effective emission control under various driving conditions

Table 1.  Development of Rotary Engine at Mazda

| Year | '75 | '80 | '85 | |
|---|---|---|---|---|
| Car | RX-3 | RX-7 | | NewRX-7 |
| Emission sys. | Thermal Reactor | 3way Catalyst | | |
| Fuel sys. | Carburetor | | EGI | |
| Charging sys. | 573cc×2 12A | 6 Port Induction | | |
| | 12A | TC | | |
| | 654cc×2 13B | Super Injection | | |
| | 13B | SI with TC | | |

By introducing these technologies, the Lean-Burn RE with catalyst system realized a fuel economy improvement of about 20% under the LA-4 mode and 30% at idling.

## High Performance RE

At the first stage of making a high performance engine, a 6-Port Induction system[6], which is a kind of variable port timing mechanism, was installed in the Lean Burn RE. This system made it possible to overcome the problems of fuel economy at low speed and output performance at high speed.

At the next stage, the carburetor was replaced by an EGI (Electric Gasoline Injection) and a turbocharger[6] was introduced. Since the RE has no exhaust valve, high exhaust gas energy can drive the impact type turbine effectively. To further improve the response and the torque characteristics over a wide range of speeds, the twin-scroll type turbo charger[7] was developed.

Together with a magnification of the displacement volume from 573 cc x 2 (Type 12A) to 654 cc x 2 (Type 13B), a new dynamic effect intake system[8], named Super Injection, has been developed.

In the process of developing the abovementioned high performance RE, several kinds of techniques to improve gas sealing, lubrication, and combustion have been put to practical use. The fundamental techniques for the RE are explained in the following chapters.

## EXISTING TECHNIQUES FOR RE

## Analysis of Mechanical Friction

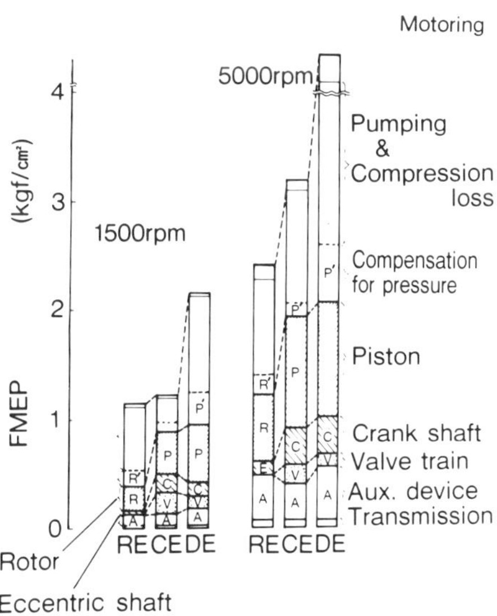


Fig. 1.  Friction Loss of RE, CE, DE

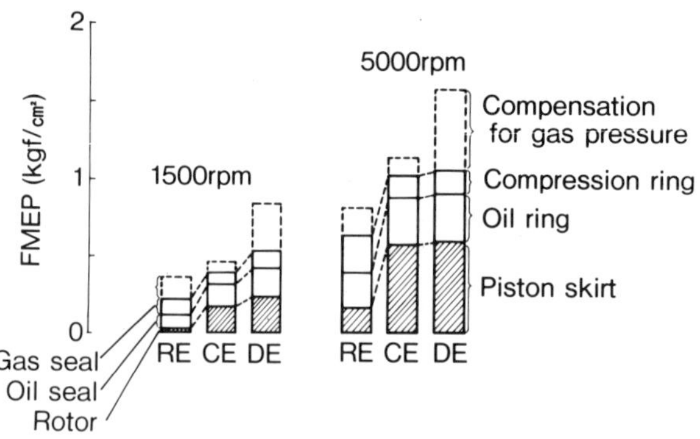



Fig. 2.  Friction Loss of Rotor and Piston

Fig. 1 [9] shows the friction loss of the RE compared with those of the reciprocating gasoline engine (CE) and the diesel engine (DE) having nearly the same displacement volume.  The friction factors were analyzed by means of the strip down method while motoring.  The friction loss of the RE is less than those of the reciprocating engines, especially in the high speed range.  The RE has such a small friction in total because it has less friction in the rotor than reciprocating engines do in the piston, and it has no valve train friction.  These results indicate that the RE is suitable for high speed use.

However, the friction of the gas seals is greater than that of the piston rings.  Moreover, as shown in Fig. 2 [9], the gas seal friction is affected significantly by gas pressure, because the gas seals are affected by the pressure under the gas seals over a larger area and longer period than are the piston rings.  Therefore, one problem is to reduce the back pressure area of the gas seals.  On the other hand, it is important to reduce the gas leakage especially in the range of low engine speed.

Improvement of Apex Seal Lubrication

Direct Supply of Lubricating Oil.  The lubricating conditions for the apex seals sliding on the trochoid surface get more and more sever by increasing the output performance.  The direct oil supply system[10], as shown in Fig. 3, enabled the lubricating oil consumption to be reduced to 1/3 - 1/5 of that of conventional manifold supply system.

Surface Treatment on Cr-Plating of Trochoid.  The inner surface of an aluminum alloy rotor housing is formed by a trochoid-shaped sheet metal with hard chromium plating on it, as shown in Fig. 4.  The Micro Channel Porous (MCP) Cr-plating[11] is the most advanced pattern.  Its surface has been made further resistant to wear by providing pinpoint pores for better oil retention and distribution.  By using the MCP Cr-plating, the wear of the trochoid wall has been reduced 50% compared with that of the conventional pin point porous plating.

To further reduce friction and wear of apex seals under poor supply of liquid lubricants, a fluorocarbon resin coating[12] was applied on the MCP chromium-plated trochoid wall.  The result was that the above coating

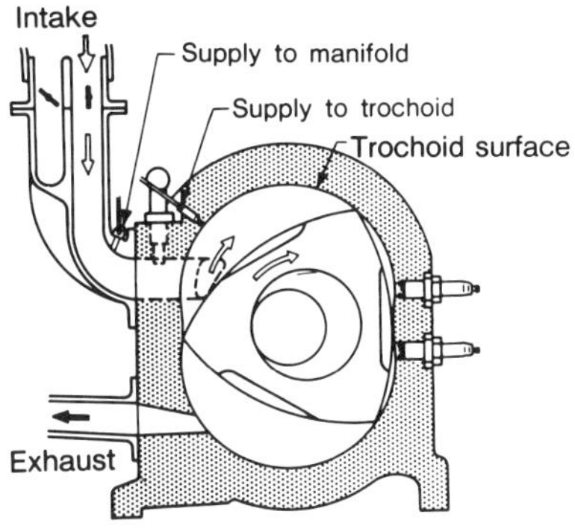



Fig. 3.   Direct Supply of Lubricating Oil

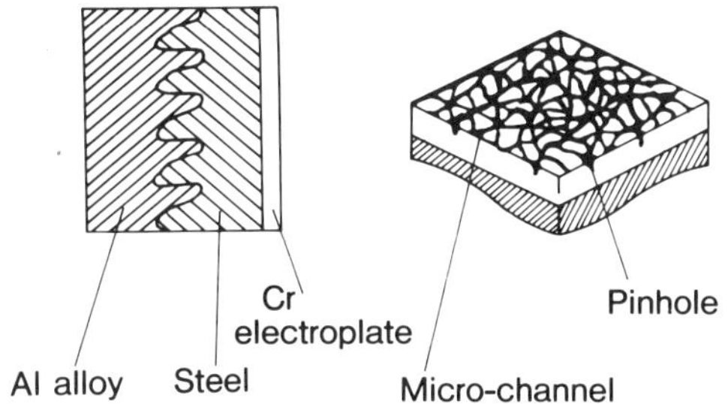



Fig. 4.   Micro Channel Porous Cr-Plating

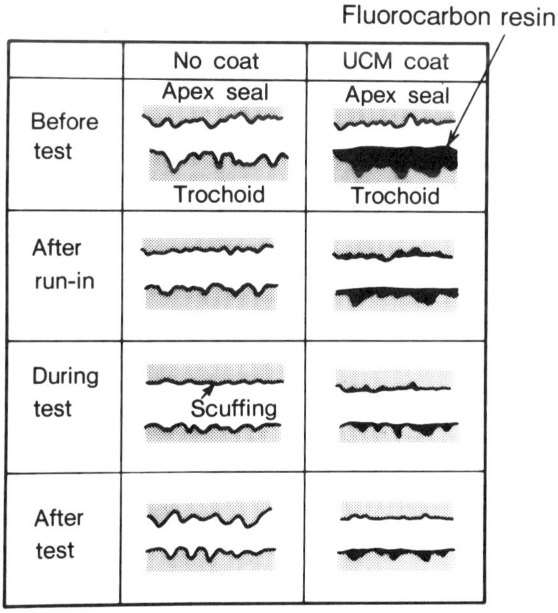

Fluorocarbon resin

| | No coat | UCM coat |
|---|---|---|
| Before test | Apex seal<br><br>Trochoid | Apex seal<br><br>Trochoid |
| After run-in | | |
| During test | Scuffing | |
| After test | | |

Fig. 5.   Mechanism of Conformability

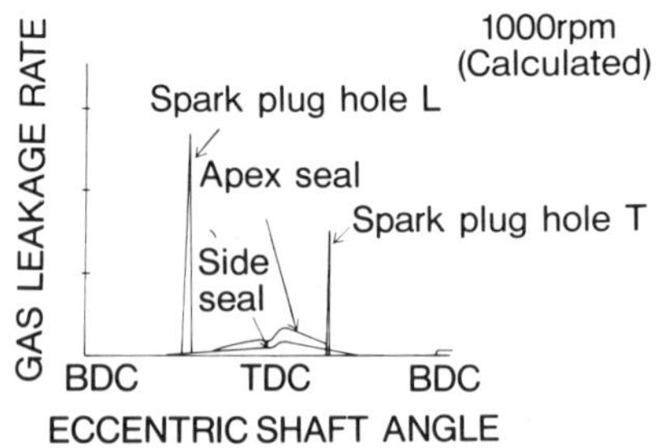



Fig. 6.  Gas Leakage Characteristics

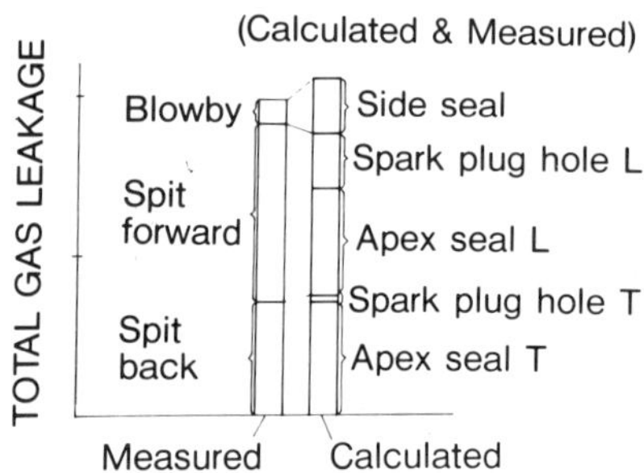



Fig. 7.  Total Gas Leakage

enabled the wear of apex seals to be reduced to 1/8 of that of the
conventional MCP Cr-plating especially under green engine conditions,
because the fluorocarbon resin, a solid lubricant, has no-sticking and
low-friction characteristics.  Fig. 5 shows a mechanism of comformability
between apex seal and Cr-plating with and without the fluorocarbon resin
coat (it is called Un-Coherence Material coat).

Improvement of Gas Seals

Fig. 6 shows the gas leakage characteristics of each factor[13], and
Fig. 7 shows the total amount of leakage[13].  The largest leakage is due
to the apex seals.  However, leakage from spark plug holes, which occurs
when the apex seals pass by the hole, is not negligible.

To restrict leakage from apex seals, it is important to reduce
clearance of the corner seals.  The development of the both types of
seals is shown in Fig. 8 [4,5].  The most progressive apex seal consists
of three pieces of reduced width, which improves their adherence on the
seal channel wall.  Owing to this, not only the leakage but also the
friction of the apex seals were reduced.

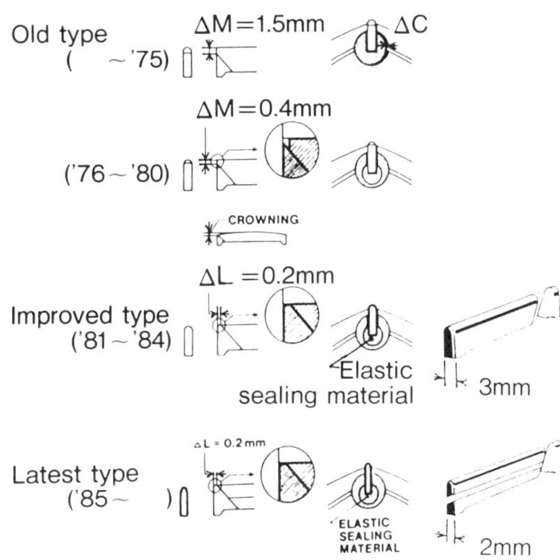



Fig. 8.  Improvement of Gas Seals

## Improvement of Combustion

Flame propagation in the 2-spark plug RE is presented schematically in Fig. 9 [14], which shows four flame fronts, TT, TL, LT and LL, generated by spark plugs, T and L, two burned gas regions and three unburned gas regions.

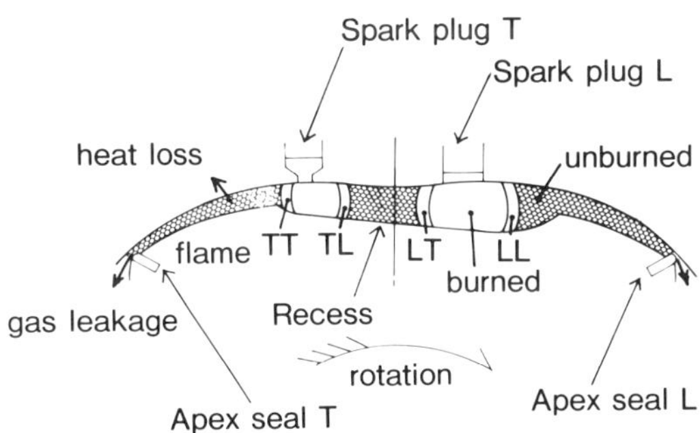



Fig. 9.  Flame Propagation Model for 2-Spark Plug Rotary Engine

Squish.  In order to improve combustion in a rotary engine, it is necessary to understand the effect of squish flow on flame propagation, because the movement of the rotor generates strong squish in the

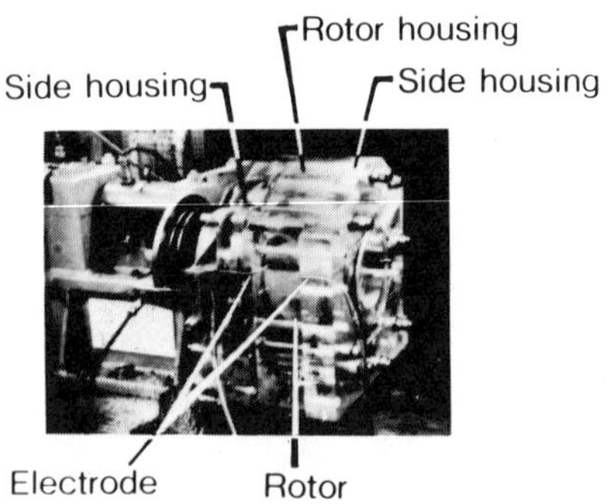


Fig. 10. Transparent Engine for Spark Tracing Technique

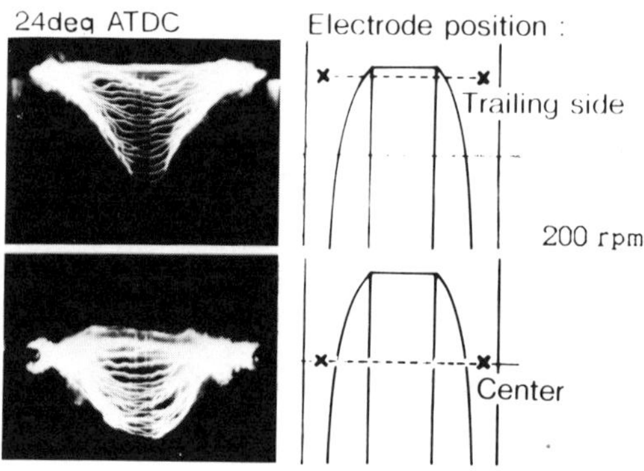


Fig. 11. Paterns of Squish Flow

direction of rotation. To this end, the patterns of squish were visualized near TDC of the compression stroke in a transparent engine by using a spark tracing technique, as shown in Fig. 10 [15].

The flow patterns have a convex shape in the direction of rotation, as shown in Fig. 11. In the vicinity of the trailing end of the rotor recess, the flow pattern is strongly affected by the shape of the recess.

Flame Propagation. The relationship between squish and flame propagation was clarified by using a high-speed photo-technique and a one-directional flame propagation model[14]. Fig. 12 shows the location of quartz windows in the region between the two spark plugs, T and L, in the housing.

The behavior of two flame fronts, TL and LT, were observed to approach each other, as shown in Fig. 13. Although the region that can be seen is limited to the abovementioned windows, the pictures show that the flame TL propagates rapidly in the leading direction, but the flame LT moves quite slowly in the trailing direction. The behavior of all

190

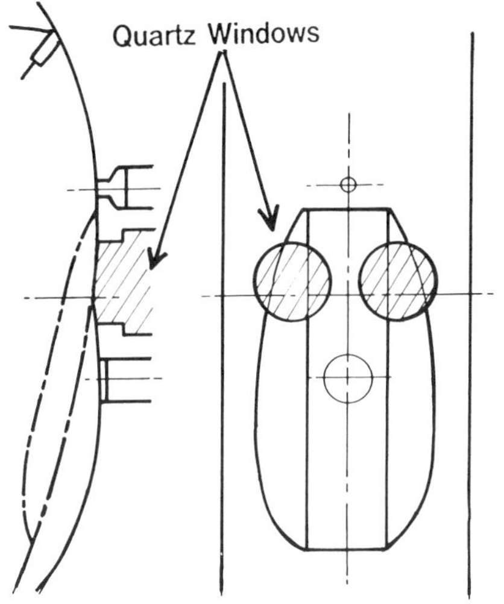


Photographic Conditions

| | |
|---|---|
| Engine Speed | 1000 rpm |
| Charging Efficiency | 70% |
| Air-Fuel Ratio | 13 |
| Camera Speed | 3000 f/s |

Fig. 12.  Quarts Windows for Photo-technique

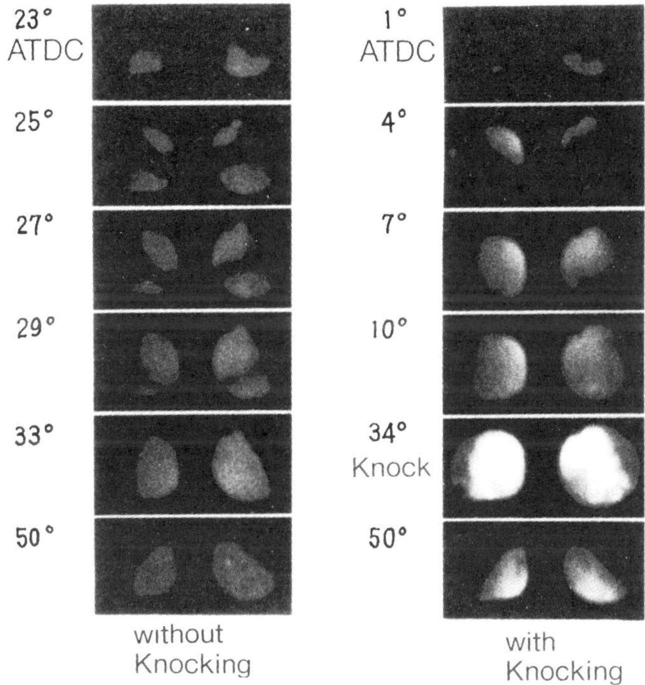


Fig. 13.  Photograph of Flame Propagation

flames in the whole combustion chamber were predicted by using a combustion model which had already been verified by measurements from the ionized gap method shown in Fig. 14.  The flame TT which propagates in the trailing direction from the trailing spark plug, presents a complicated phenomena, because the flame is strongly affected by the squish.

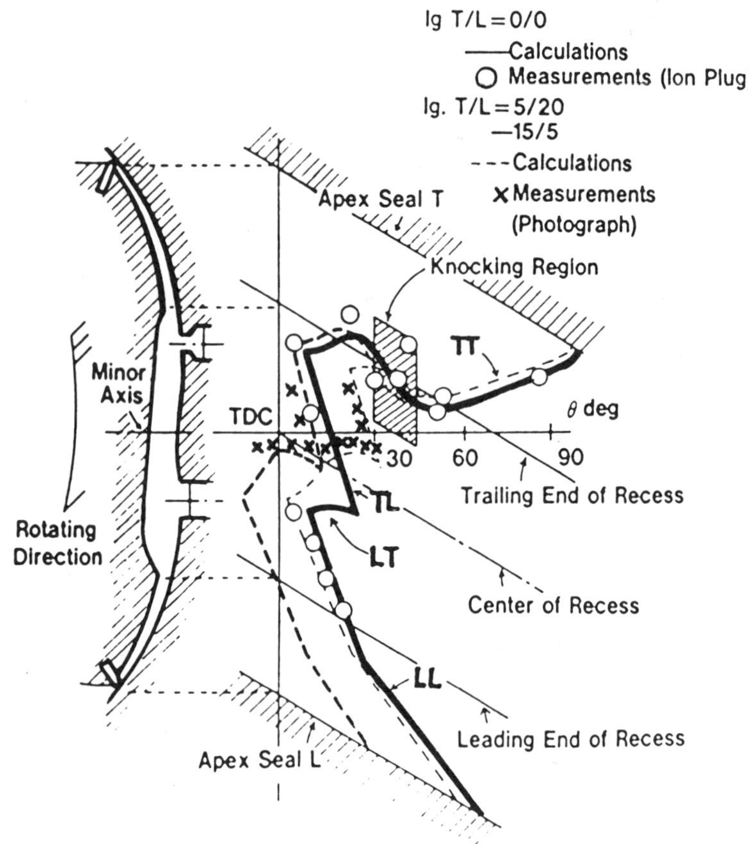


Fig. 14.  Flame Propagation in 2-Spark Plug RE

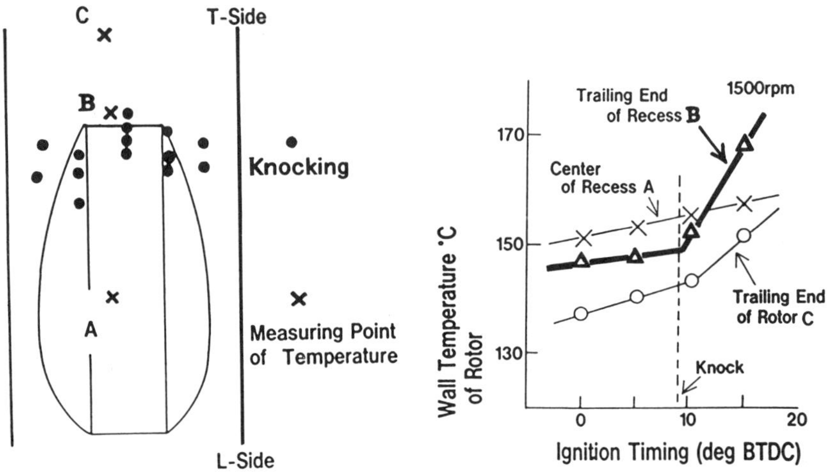


Fig. 15.  Location of Knocking, and Wall Temperature of Rotor

A twin-spark plug combustion is effective to improve thermal
efficiency.  In the RE, it is important to assist combustion by using the
trailing side spark plug, because the flame in the trailing region is
cooled by strong squish.

Knocking.  Fig. 15 [14] shows where knocking occurs in the RE.
Knocking occurs in the vicinity of the trailing end of the rotor recess
when strong squish is generated, as shown in Fig. 14.  In the flame
pictures, the knocking was observed as a very bright flame which spreaded
abruptly.  In order to increase the compression ratio or the charge
pressure to get heigher performance, it is important to improve the
combustion and cooling characteristics in the region where knocking
occurs.

## Improvement of Charging Efficiency

Fuel Injector with Mixing-plate Socket.  Although the electronic
fuel injection system has some benefit in charging efficiency when it is
compared with the conventional carburetor system, it is still necessary
to improve the atomization of the spray from the injector.
Accordingly, a new injector with a mixing-plate socket[6] has been
developed, which promotes air and fuel mixing by atomizing air directed
to the nozzle receptacle at low engine speeds, and by a mixing-plate
having a number of small holes in an open-sided plastic tube socket at
higher speeds, as shown in Fig. 16.

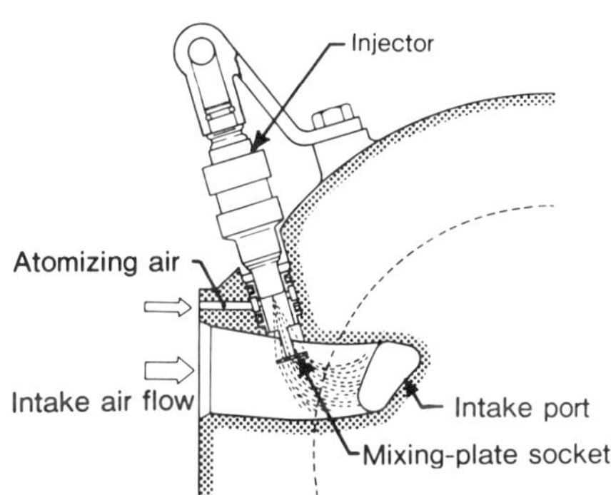


Fig. 16.  Fuel Injector with Mixing-plate Socket

6-Port Variable Induction System.  The 6-port variable induction
system[6] has three intake ports per one rotor, that is, the primary
port, the secondary port, and the auxiliary power port, as shown in Fig.
17.  The auxiliary power port is controlled by a cylindrical valve which
is actuated by exhaust pressure in accordance with engine operation.  The
primary port has a large intake gas speed and a short period where the
openings of the intake-exhaust ports overlap to improve combustion and
low-end torque.  The auxiliary power port increases the intake port
opening area and duration to improve top-end power.

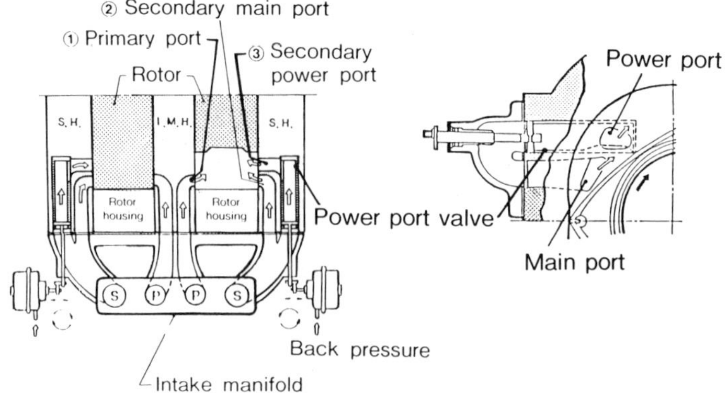



Fig. 17.   6-Port Variable Induction System

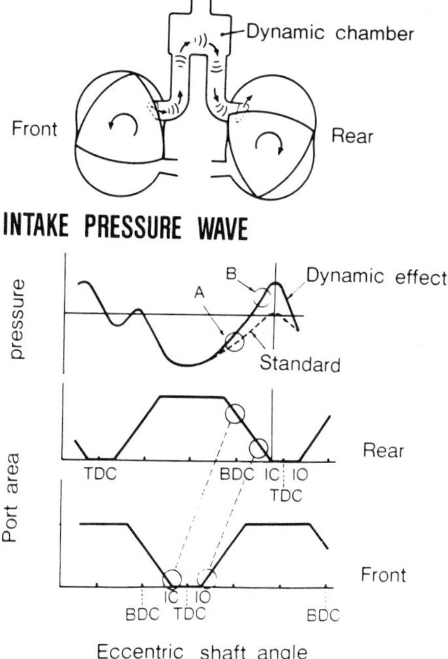



Fig. 18.   Principle of Dynamic Effect on
Volumetric Efficiency

Dynamic Effect Intake System. The dynamic effect intake system[6,8] consists of comparatively long intake passages which connect to a dynamic chamber. The system uses the effect of the pressure wave interference between the two rotors to increase charging efficiency, as shown in Fig. 18. As the intake port closes, a compressed pressure wave is generated by its own inertia in the intake passage. As it begins to open, another pressure wave occurs owing to the counter flow of residual gas from the combustion chamber into the intake passage. These two pressure waves feed more air into the other port, owing to both the quick opening-closing characteristics and the long duration intake-overlap characteristics of RE's intake ports.

Twin-Scroll Turbocharger. In general, the use of a small turbocharger with a small A/R (where A is the sectional area of the scroll, and R is the distance between the scroll center and the turbine center) improves low-end torque and acceleration response, but remarkably drops the supercharging efficiency at higher engine speeds.

To improve quick response and performance over the whole range of

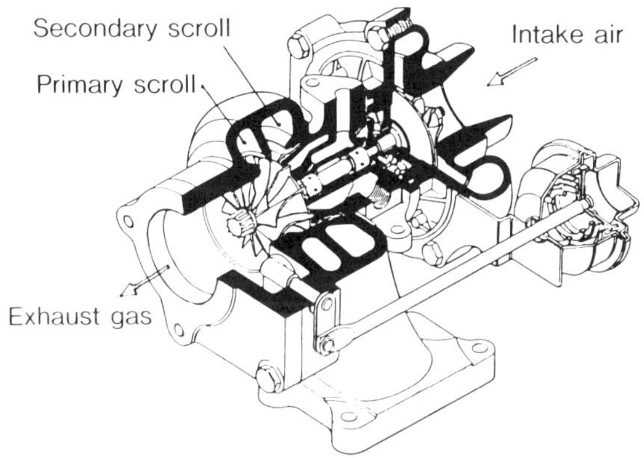


Fig. 19.  Twin-Scroll Turbocharger

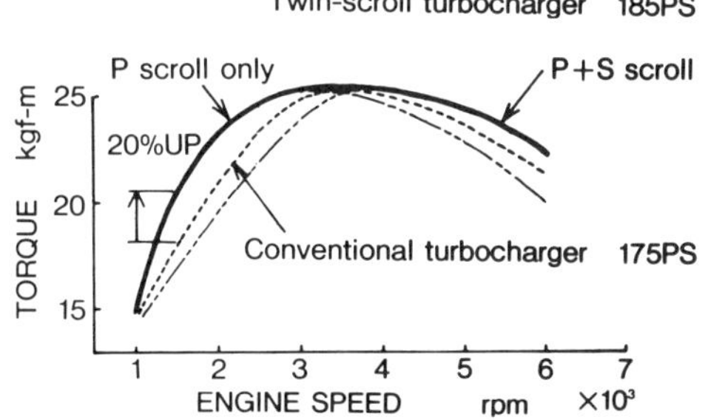


Fig. 20.  Performance of Twin-scroll Turbocharged RE

speeds, the twin-scroll turbocharger[7] has been introduced.  Fig. 19
gives a cutaway view of the twin-scroll.  By switching the scrolls to
operate in accordance with engine operation, not only a small A/R (=0.4)
at low engine speeds but also larger A/R (=1.0) at higher speeds can be
obtained to realize excellent performance over a wide range of speeds, as
shown in Fig. 20.


NEW RX-7's RE

The new RX-7 is a state of the art sports car which was developed
with the intention to realize a high level combination of styling,
handling, and maneuverability.  The compactness and lightness of the RE
allowed the low-hood profile and front-midship layout which are important
factors in the superior handling of the RX-7.

Fig. 21 shows the 13B-SI (Super Injection) engine and the 13B-TC
(Turbo-charged) engine for the New RX-7 [7].
The major techniques incorporated into these engines to realize these
demands are shown in Table 2.  The improved dynamic effect intake system,
the twin-scroll turbocharger and a direct inter cooler were incorporated
for higher output.  The dual injector, a very short intercooler passage
(therefor it is called direct intercooler) and a 8-bit, one-tip
microcomputor total electronic management system were adopted for quick
response.  In addition, improved gass sealing and control of rotor
cooling oil flow contribute to reduce fuel consumption.  The
dual-silencer was adopted for quietness.  Improvement in the rotor
housing surface lubrication, reduction of rotor weight, modification of
the stationary gear fixing pins, control of the rotor cooling oil flow by
a thermovalve, and the employment of water cooled turbocharger contribute
to ensure durability.  Because of these techniques, engine output is
greatly increased and fuel economy under normal driving conditions is
improved, as shown in Fig. 22.


Table 2.  Major Techniques Incorporated into
New RX-7's RE

|  | 13B-SI | 13B-TC |
| --- | --- | --- |
| Improved 6-Port Variable Induction System | o | – |
| Improved Dynamic Effect Intake System | o | o |
| Dual Injection | o | o |
| Twin-Scroll Turbocharger | – | o |
| Direct Inter Cooler | – | o |
| Total Electronic Management System | o | o |
| Improved Gas Seals | o | o |
| Improved Rotor Cooling System | o | o |

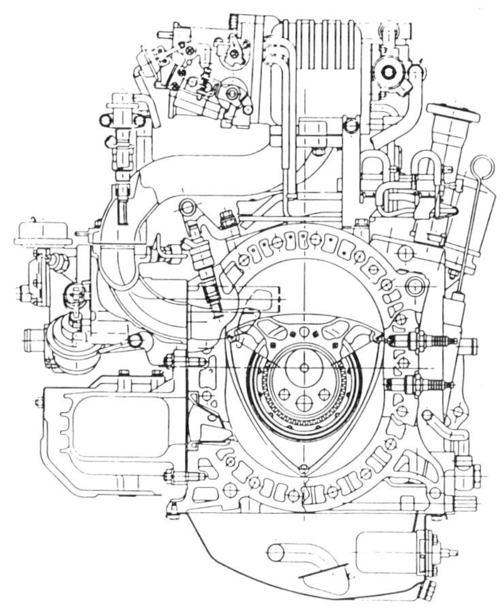

**13B-SI**
Dynamic effect intake system

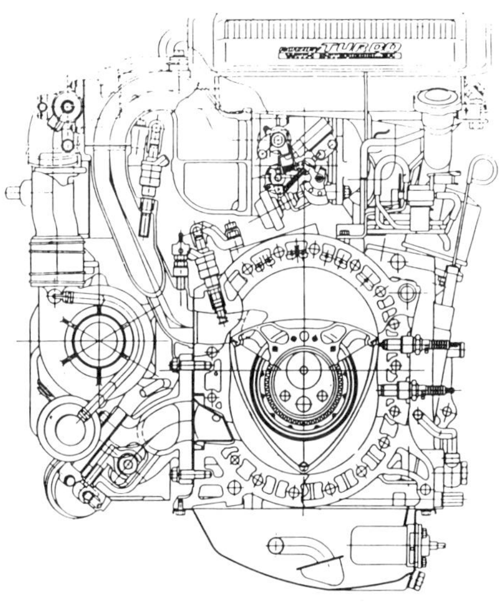

**13B-TC**
Turbo-charged system

Fig. 21.  New RX-7's RE

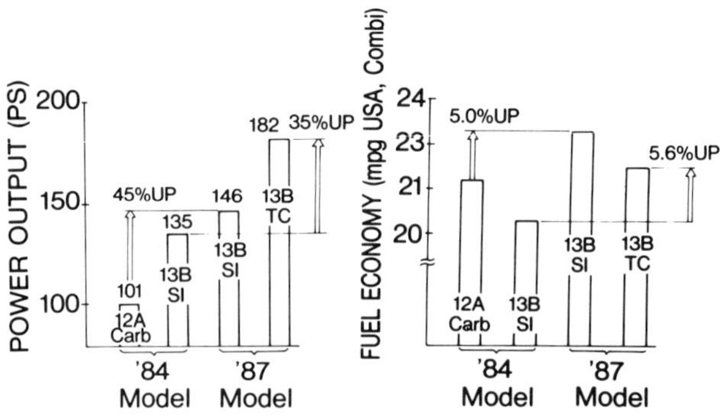


Fig. 22.  Performance of New RX-7's RE

## THE FUTURE OF THE RE

Because of the increasing demand to diversify automobile engines, light and compact engine designs, which will be able to contribute to imaginative styling of cars by increasing the design flexibility, will become of great importance in addition to the high performance technology for high-level drivability.  For energy saving, pursuit of thermal efficiency improvement is an eternal subject, with multi-fuel combustion technologies become even more important in the future.

To fulfill the above demands, research and development programs for the following future technologies are on-going now.

1)   High efficiency mechanical supercharging.
      Multi-rotor; higher engine speed.
2)   Reduction of weight and size; Extention of designing freedom.
3)   Reduction of friction loss, pumping loss and cooling loss.
4)   Multi-fuel stratified charge combustion.
      Improvement of ignitability.

Some of these are introduced below.

### TISC (Timed Induction Super Charging)

The RE has great design freedom for intake port layout.  TISC makes the most of this feature, as shown in Fig. 23.  Part of the intake air is supercharged by a small, and thus low friction, air pump.  Timing of the supercharging is controlled by a rotary valve synchronized with the output shaft.  The performance improvements for the speed range for normal driving are significant, as shown in Fig. 24, and thus it is expected to be a potential super charging system in the near future.

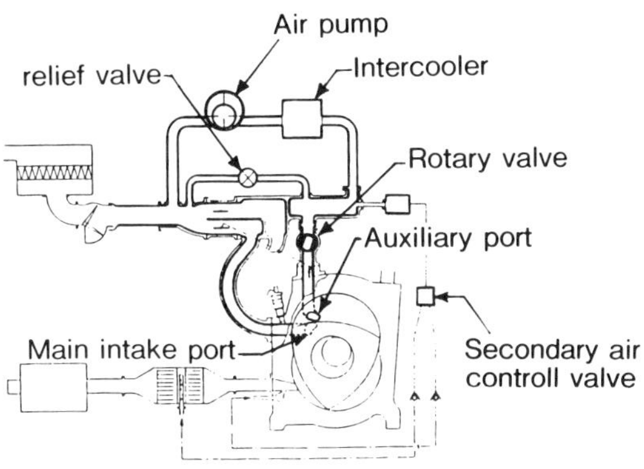


Fig. 23.  TISC

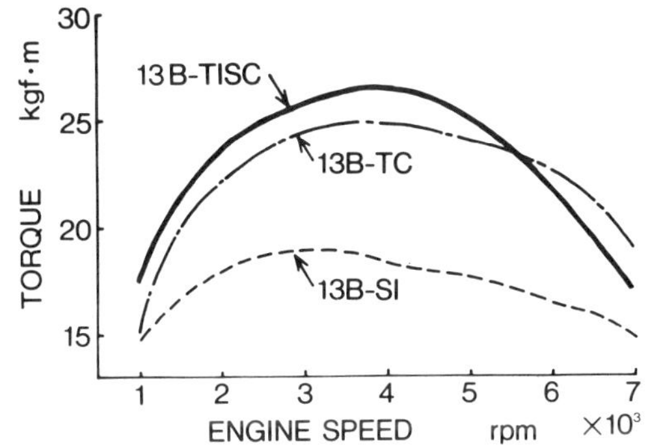


Fig. 24.  Performance of TISC

## Three Rotor RE

For the realization of a high grade engine which has higher output
and torque, and lower NVH and thmooser revolution, a three rotor engine
is now under development.  As shown in Fig. 25, the output shaft has a
two-piece configuration by simple conical coupling.  By use of this
coupling technique, a four or more rotor configuration will become
possible.

## Low Pumping Loss RE

Reduction of pumping loss is an important subject for gasoline
engines with intake air throttling.  By the delayed-inlet-close cycle,
which has a connecting port between the two intake chambers of the rotor

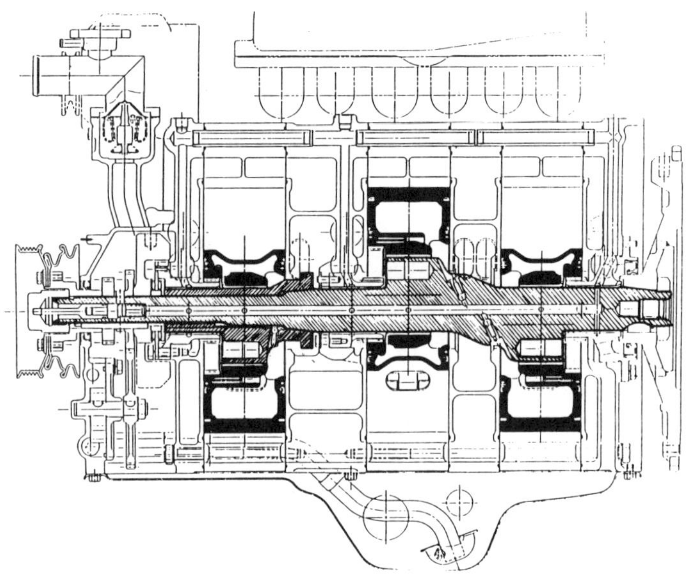

Fig. 25. Three Rotor Engine

housings, pumping loss can be reduced, owing to the special configuration
of the RE. As compression pressure is reduced by this technique, the
friction of the gas seal, which is of great significance for the RE, is
also reduced.

## Direct Injection Stratified Charge RE

The Direct Injection Stratified Charge (DISC) engine, like the
diesel engine, has no pumping loss due to intake throttling. In this
engine, mixture from the fuel spray is stratified in the vicinity of the
ignition plug with the high-pressure fuel injection system directly
attached to the combustion chamber.

The DISC engine not only increases the thermal efficiency
dramatically but also improves combustion over a wide range of fuels, as
compared with conventional spark ignition engines.

CURTISS-WRIGHT has actively tested and improved the DISC RE, and
clarified that the RE's geometry could readily provide the requisit air
motion for a stratified charge, with no ill effects to volumetric
efficiency and pumping work[16]. As it will further improve combustion,
the adiabatic engine technology will increase the potential benefits of
DISC RE.[17]

Therefore, the unthrottled, low heat rejection, multi-fuel DISC RE
is expected to become one of the most advanced internal combustion
engines.


CONCLUSION

The fundamental technologies of the RE, the latest techniques of the
New RX-7's engine, and the future of the Mazda Rotary Engine were
described. From now on, research and development on the techniques that
make the most of the rotary engine's inherent features and the techniques
that will meet diversified demands will be positively carried out.

Regarding its automotive use in the future, we think the abovementioned
RE will be a tool to realize the dreams of our individual customers.


REFERENCES

1. Phillip R. Meng, William F. Hady, and Richard F. Barrows,
   An Overview of the NASA Rotary Engine Research Program,
   SAE Paper 841018 (1984).
2. T.N. Chen, R.N. Alford, and S.S. Kim, Detonation
   Characteristics of Industrial Natural Gas Rotary Engines,
   SAE Paper 860563 (1986).
3. H. Ohzeki and T. Yamaguchi, President and Future of Rotary
   Engine Technology, JSAE Review, March (1982), p.9.
4. Y. Tatsutomi, H. Ohzeki, T. Tadokoro, and H. Okimoto,
   Present and Future of Rotary Engine Technology, Journal
   of the Society of Automotive Engineers of Japan, Vol. 40,
   No. 1 (1986), p.67. (in Japanese)
5. K. Shimamura and T. Tadokoro, Fuel Economy Improvement of
   Rotary Engine by Using Catalyst System, SAE Paper 810277
   (1981).
6. T. Muroki, Recent Technology Development of High-Powered Rotary
   Engine at Mazda, SAE Paper 841017 (1984).
7. T. Tadokoro, Y. Fujimoto, I. Matsuda, and M. Nakao,
   Turbocharged 13B Rotary Engine for New Savanna RX-7,
   Internal Combustion Engine, Vol. 24, No. 313 (1985),
   p.36. (in Japanese)
8. H. Okimoto, Improvement of Rotary Engine Performance by New
   Induction System, SAE Paper 831010 (1983).
9. A. Nagao and K. Tanaka, Friction Analysis of Three Types of
   Automotive Engine, Journal of the Society of Automotive
   Engineers of Japan, Vol. 38, No. 9 (1984), p.1094. (in
   Japanese)
10. H. Ohzeki, N. Kurio, and Y. Fujimoto, Wear Prevention
    Technology of High-power Rotary Engine, Journal of the
    Society of Automotive Engineers of Japan, Vol. 39, No. 4
    (1985), p.375. (in Japanese)
11. T. Muroki and J. Miyata, Material Technology Development
    Applied to Rotary Engine at Mazda, SAE Paper 860560
    (1986).
12. Y. Shidahara, Y. Murata, Y. Tanita, and Y. Fujimoto,
    Development of Sliding Surface Material for Combustion
    Chamber of High-Output Rotary Engine, SAE Paper 852176
    (1985).
13. A. Nagao, S. Yoshioka, S. Kariyama, K. Onishi, and
    J. Funamoto, Analysis of Flame Propagation and Knocking
    in a Rotary Engine, Mazda Technical Review, No. 4 (1986),
    p.69. (in Japanese)
14. A. Nagao, S. Yoshioka, K. Ohnishi, and K. Tanaka,
    Flame Propagation and Knocking in Wankel Rotary Engines,
    Proc. of Combustion Symposium of JSME, December (1985),
    p.91. (in Japanese)
15. S. Yoshioka, A. Nagao, S. Kariyama, K. Ohnishi, and K. Tanaka,
    Visualizing Study on Squish Flow and Flame Propagation in
    Rotary Engine, Journal of the Flow Visualization Society
    of Japan, Vol. 6 No. 22 (1986), p.3. (in Japanese)
16. Charles Jones, Advanced Development of Rotary Stratified
    Charge 750 and 1500 HP Military Multi-Fuel Engines at
    Curtiss-Wright, SAE Paper 840460 (1984).
17. R. Kamo, R.M. Kakwani, and W. Hady, Adiabatic Wankel Type
    Rotary Engine, SAE Paper 860616 (1986).

THE STRATIFIED CHARGE ROTARY ENGINE


James W. Walker and
Robert E. Mount

Rotary Engine Division
John Deere Technologies Int'l Inc.
Woodridge, N.J.

ABSTRACT

A brief history of the Rotary engine leading up to
the development of the turbocharged stratified
charge rotary engine is presented. The dual nozzle
stratified charge concept is discussed as are some
of the design and development techniques used to
expedite the design, analysis and optimization of
the engine. Various market segments for the application of the rotary engine to power generation and
propulsion uses for ground vehicles, aircraft and
marine are discussed.

INTRODUCTION

The emphasis on recent developments in Internal Combustion
Engines has been toward compact fuel efficient engines. The
major thrust for this push had its focus in the energy shortage of the 70's and the subsequent push for improved fuel
consumption for the automobile.

Smaller, lighter cars and smaller engines were the initial
primary result of this push. In the mid eighties, we have
seen the retention of the smaller engines, but there continues
to be a significant demand on the part of the customer for
increased performance from the engines. The increased power
density from the engines is an inevitable result of this
customer pressure. The technical publications contain much
information on how the engine industry has responded to the
above stimulus.

The Stratified Charged Rotary Engine represents a power plant
that responds to the customer's desire for increased power,
while retaining the desired small package size, light weight,
and improved fuel efficiency. In addition the SCORE design
engine has a significant plus in its ability to burn a wide
variety of fuels. John Deere's Rotary Engine trademark is
SCORE™ Stratified Charged Omnivorous Rotary Engine.

This presentation will deal with some of the history toward
the development of the Stratified Charged Omnivorous Rotary
Engine, some of our more recent development information, and
a discussion of some market developments to date.


HISTORY


A number of very good books have been written which discuss
the invention, introduction and development of the 'Wankel'
based rotary engine configurations (1)(2)(3)(4). The basic
concept involves a rotating, triangular shaped rotor inside a
trochoid shaped housing.

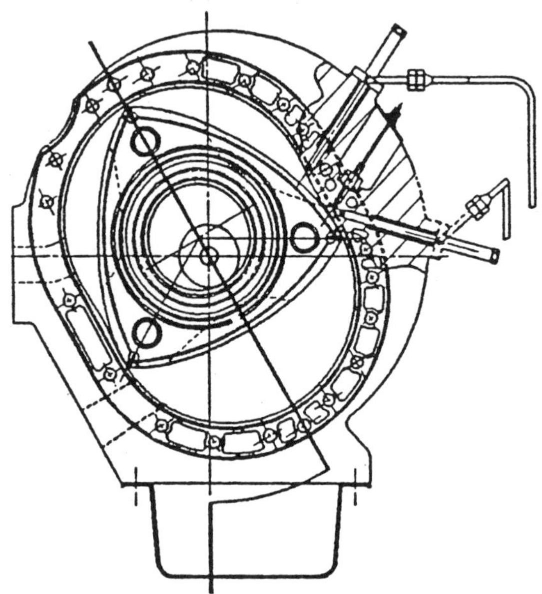

FIG 1 Cross section of Rotary engine showing trochoid shaped
rotor housing and triangular shaped rotor.


In the first engines which Felix Wankel ran, the shaft was
stationary, and the rotor and trochoid housing rotated.

A number of development difficulties (getting ignition to
moving spark plugs, engine idling characteristics, dynamics
of the moving housing, and power extractions, etc.) resulted
in subsequent designs evolving to the basic configuration of
today's Rotary engine.

A number of engine prototypes were made by the various
organizations to whom Wankel sold specific 'rights'. Various
prototype applications, include; autos, motorcycles, boats,
snowmobiles, airplanes, trucks, lawnmowers and numerous
others. The early enthusiasm for this new engine concept was
one of the factors in sometimes rushing into production
designs which had not had sufficient development effort to
work out the various design bugs.

Early problems include; rough idling characteristics, high oil consumption, excessive leakage past the combustion gas seals, and excessive wear of apex seals.

Most of these early engines were precharged (carbureted) and spark ignited, although there was some development done on a compression ignition version of the engine.(7) The carbureted engines tended to have higher fuel consumption than their reciprocating counter parts, in part due to the high surface to volume characteristics of the combustion chamber. This high surface to volume ratio, also meant that there was potential for more of the partially burned fuel/air mixture at the boundary layer surfaces to show up in the exhaust as unburned hydrocarbons.

The rotary engine as applied to automobiles in today's Mazda cars, and its popularity indicates that these early engine problems have been satisfactorily solved, and applied to production versions of the engine.

THE STRATIFIED CHARGE COMBUSTION CONCEPT

As development of the engine proceeded at Curtiss-Wright, it was felt that if the engine could operate unthrottled as does the diesel engine, and if the fuel could be injected directly into the compressed air charge at or near 'TDC', rather than introduced as a fuel air mixture, engine fuel consumption could be improved and emissions decreased. The improved fuel consumption would come from a combination of the unthrottled intake system, and the introduction of fuel only as required rather than trying to 'fill the entire combustion chamber' with an ignitable fuel/air mixture.

Furthermore, if the fuel which was introduced could be ignited by an ignition source, rather than relying on the self ignition characteristics of the fuel (as in diesel-compression ignition) then the engine would have a wider tolerance to fuel characteristics. Successful development of such a combustion system would potentially enable the engine to operate as a true "multi-fuel" engine.

Early efforts at Curtiss-Wright to develop the stratified combustion concept involved the use of a multihole nozzle with one of its sprays directed toward the spark plug. This engine, as did most efforts to stratify reciprocating "diesel" engines resulted in engines that ran well under rather narrow operating conditions of speed and load. (The primary difficulty was one of maintaining the correct fuel/air ratio in the vicinity of the spark plug. Under specific operating conditions, the 'fuzz' from the injector spray would be 'just right', and the engine ran fine, having the smoothness, and fuel economy desired. Once the spray penetration characteristics changed as the amount of fuel injected either increased or decreased, the optimum conditions for ignition of the fuel spray that existed by the spark plug changed, resulting in poorer combustion characteristics.)

Various schemes of the single nozzle stratified charged combustion concept were evaluated with varying degrees of success.(6)(See fig. 2&3) As the Curtiss-Wright engineers worked with the combustion system, they developed the two nozzle stratified charged concept. See Figure (4). This concept involves a single orifice 'pilot' nozzle which essentially injects a constant quantity of fuel optimized for producing an ignitable mixture in the vicinity of the spark plug. A second nozzle which incorporates multiple orifices, serves as the main fuel source for the combustion process, and is designed to inject fuel as it is required for controlled combustion.

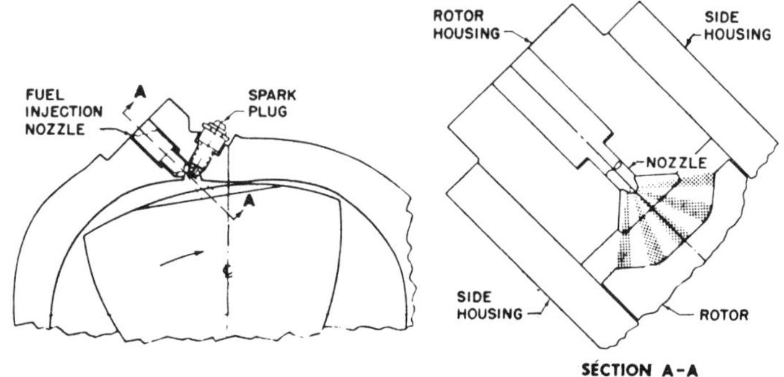


FIG 2   CO-PLANAR SINGLE NOZZLE SC DESIGN.

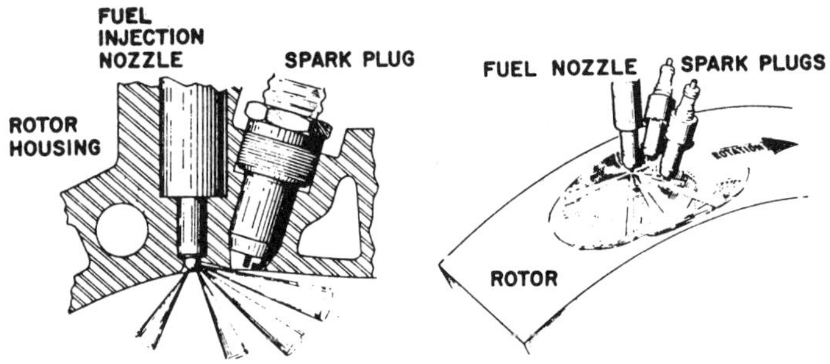


FIG 3 SHOWER HEAD SC NOZZLE DESIGN.

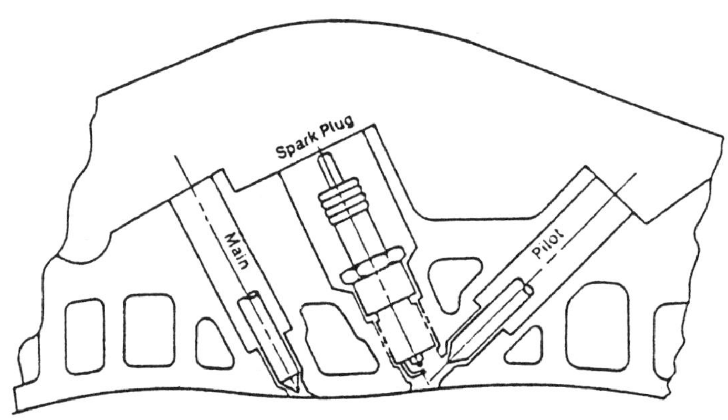


FIG 4 DUAL NOZZLE STRATIFIED CHARGED CONFIGURATION.

The quantity of fuel injected is determined by the load requirements on the engine.

This dual nozzle stratified combustion system has lived up to it expectations. It has successfully demonstrated its ability to run on gasoline, methanol, diesel fuel, and jet fuel without any engine adjustments. Further details of the performance results obtained during the development program on the Stratified Charged Combustion System are presented in reference(6).

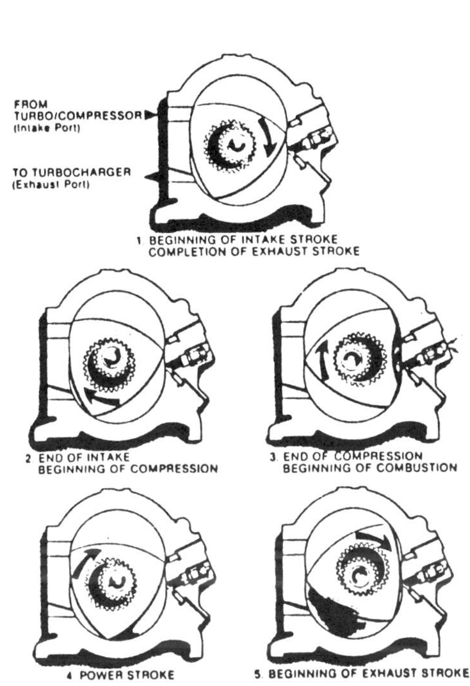


The turbocharged rotary engine uses a rotor with three combustion faces. These faces (which are equivalent to pistons in a reciprocating engine) provide for a power impulse (stroke) during each crank revolution. The rotor fits closely around the crank eccentric, but turns at 1/3 of its speed. Producing rotary motion directly eliminates all those parts needed in a reciprocating engine to convert the up and down motion to rotary motion.

The cutaway drawings (shown at right) demonstrate a typical rotary cycle. (For this explanation, we will follow only one of the three rotor faces.)

1. The operation begins when the APEX seal uncovers the intake port and unthrottled air from the turbocharger compressor enters into the combustion chamber.

2. Air continues to enter the chamber until the trailing APEX seal closes the port. Air is compressed as the rotor continues its rotation.

3. As the air is compressed to its minimum volume the pilot fuel charge is ignited.

4. Combustion is stratified. This is controlled by the main injector and air motion resulting from a specially designed rotor pocket. The power stroke is completed when the exhaust port is uncovered.

5. High temperature gases then exit through the exhaust port. The exhaust turns the turbocharger turbine, which in turn powers the turbocharger compressor.

FIG 5  BASIC GEOMETRY AND OPERATION CYCLE STRATIFIED CHARGE ROTARY ENGINE.

TURBOCHARGING

One of the distinct advantages of the rotary engine, is the elimination of the intake and exhaust valves required of the typical reciprocating engine. The elimination of the valves and their accompanying flow restrictions and dynamics limitations, as in the Rotary engine, expands the engine maximum speed potential, and increases the exhaust energy available.

Turbocharging to use this available exhaust energy is a logical and viable step in the evolution of the Rotary engine, and especially the Stratified Charged Rotary engine. (Since the fuel is injected as required rather than precharged, the engine can be boosted to a higher level with out concern for preignition.)

Some turbocharging performance was done on the Stratified
Charged Rotary Engine in the early 80's.(7) Much of this
work was done on the engine designed primarily to run as a
naturally aspirated engine.

As the result of an extensive study funded by NASA, which
pinpointed the Advanced Turbocharged Stratified Charged
Rotary Engine as being the most promising small aircraft
power plant of the future. A Technology Enablement Program
was funded by NASA to conduct tests on a 40 cubic inch (0.7
Liters) single chamber Stratified Charged Rotary research
engine.

Early results of these tests are reported in (12). The
engine which has been built and run by John Deere was
designed to withstand the higher operating pressures
resulting from turbocharging and the higher power outputs.

FIG 6   SCORE[T.M.] 70 SINGLE ROTOR RESEARCH RIG ENGINE.

   This single rotor engine has successfully run at over 75
kW   (100 hp) and up to 8000 rpm. Fuel injection system
limitations precluded running at higher power levels.

As a result of our initial running of the NASA research rig
engine, we have embarked upon the second phase of the
program. This phase includes evaluation of an advanced
injection system capable of delivering controlled fuel
injections over the speed and load range capabilities of the
engine. A electronically controlled, hydraulically actuated
unit injector fuel injection system has demonstrated its
capabilities to satisfactorily inject and control fuel

208

quantities up to 100 cubic millimeters per injection at speeds in excess of 10,000 injections per minute. This system is scheduled for engine testing starting in August 1986.

We have used this research rig engine to evaluate the benefits of various engine porting arrangements, nozzle configurations, turbocharger trims, and intercooler conditions.

Additional turbocharging performance has been conducted on the SCORE [T.M.] 580 engine (a 5.8 liters per rotor engine being currently run in a single rotor research configuration, and a two rotor demonstration engine ) to in part verify our Finite Element predictions and to evaluate some of the design changes required to modify the basic naturally aspirated version of the engine.

Extensive finite element analysis of the major structural components led to significant design changes as compared to the engine as designed for naturally aspirated operation. (See figures 7 & 8)

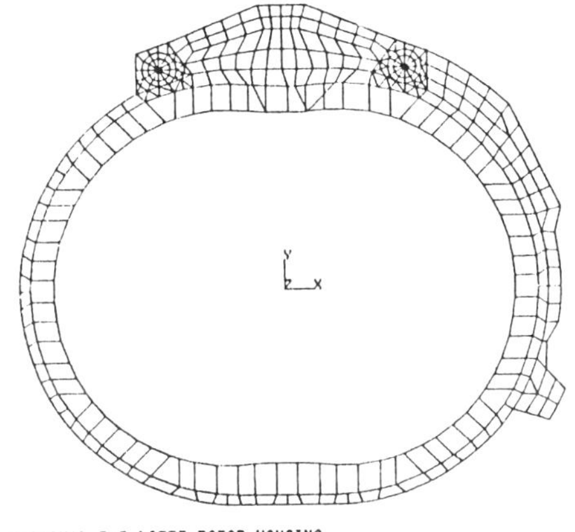


FIG 7 NATURALLY ASPIRATED
VERSION OF ROTOR HOUSING.

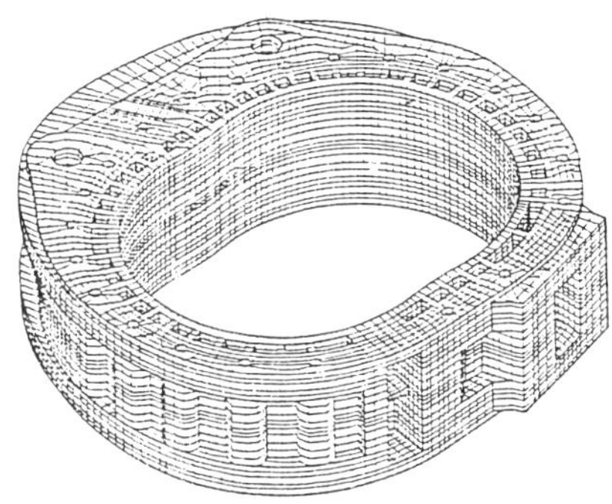

FIG 8 FEM OF ROTOR HOUSING
MODIFIED FOR TURBOCHARGING.

In order to accommodate the higher firing pressures due to turbocharged operation, the rotor was beefed up by incorporating additional support ribs. The rotor housing has been modified to change the bolt locations, and to provide extra support in the combustion chamber portion of the housing.

In addition to rotor housing modifications, a center main bearing was added to the two rotor version of the engine.

Additional factors which needed to be considered as the result of turbocharging to obtain additional engine output, were housing vibration levels, excitation forces on the apex seal, crankshaft configuration, rotor and housing surface temperatures.

Based upon the results of preliminary turbocharger testing, turbochargers were selected for our engines operating at a rated power of 280 kW (375 hp) per rotor at 3600 rpm.

Figure (9) shows some initial performance results.

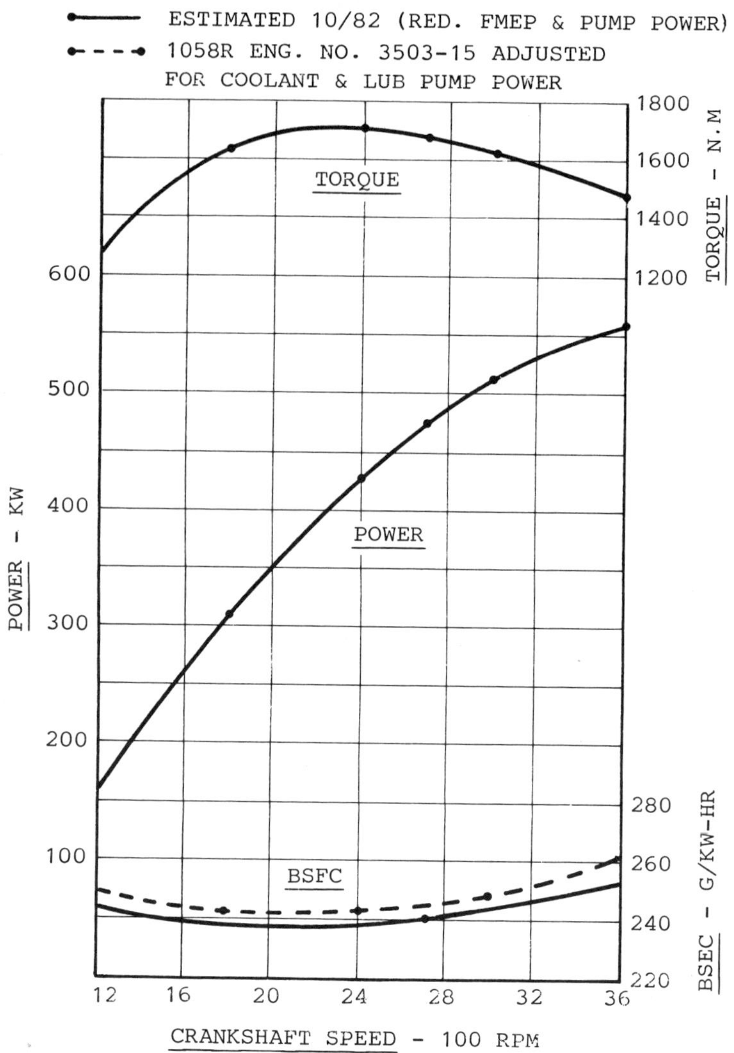


FIG 9 CURVE SHOWING SOME PERFORMANCE RESULTS FROM TURBOCHARGED 580 TWO ROTOR ENGINE.

The engine as designed has significant power growth potential. As we gain experience at the current rating, and verify the validity of the redesigns (based upon FEM studies) we will expect to increase the output per rotor to 375 kW (500 hp) or more.

FIG 10   SCORE 580 EARLY CONFIGURATION


VALUE CONCEPTS STUDY

There has been significant market interest in the Stratified
Charged Rotary engine, primarily because of the power density
and fuel savings advantages to its potential users. In order
to determine if this engine concept could cost effectively
compete in the market, John Deere conducted an extensive
Value Concepts study early in the SCORE T.M. design program.

The purpose of such a study was to get input from experts of
various pertinent disciplines to assess such factors as,
optimum design and function of engine subsystems,
manufacturability, potential target markets and market size,
and production costs. Input was also obtained on where we
should put emphasis for our Engineering Development and
Research programs.

The results of this study indicated, the market areas on
which we should concentrate for initial production units. In
addition the cost data indicated that the engine could be
produced to compete effectively with the light, medium and
heavy duty diesel market, as well as the small gas turbine
market. While much of the data generated must necessarily be
considered proprietary, the technique used for the Value
Concepts study are discussed in reference (10).

The system approach to value analysis not only provided
direction on potential significant cost savings, but provided
information where we should apply extra effort to assure the
appropriate level of reliability and durability.

FIG 11 MOCK-UP OF ENGINE CONFIGURATION RESULTING
FROM VALUE CONCEPTS STUDY.


MARKET DEVELOPMENT

The high level of power output per unit volume available in
the Stratified Charge Rotary Engine is of particular interest
in commercial and military applications where space claim is
critical. Automotive applications are advanced small,
aerodynamially clean automobiles, a variety of tracked and
wheeled military vehicles. Other applications are high
mobility military generator sets, shipboard generator systems
and airborne auxiliary power units. The engines small size
and weight, combined with the ability to burn jet fuel also
makes the engine attractive for propulsion, in fixed wing and
rotary wing aircraft, both commercial and military.

To address these market demands, John Deere has planned late
1980's, early 1990's production of three families of engines,
covering a wide power range from 60 to 1680 kw (80 to 2250
bhp), Figure 12.

The three families are identified as :

o SCORE 70 (0.67 liters/rotor or  40 cu. inch/rotor)

o SCORE 170 (1.72 liters/rotor or 105 cu. inch/rotor)

o SCORE 580 (5.8 liters/rotor or  350 cu. inch/rotor)

Within the three families of engines, current activities
involve a variety of end applications.

## SCORE FAMILIES OF ENGINES
## BASIC PERFORMANCE AND BRIEF SPECIFICATIONS

### SCORE 70 SERIES

| Model | No. of Rotors | Power kW(bhp) | Displacement 1(in³) | Height mm(in) | Width mm(in) | Length mm(in) | Weight kg(lbs) |
|---|---|---|---|---|---|---|---|
| 1007R | 1 | 60 (80) | .7 (40) | 505 (20) | 530 (21) | 560 (22) | 90 (198) |
| 2013R | 2 | 120 (160) | 1.3 (80) | 505 (20) | 530 (21) | 685 (27) | 115 (253) |
| 3020R | 3 | 180 (240) | 2.0 (120) | 505 (20) | 530 (21) | 840 (33) | ·166 (366) |
| 4026R | 4 | 240 (320) | 2.6 (160) | 505 (20) | 530 (21) | 965 (38) | 195 (430) |

### SCORE 170 SERIES

| Model | No. of Rotors | Power kW(bhp) | Displacement 1(in³) | Height mm(in) | Width mm(in) | Length mm(in) | Weight kg(lbs) |
|---|---|---|---|---|---|---|---|
| 1017R | 1 | 150 (200) | 1.7 (105) | 508 (20) | 661 (29) | 585 (23) | 147 (325) |
| 2034R | 2 | 300 (400) | 3.4 (210) | 508 (20) | 661 (29) | 745 (29) | 223 (490) |
| 3051R | 3 | 450 (600) | 5.1 (315) | 508 (20) | 661 (29) | 880 (35) | 291 (640) |
| 4068R | 4 | 600 (800) | 6.8 (420) | 508 (20) | 661 (29) | 1020 (40) | 364 (800) |
| 6102R | 6 | 900 (1200) | 10.2 (630) | 508 (20) | 661 (29) | 1460 (57) | 514 (1130) |

### SCORE 580 SERIES

| Model | No. of Rotors | Power kW(bhp) | Displacement 1(in³) | Height mm(in) | Width mm(in) | Length mm(in) | Weight kg(lbs) |
|---|---|---|---|---|---|---|---|
| 1058R | 1 | 280 (375) | 5.8 (350) | 760 (30) | 990 (39) | 610 (24) | 385 (848) |
| 2116R | 2 | 560 (750) | 11.6 (700) | 760 (30) | 990 (39) | 1118 (44) | 615 (1354) |
| 3174R | 3 | 840 (1125) | 17.4 (1050) | 760 (30) | 990 (39) | 1320 (52) | 815 (1795 |
| 4231R | 4 | 1120 (1500) | 23.1 (1400) | 760 (30) | 990 (39) | 1525 (60) | 1020 (2246) |
| 6347R | 6 | 1680 (2250) | 34.7 (2100) | 760 (30) | 990 (39) | 2185 (86) | 1450 (3190) |

NOTE: BASIC DIMENSIONS AND PERFORMANCE CAN VARY
IN PACKAGING FOR SPECIFIC APPLICATIONS.

FIGURE 12

SCORE 70 SERIES

An envelope drawing for a single rotor, SCORE 70, for advanced automotive application is shown in Figure 13. The engine height of 16.6 inches is very attractive from a vehicle packaging standpoint in advanced design, smaller and lighter cars. The engine performance is shown in Figures 14 & 15, reflecting a relatively low level of specific fuel consumption over a wide speed and load range and a reasonable torque curve.

A two rotor version of the SCORE 70 series engine, designated 2013R, is currently in the design phase at John Deere with prototypes scheduled for mid 1987. This engine, with 150-200 HP range for the near term is applicable to light trucks and minivans (in the automotive field). It is also of interest in high mobility military generator sets, airborne auxiliary power units and other military airborne applications, Figure 16. A four rotor, uprated version at 550 HP has been defined for studies toward automotive vehicular application in the military's Small Integrated Propulsion Systems (SIPS).

SCORE 170 SERIES

A license agreement is in place between John Deere Technologies International, Inc. and AVCO Lycoming, Williamsport Division for development of a two rotor, SCORE 170 engine for general aviation.

The joint development program was initiated in July 1985. The purpose of the joint program is to obtain FAA Certification and prepare for production in early 1990 of a 300 kw (400 HP) twin rotor general aviation engine. A basic preliminary specification for the engine is presented in Figure 17. The designation L2A-400-A1 is AVCO's designation for 2 rotors of 3.4 liters total displacement (210 cu. in.), 400 BHP and accessory arrangement designation A1.

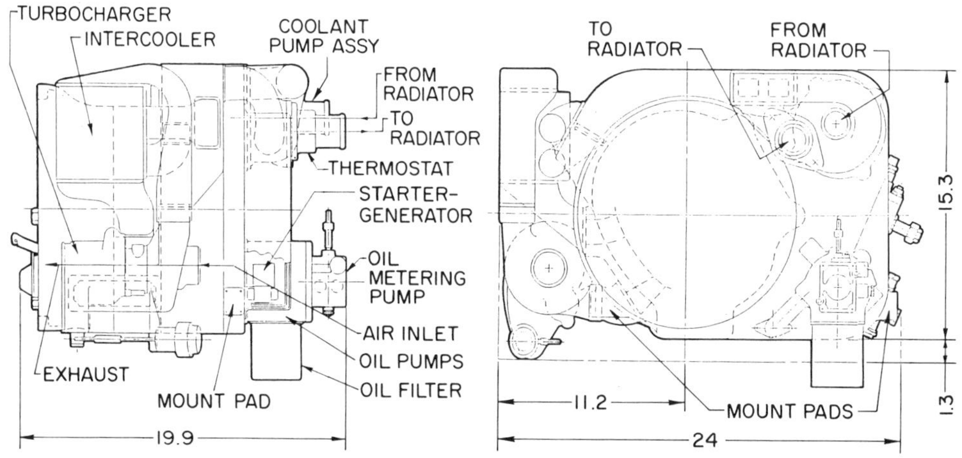


FIGURE 13 1007R SCORE 70 ENGINE-AUTOMOTIVE APPLICATION 100 HP.

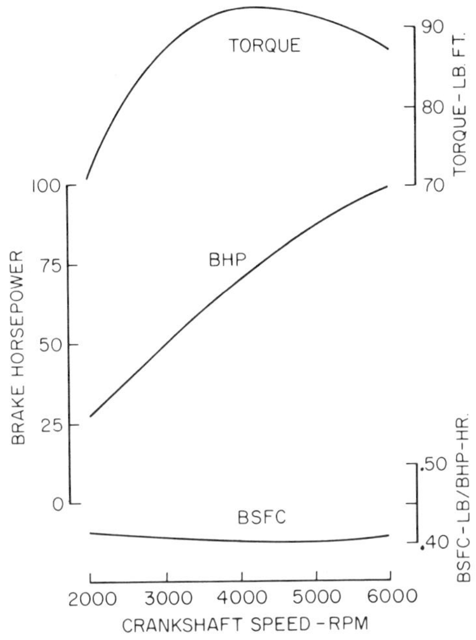



FIGURE 14 SCORE 70 SERIES 1007R ESTIMATED FULL LOAD PERFORMANCE.

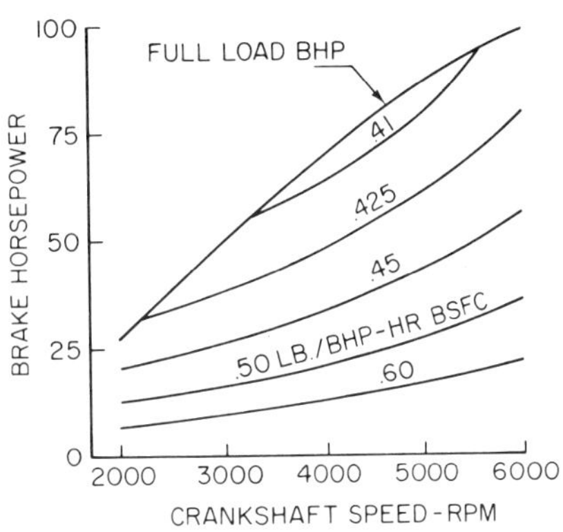



FIGURE 15 SCORE 70 SERIES 1007R ESTIMATED PART LOAD PERFORMANCE.

2 rotors of 0.66 liters (40 cu. in.) each, 1.3 liters total displacement

FIGURE 16 2013R GENERAL AVIATION PRELIMINARY ARRANGEMENT.

**L2A-400-A1 ENGINE**

**FUEL INJECTED, TURBOCHARGED, LIQUID COOLED, GEARED, INTERCOOLED, WITH CABIN BLEED AIR PROVISIONS**

| | |
|---|---|
| Take-off Power | 350kW (400hp) |
| Take-off Crankshaft Speed | 5800 RPM |
| Take-off Altitude Capability | Sea level to 6000m (20.000 Ft.) |
| Cruise Power (75% Cruise) | 225kW (300HP) |
| Cruise Crankshaft Speed | 4350 RPM |
| Cruise Altitude Capability | To 7500m (25,000 Ft.) |
| Engine Weight Goal | 228kg (506 Lbs.) |
| BSFC at Take-off | 243-255kg/kW-Hr. (.40-.42 Lbs/BHP-Hr.) |
| BSFC at 75% Cruise | 231-249kg/kW-Hr. (.38-.41 Lbs./BHP-Hr.) |
| TBO | 2,000 Hours |
| Fuel, Aviation Grade | Jet-A |
| Oil Consumption (Based on 75% Cruise) | 27kg/Hr. (0.6 Lbs./Hr.) |
| Production Target | Early 1990 |

FIGURE 17 PRELIMINARY ENGINE SPECIFICATION.

JDTI, is performing the primary design for the core power section. AVCO is performing the primary design for the accessory drive gear box, prop shaft and reduction gear and the integration of the package. Coordination with airframe manufacturers toward general requirements and flight test activities is in progress by AVCO. Also, FAA involvement has been initiated.

Automotive versions have been defined for the SCORE 170 family for consideration in the anticipated U.S. Army Armored Family of Vehicles (AFV) as shown in Figure 18. Ratings for 1989 and growth ratings for 1994 are listed. Figure 19 outlines general arrangement and installation features for a three rotor version of the SCORE 170 series, designated 3051R, rated at 600 HP in 1989 and 750 in 1994. Full load performance for the engine is shown in Figure 20.

SCORE 170 SERIES FAMILY FOR AFV

| MODEL NO. | NO. OF ROTORS | RATED BHP (1989) | GROWTH BHP (1994) | LxWxH, INCHES | WT., LBS. |
|---|---|---|---|---|---|
| 1017R | 1 | 200 | 250 | 23 x 29 x 20 | 325 |
| 2034R | 2 | 400 | 500 | 29 x 29 x 20 | 490 |
| 3051R | 3 | 600 | 750 | 35 x 29 x 20 | 640 |
| 4068R | 4 | 800 | 1000 | 40 x 29 x 20 | 800 |
| 6102R | 6 | 1200 | 1500 | 57 x 29 x 20 | 1,130 |

FUEL CONSUMPTION (LB./BHP-HR.)

| | RATED POWER | 65% POWER |
|---|---|---|
| 1989 | 0.42 | 0.40 |
| 1994 | 0.39 | 0.37 |
| 1994 w/TURBOCOMPOUNDING | 0.35 | 0.34 |

FIGURE 18

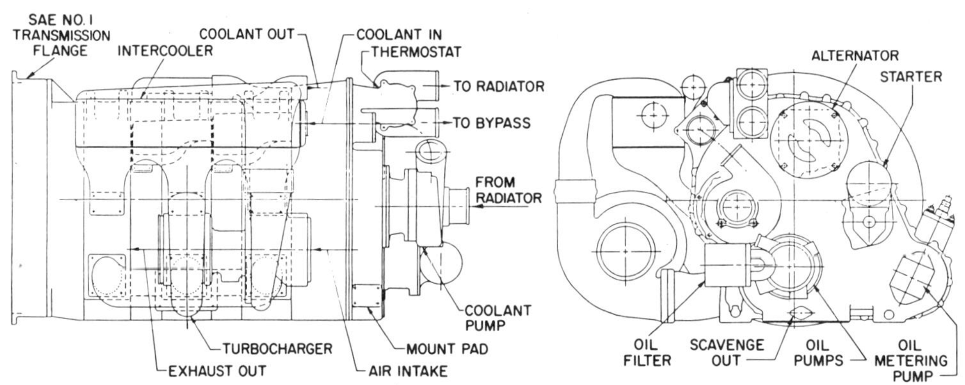


FIGURE 19  3051R SCORE 170 SERIES.

SCORE 580 SERIES

Contractural efforts with the USMC are in progress for the Demonstration and Validation of a two rotor, 5.8 liters/rotor, 2116R engine rated at 750 BHP/3600 RPM for amphibious, tracked vehicle application. Higher output engines, using a modular approach of coupling of two rotor or three rotor sections, are under consideration for a variety of automotive vehicle applications and ship electrical service generators.

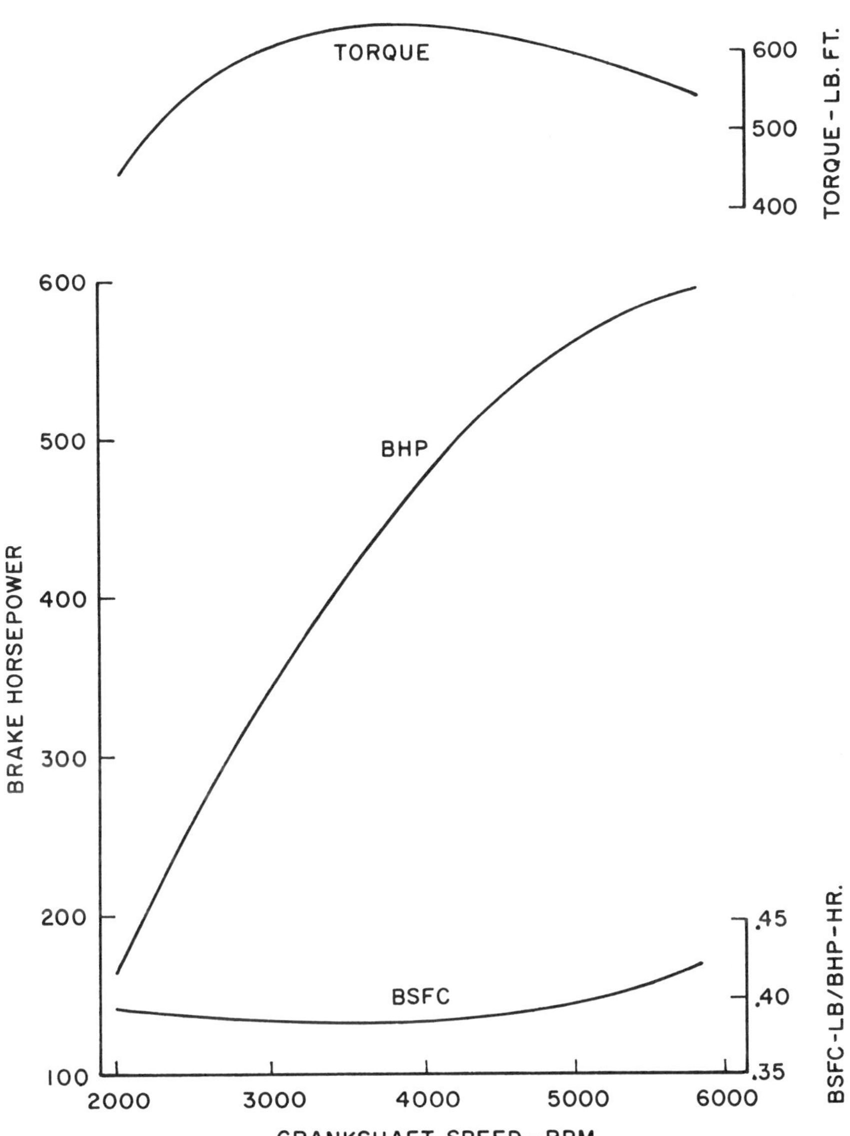


FIGURE 20 SCORE 170 SERIES 3051R ESTIMATED FULL LOAD PERFORMANCE.

REFERENCES

(1)     Norbye, Jan P.  *The Wankel Engine* Chilton Book Co., Philiadelphia 1971.

(2)  Yamamoto, Kenichi, *The Rotary Engine*, Toyo Kogyo Co. Ltd. Horoshima, Japan 1969.

(3)  Wankel, Felix, *Rotary Piston Machines*, London ILIFFE Books Ltd, London 1965.

(4)  Ansdale, R.F., *The Wankel RC Engine*, London ILIFFE Books Ltd, London 1968.

(5)  Faith, Nicholas, *Wankel*, Stein & Day, Briarcliff, NY,1975.

(6)  Jones, C., Lamping, H.D., Myers, D.M., and Loyd,R.W.; "An Update of the Direct Injected Stratified Charge Rotary Combustion Engine Developments at Curtiss-Wright", SAE Paper 77044,Feb. 1977.

(7) Jones,C., "An Update of Applicable Automotive Engine Rotary Stratified Charge Developments", SAE Paper 820347, Feb. 1982.

(8) Jones, C. "Advanced Development of Rotary Stratified Charge 750 and 1500 HP Military Multi-Fuel Engines at Curtiss-Wright", SAE Paper 840460, Feb. 1984.

(9)    Jones,C., Mack, J.R. & Griffith, M.J., "Advanced Rotary Engine Developments for Naval Applications", SAE Paper 851243, May 1985.

(10) Jain,A.,"Challenge and Opportunity For the Manager of The 80's", SAE Paper 851578, Sep. 1985.

(11) Kulina, M.R., "Experimental Evaluation of Rotary Engine Timing Gear Loads", SAE Paper 860562, Feb. 1986.

(12)
  Mount, R.E., and Greiner, W.L.,"High Performance, Stratified Charge Rotary Engines for General Aviation", AIAA-86-1553, Jun 1986.

(13)
  Jones, C, "A Survey of Curtiss-Wright's 1958-1971 Rotating Combustion Engine Technological Developments" SAE Paper 20468 May 1972.

# TURBO-COMPOUND DIESEL ENGINES

F.J. Wallace

School of Engineering
University of Bath
Bath, BA2 7AY, U.K.

## INTRODUCTION

The Diesel Engine is still, despite all claims to the contrary by the
protagonists of the gas turbine and of the Stirling engine, the most
efficient prime mover at our disposal.  Furthermore, unlike the gas turbine
which is in difficulty at the lower end of the power spectrum because of
inherent aerodynamic losses, and the Stirling engine which would seem to
suffer from a severe weight disadvantage, the Diesel engine successfully
covers the full automotive spectrum and indeed has been successfully
developed in marine form to cover power requirements as high as 60,000 h.p.
in a single unit.  In this latter form, where it is generally designed as
a low speed, uniflow scavenged two stroke engine, it has also achieved
astonishingly high efficiencies, in excess of 50%.

Nevertheless, the search for even better efficiency continues, not
least in the automotive field where both the private car owner and the
operator of large fleets of public service or goods vehicles, is still
looking for possible economies, and this in spite of the slump in the price
of oil.  It is in this context that the turbo compounded Diesel Engine is
of particular importance.

## BASIC THERMODYNAMIC CONSIDERATIONS

First Law analysis of the Diesel cycle operating in conjunction with
compressors and turbines (1) shows that, making certain simplifying
assumptions, it is possible to calculate, on a basis of common air through-
put, the power flows through engine, compressor and turbine for any system
such as those shown in figs. 1a to 1d representing the basic combinations
of the Diesel engine with turbomachinery in turbocharged, gas generator or
compounded form.

Fig. 1d shows a particular compounding arrangement pioneered by
Cummins,        in which the free running turbocharger is retained, with a
geared power turbine operating in series with the turbocharger turbine, so
that the total output of the system becomes the sum of engine and geared
turbine power.  With fixed turbine and compressor efficiencies, and at a
fixed exhaust temperature of 600C, (implying an approximately constant A/F
ratio), it is possible to predict, on a base of boost pressure ratio, the
likelihood of any surplus of turbine over compressor power which would

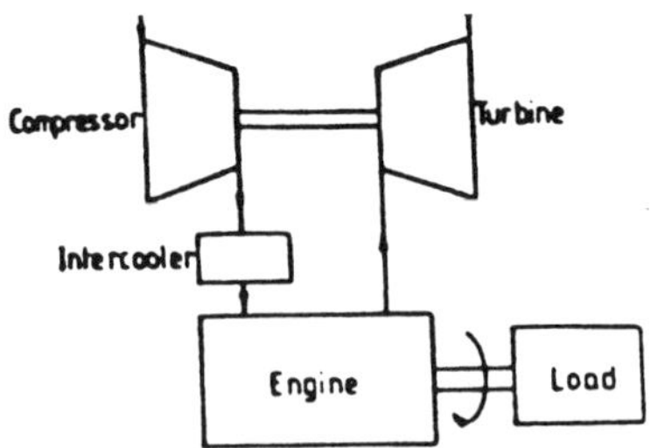



Fig. 1(a)  Turbocharged engine

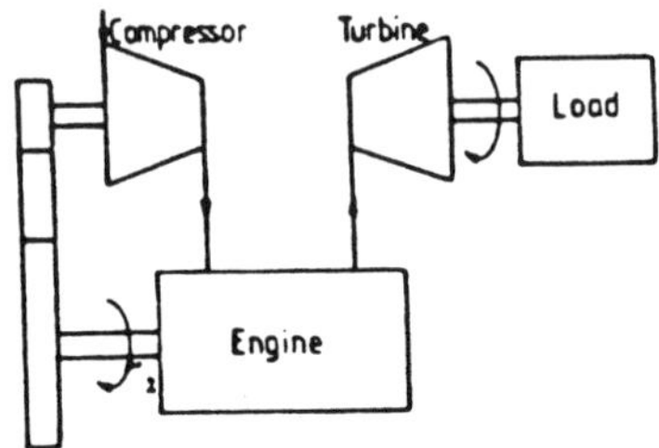



Fig. 1(b)  Gas generator

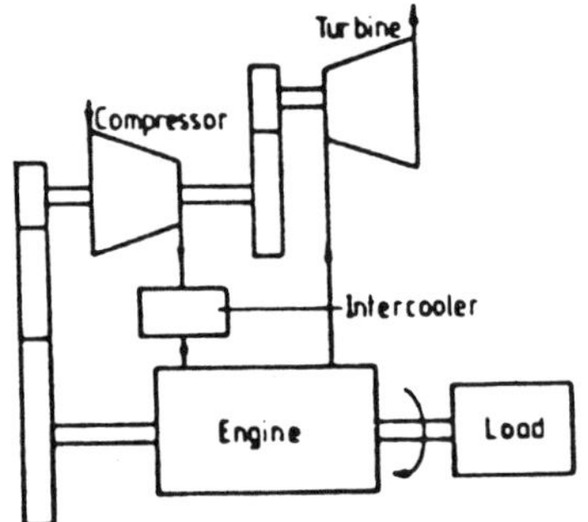



Fig. 1(c)  Fixed gear ratio compound engine

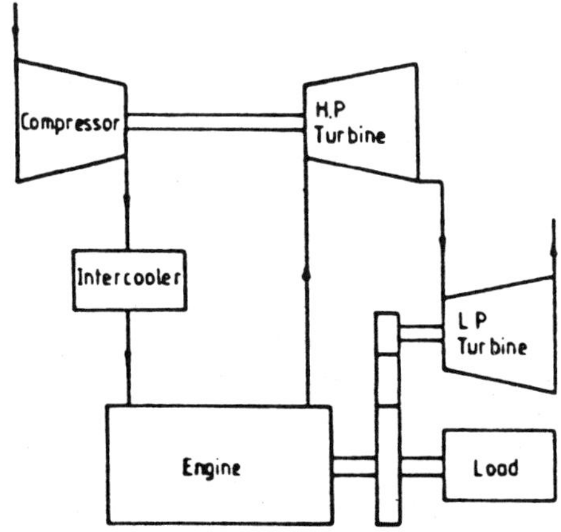



Fig. 1(d)  Cummins compound engine scheme

indicate favourable conditions for compounding, or, alternatively, for
the gas generator concept, where turbine power may exceed engine power.
Fig. 2 shows that indeed compounding can be implemented with advantage
over almost the full boost range, whereas gas generation is attractive
only at extremely high boost ratios.  However, the analysis of fig. 2
is extremely superficial, and a deeper Second Law analysis is called for.
This Second Law analysis is particularly valuable in that, through the
concept of availability (2,4), i.e. the max. useful work which can be
extracted from a thermodynamic system operating between a max. energy state
(usually the peak cylinder conditions) and the state of the surrounding
atmosphere, may readily be calculated.  At the same time, the effect of
individual system losses on overall performance may be quantified.

Sophisticated cycle simulation programs are now widely used.  By
extending them to include availability functions for both closed and gas
system processes, we can draw up a First and Second Law analysis side by
side.  Such an analysis was carried out by Primus et al (3) who drew up
energy and availability balances for a 14ℓ direct injection Diesel engine
operating in the following modes :-

1) Turbocharged, non aftercooled            (fig. 1a without cooler)
2) Turbocharged, aftercooled                (fig. 1a)
3) In 'Cummins' compounded form             (fig. 1d)
4) In 'Cummins' compounded form, with thermal
   insulation of the combustion chamber.    (fig. 1d)

Table 1 summarises the operating conditions and by implication also
the energy balances, while Table 2 gives the Second Law availability
balance for these four arrangements which represent a progression from
'conventional' to very advanced practice for modern heavy truck power plant.
A glance at Table 1 shows a progressive increase in power from 294 kW for
the non intercooled turbocharged engine to 270+38 = 308 kW for the 'conventional' compound and (274+46) = 320 kW for the insulated compound engine,
an increase of 8.84%, for a fixed fuel input of 16.92 g/sec. and a fixed
trapped air fuel ratio of 30:1.  The corresponding increase in brake
thermal efficiency is from 40.06% to 43.60%.  Bearing in mind the greatly
increased plant complexity, these gains must be considered modest.

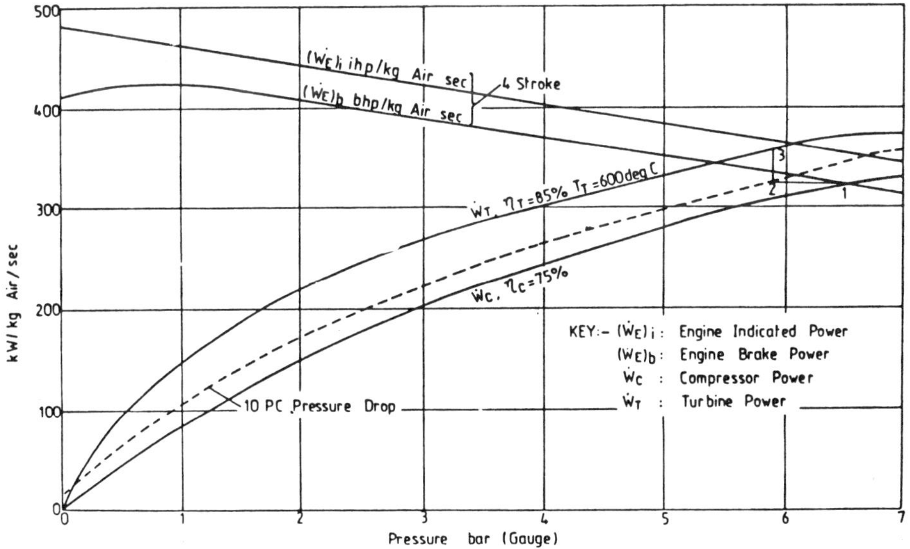


Fig. 2.  Compressor, Turbine and Engine power curves

Table 1.  First Law Analysis of Turbocharged
and Compound Systems [A/F = 30]

|  | Turbocharged | Turbocharged & Aftercooled | Compound | Compound Insulated |
|---|---|---|---|---|
| Engine Speed (rev/min) | 2100 | 2100 | 2100 | 2100 |
| Fuel rate (g/sec) | 16.92 | 16.92 | 16.92 | 16.92 |
| Cylinder power (kW) | 294 | 299 | 270 | 274 |
| Power turbine (kW) | – | – | 38 | 46 |
| Cylinder heat transfer (kcal/min) | 2056 | 1858 | 1967 | 772 |
| Cylinder Exh. energy (kcal/min) | 3982 | 4012 | 4024 | 4975 |
| Aftercooler heat transfer (kcal/min) | – | 304 | 442 | 528 |
| Inlet manifold temperature (K) | 429 | 370 | 370 | 370 |
| Inlet minifold pressure (bar) | 2.562 | 2.03 | 3.35 | 3.18 |
| Exhaust manifold temp (K) | 870 | 832 | 861 | 995 |
| Exhaust manifold pressure (bar) | 2.186 | 2.03 | 3.35 | 3.18 |
| Peak cylinder pressure (bar) | 125 | 125 | 125 | 125 |
| Comp. ratio | 13.90 | 14.50 | 13.70 | 12.50 |
| Comp. P.R. | 2.57 | 2.28 | 2.59 | 2.79 |
| Comp. efficiency | 0.72 | 0.72 | 0.72 | 0.72 |
| Turbine P.R. | 2.18 | 2.03 | 2.23 | 2.12 |
| Turbine efficiency | 0.70 | 0.70 | 0.70 | 0.70 |
| Power turbine P.R. | – | – | 1.50 | 1.50 |
| Power turbine efficiency | – | – | 0.8 | 0.8 |
| Brake thermal efficiency | 40.06 | 41.09 | 42.13 | 43.60 |

Table 2 summarises the availability analysis, in which all items
except that labelled 'indicated work + power turbine' must be regarded as
losses.  The combustion loss is the largest single item and represents
thermodynamic irreversibility, mainly due to mixing of cold excess air with
stoichiometric products of combustion.  It is least for the insulated
compound engine, as is cylinder heat transfer.  The next biggest loss is
'exhaust to ambient' which is least for the 'conventional' compound engine
due to energy abstraction in the power turbine and greatest for the insulated
engine with its high exhaust temperature.  The various throttling and turbomachinery losses are all of a smaller order, and little can be done to reduce
them further.

PARTICULAR COMPOUND SYSTEMS

Although the gains achievable by compounding are only limited, nevertheless, they are important.  Two units will now be described in some detail,
viz.
   a) the Cummins Compound Engine
   b) the Wallace Differential Compound engine (DCE).

The Cummins Compound Engine (fig. 1d)

The Cummins Compound Scheme has been extensively reported on (3,5).


Table 2.  Second Law Analysis of Turbocharged
and Compound Systems

| | Turbocharged | Turbocharged &<br>Aftercooled | Compound | Compound<br>Insulated |
|---|---|---|---|---|
| Indicated work plus<br>power turbine | 43.14 | 44.41 | 45.40 | 47.08 |
| | | no power turbine | | |
| Combustion loss | 19.11 | 20.15 | 20.35 | 17.68 |
| Cylinder heat<br>transfer | 15.05 | 13.70 | 14.23 | 8.24 |
| Inlet throttling | .85 | 0.94 | 0.82 | 1.18 |
| Exhaust throttling | 2.49 | 2.69 | 1.32 | 1.70 |
| Charge cooler ht.<br>transfer | – | 0.67 | 1.06 | 1.31 |
| Compressor loss | 1.62 | 1.61 | 1.68 | 1.81 |
| T/C turbine loss | 1.89 | 1.65 | 1.95 | 1.73 |
| Power turbine loss | – | – | 1.13 | 1.30 |
| Exhaust to ambient<br>loss | 15.85 | 14.35 | 12.07 | 17.97 |
| | 100.00 | 100.00 | 100.00 | 100.00 |

(All figures expressed as percentages of ideal fuel availability)

It represents a transition from turbocharging to compounding with minimal
additional complication, as it retains the conventional turbocharger, and
merely adds to the system a low pressure ratio geared power turbine operating in series with the high pressure ratio turbocharger turbine.  This
system has operated successfully, in non insulated form, as a prototype
truck engine while, in insulated form, Komatsu (6) are offering a limited
production version of an 11ℓ engine for use in earth moving vehicles.  In
the following the results of studies at Bath University on system optimization, based on the use of the 8.2ℓ Leyland 520, 6 cylinder D.I. high
output Diesel engine, are briefly reported(7).[This engine is also used in
the latest version of the Bath Differential Compound Engine (DCE)].

In order to achieve the highest possible efficiencies, the compressor
map, fig. 3, was adjusted to give a peak efficiency of 82% as against the
actual best value of 76.5%.  Likewise, the turbine maps, based on the turbocharger turbine of the standard Leyland engine, were adjusted to give best
efficiencies of 80%.  Separate optimization studies were carried out at the
design point on

    i) the relative sizes of the H.P. and L.P. turbines, and
   ii) the transmission gear ratio for the power turbine

in order to achieve the best possible performance.  Tables 3 & 4 summarize
the results of these optimizations.

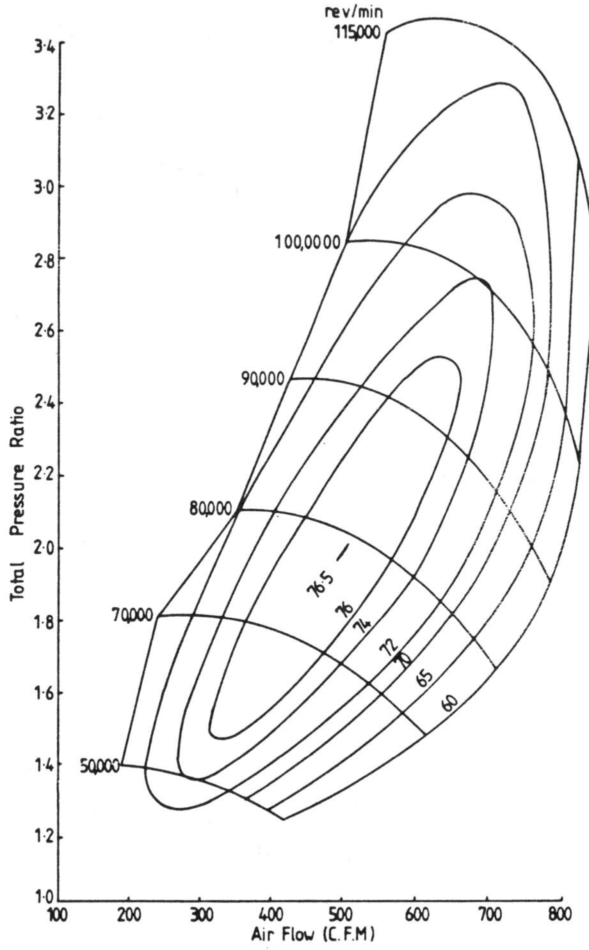


Fig. 3.  Compressor map of the HOLSET H2C

Table 3.  Optimization of Turbocharger and Power Turbine Size

| | High Pressure Turbine Size | Low Pressure Turbine Size | Piston Power (kW) | Low Pressure Turbine Power (kW) | Brake Thermal Efficiency | Overall Thermal Compound Efficiency |
|---|---|---|---|---|---|---|
| Run A | S | S | 225.9 | 39.65 | 0.3444 | 0.4018 |
| Run B | S | S+5% | 229.7 | 38.05 | 0.3501 | 0.4052 |
| Run C | S | S+10% | 231.7 | 36 | 0.3532 | 0.4053 |
| Run D | S-10% | S+5% | 229.0 | 40.22 | 0.3491 | 0.4075 |
| Run E | S-10% | S+10% | 230.1 | 37.46 | 0.3508 | 0.4050 |
| Run F | S-15% | S+5% | 226.8 | 41.07 | 0.3457 | 0.4051 |
| Run G | S-15% | S+10% | 227.5 | 38.17 | 0.3468 | 0.4021 |

Table 4.  Optimization of Transmission Gear Ratio

| | tgr | Piston Power (kW) | Low Pressure Turbine Power (kW) | Brake Thermal Efficiency | Overall Thermal Compound Efficiency |
|---|---|---|---|---|---|
| Run G | 25 | 230.1 | 37.73 | 0.3507 | 0.4053 |
| Run H | 27 | 229.7 | 38.93 | 0.3502 | 0.4065 |
| Run I | 30 | 229 | 40.42 | 0.3491 | 0.4077 |
| Run J | 34 | 228 | 40.69 | 0.3475 | 0.4064 |
| Run K | 38 | 226.5 | 40.06 | 0.3453 | 0.4033 |
| Run L | 40 | 225.7 | 39.45 | 0.3440 | 0.4011 |

Table 3 shows that run D, with a 10% smaller turbocharger turbine and 5% larger power turbine, (compared with the standard turbocharger turbine), gave the best overall efficiency of 40.75% while Table 4 shows that while the influence of turbine gear ratio is relatively small, a ratio of 30:1 gives the best overall efficiency.

Fig. 4a is the system performance map with contours of overall system efficiency showing the very substantial torque back up which can be achieved, with a peak system efficiency of 42.3%, in line with the predictions of Table 1.

Fig. 4b gives the corresponding map for the turbocharged engine with aftercooling.  While the limiting torque envelope is similar, the efficiency contours show a somewhat different trend, as well as generally rather lower values.  Nevertheless, these differences are relatively small, and once again underline the marginal nature of the gains achieved.

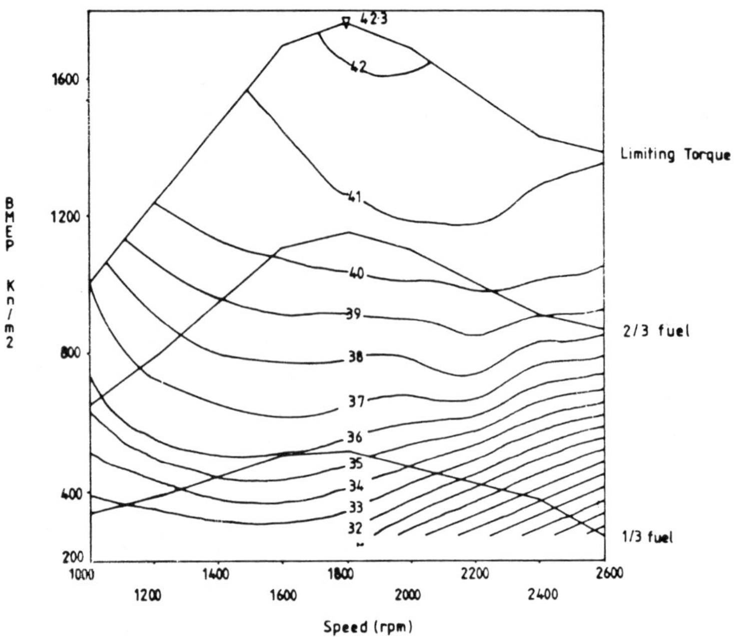


Fig. 4(a)   Brake mean effective pressure vs engine speed (for constant efficiency contours)

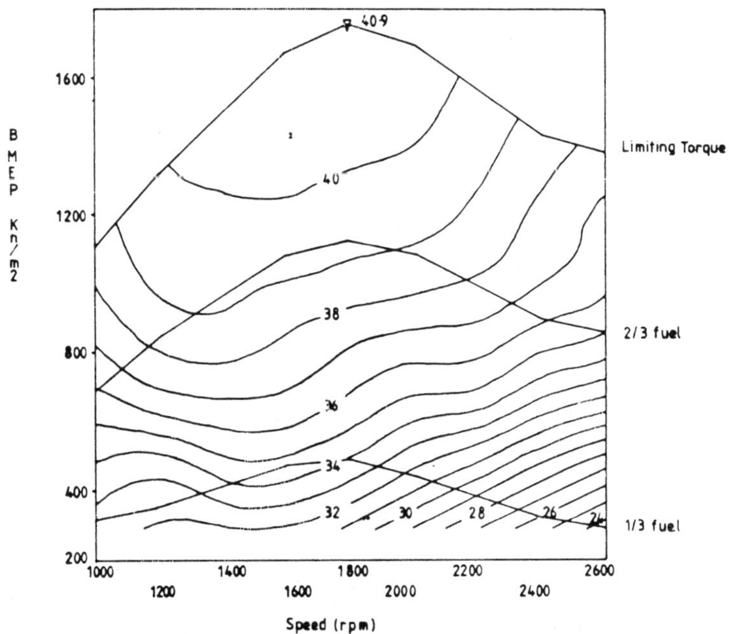


Fig. 4(b)   Brake mean effective pressure vs engine speed (for constant efficiency contours)

226

Fig. 5 is the system power map, with contours of power turbine contribution, and illustrating the rapid reduction in power turbine input as engine speed and load are reduced.

An interesting aspect of most compounded systems is that the engine tends to operate with adverse pressure gradients, i.e. exhaust manifold pressure exceeds inlet manifold pressure. This leads to negative pumping work and somewhat reduces engine efficiency. However, this loss is more than made good by the additional power recovered from the geared turbine (fig. 5). Table 5 gives values of these pressures on the limiting torque curve;  showing a rapid increase in the adverse pressure gradient with increase in engine speed.

Finally, it has been shown both theoretically and in practice that thermal insulation of the combustion chamber can lead to further significant improvements in efficiency (see Table 1). Computer predictions at Bath (7) indicate a further increase of peak system efficiency from 42.3% to 44.7%, for the uninsulated and insulated compound system respectively, very much in line with the values quoted in Table 1.

It is interesting that the published experimental figures for the Komatsu engine (6) show fairly similar trends.

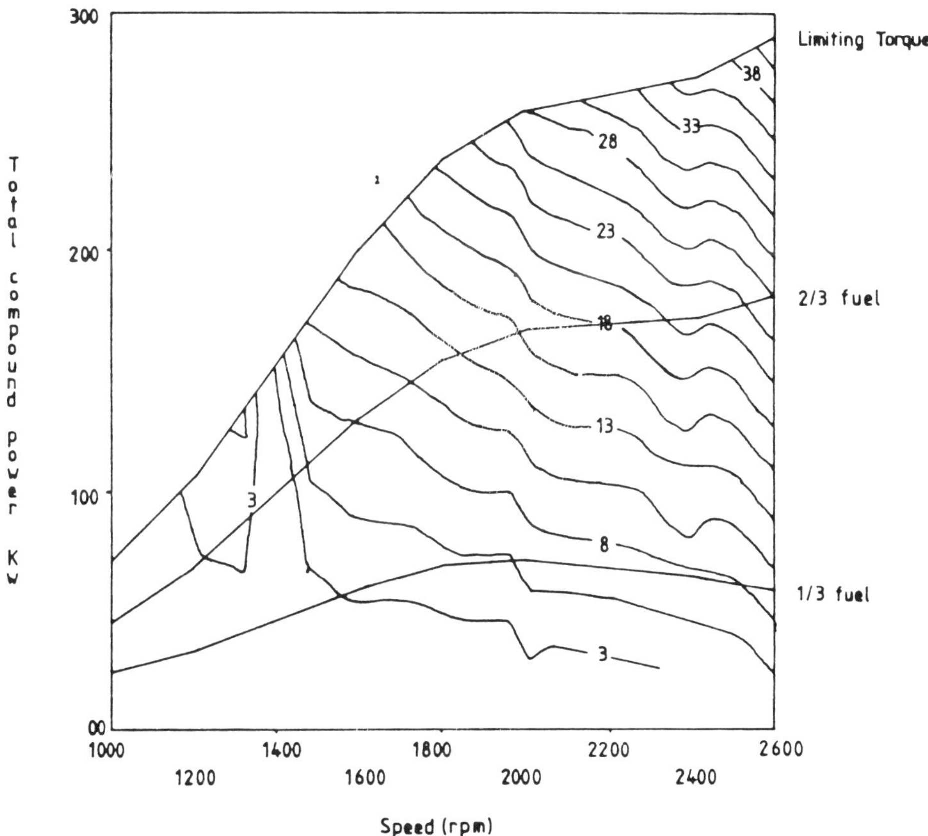


Fig. 5.  Total compound power vs engine speed (for constant turbine recovery power contours)

| Engine speed (rev/min.) | 1000 | 1200 | 1600 | 1800 | 2000 | 2400 | 2600 |
|---|---|---|---|---|---|---|---|
| Inlet manifold pressure (bar) | 1.348 | 1.633 | 2.144 | 2.284 | 2.335 | 2.263 | 2.261 |
| Exhaust manifold pressure (bar) | 1.491 | 1.692 | 2.308 | 2.587 | 2.777 | 2.987 | 3.160 |

## The Differential Compound Engine (DCE) (8, 9,10)

_General_ -It is clear from the foregoing that the gains achieved
with Cummins type compounding are relatively small, considering the added
cost and complication of the power turbine with its high speed gear train
and torsional oscillation damper, usually in the form of a fluid coupling.

The concept underlying the DCE is the simple one of trying to maximise
the benefit of the gearing implied by compounding by using it not only to
achieve modest efficiency gains but also to perform a transmission function,
preferably by achieving continuously variable transmission (CVT) characteristics.  Since many CVT's incorporate a fully floating epicyclic gear train,
the DCE scheme incorporates such a train between engine, compressor and
output shaft (fig. 6), with the power turbine geared down to the output
shaft, preferably through an auxiliary CVT which allows the turbine always

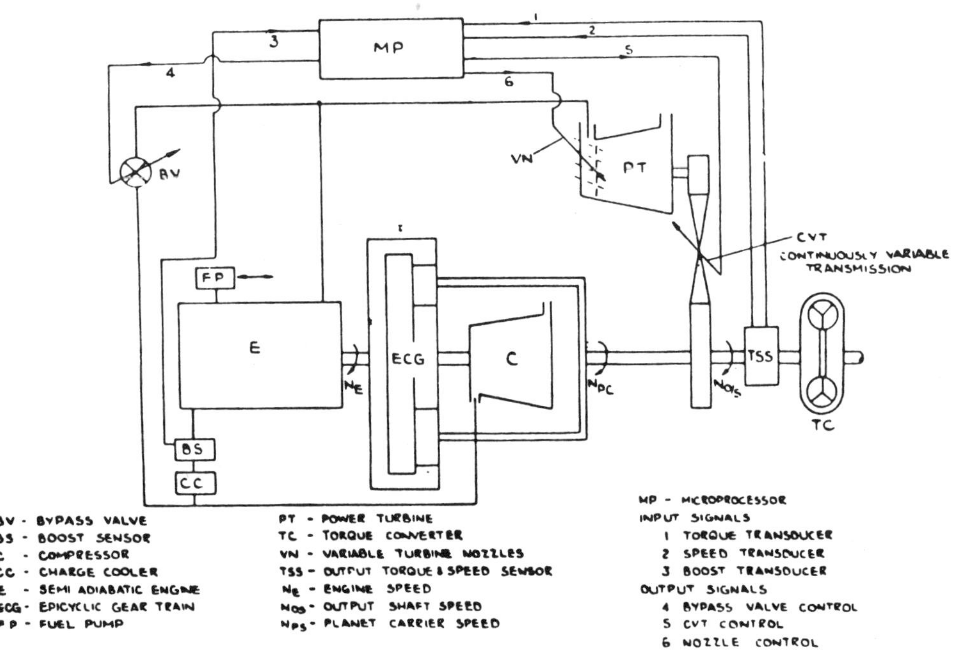


Fig. 6   DCE layout - final version

to operate at its most efficient speed.  The gear train acts as a differential, so that any reduction in output shaft speed, at fixed engine speed,
automatically leads to increased compressor speed and hence increased
availability of air to the engine.  The result is that, with the large torque
contribution made by the turbine, particularly at low output shaft speeds,
the system as a whole operates on a steeply and continuously rising torque
characteristic.  The laboratory plant at Bath University, as already stated,
is based on the 8.2ℓ Leyland 520, D.I. Diesel engine and is currently undergoing its commissioning tests.

Performance characteristics - Figs. 7a, b and c show the remarkable
output shaft characteristics as torque vs. speed with contours as follows :-

    Fig. 7a:  Overall system efficiency
    Fig. 7b:  Boost pressure ratio with engine BMEP in parenthesis
    Fig. 7c:  Turbine CVT ratio

The engine BMEP ratings are very similar to those adopted in the
earlier calculations, with the same engine, for the Cummins compound scheme,
but the compressor and turbine efficiencies are lower (and more realistic)
with best values of 75% for the compressor and 77.6% for the turbine, and

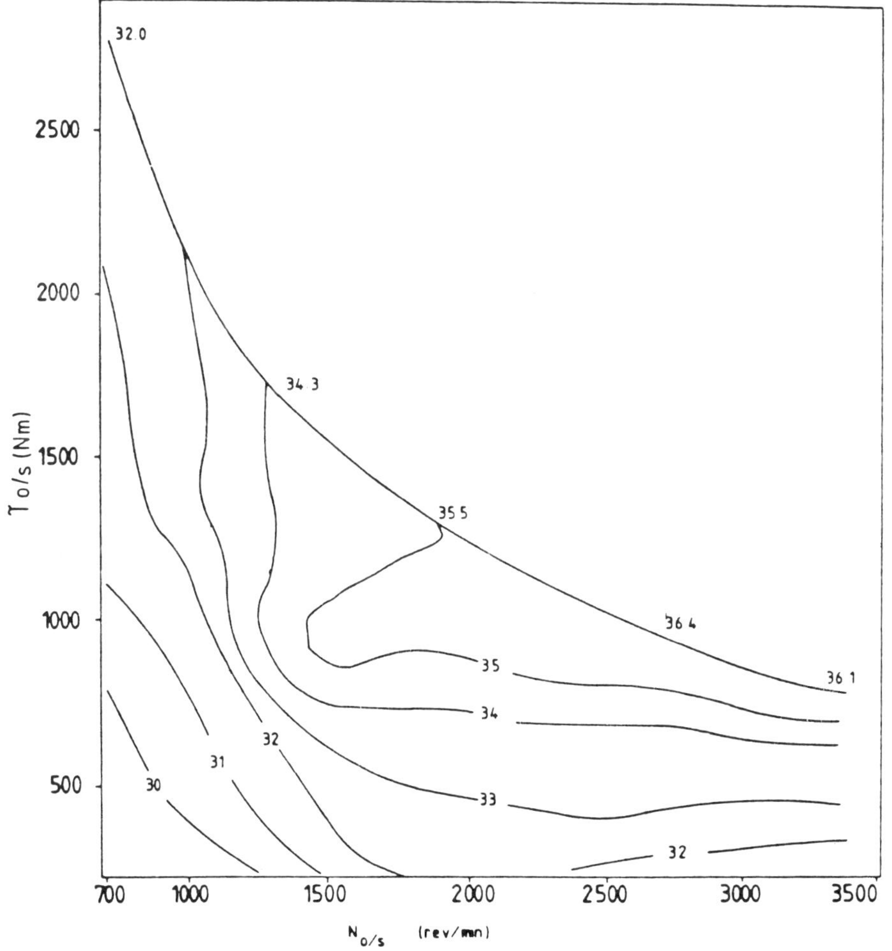


Fig. 7(a)  Output shaft thermal efficiency contours

229

more generous allowance having been made for gear losses.  The resultant
system efficiencies in fig. 7a show a best value of 36.4% at 80% of full
speed, and max. torque.   The output shaft speed range is from 3409 rev/min.
(rated) to 682 rev/min. (min), below which the transmission torque converter (fig. 6) comes into operation to give a very high level of stall torque.
The output shaft torque values rise from 769 Nm ($\equiv$ 274.4 kW) at rated speed,
to 2718 Nm at 682 rev/min. ($\equiv$ 194.0 kW), and finally to 9785 Nm at stall
with the torque converter in operation, giving an overall torque ratio of
$\frac{9785}{769}$ = 12.73 which should satisfy most truck applications.

[Stall gradability for a 38 ton truck = 25% for a max. vehicle speed of
140 km/hr].   The system thus achieves stepless operating characteristics
with very acceptable overall efficiencies which could be further improved
by the adoption of insulated engine technology (see Table 1).

   Fig. 7b shows the boost and BMEP contours and fig. 7c required turbine
CVT ratios to allow turbine operation at best efficiency for all conditions.

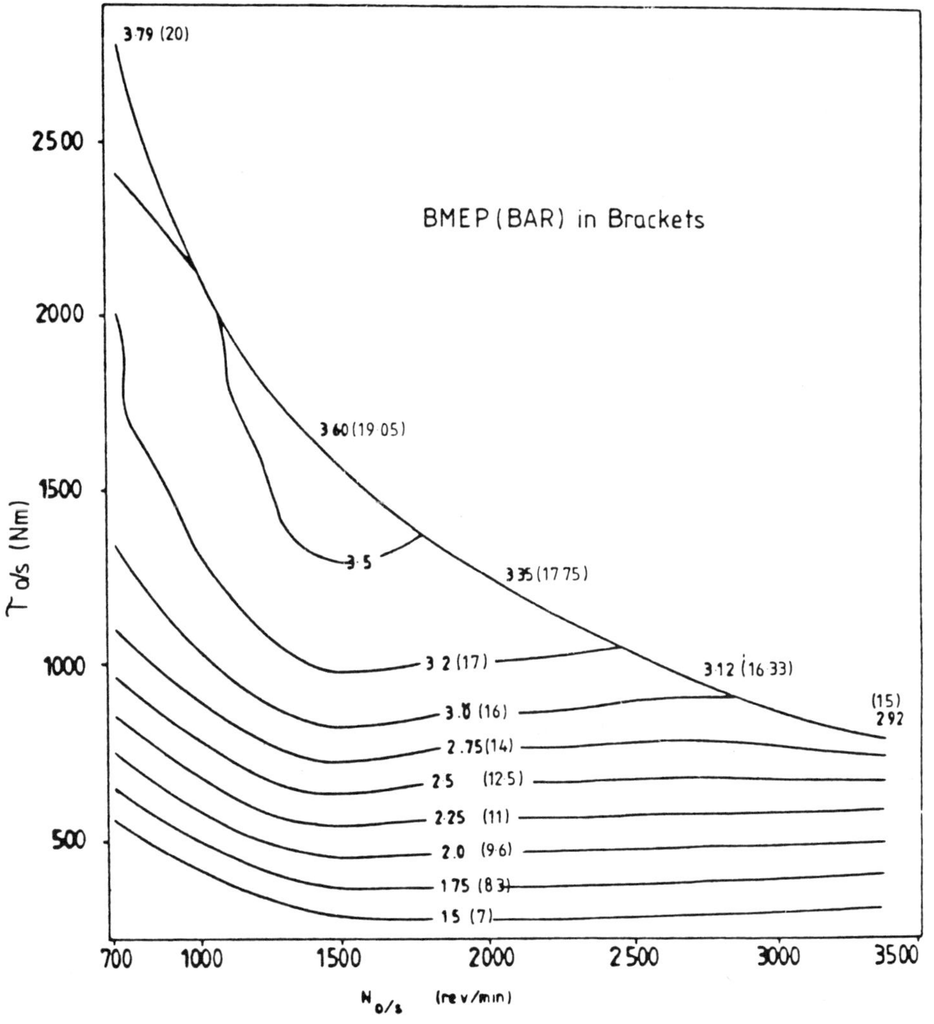



Fig. 7(b)  Boost pressure ratio contours

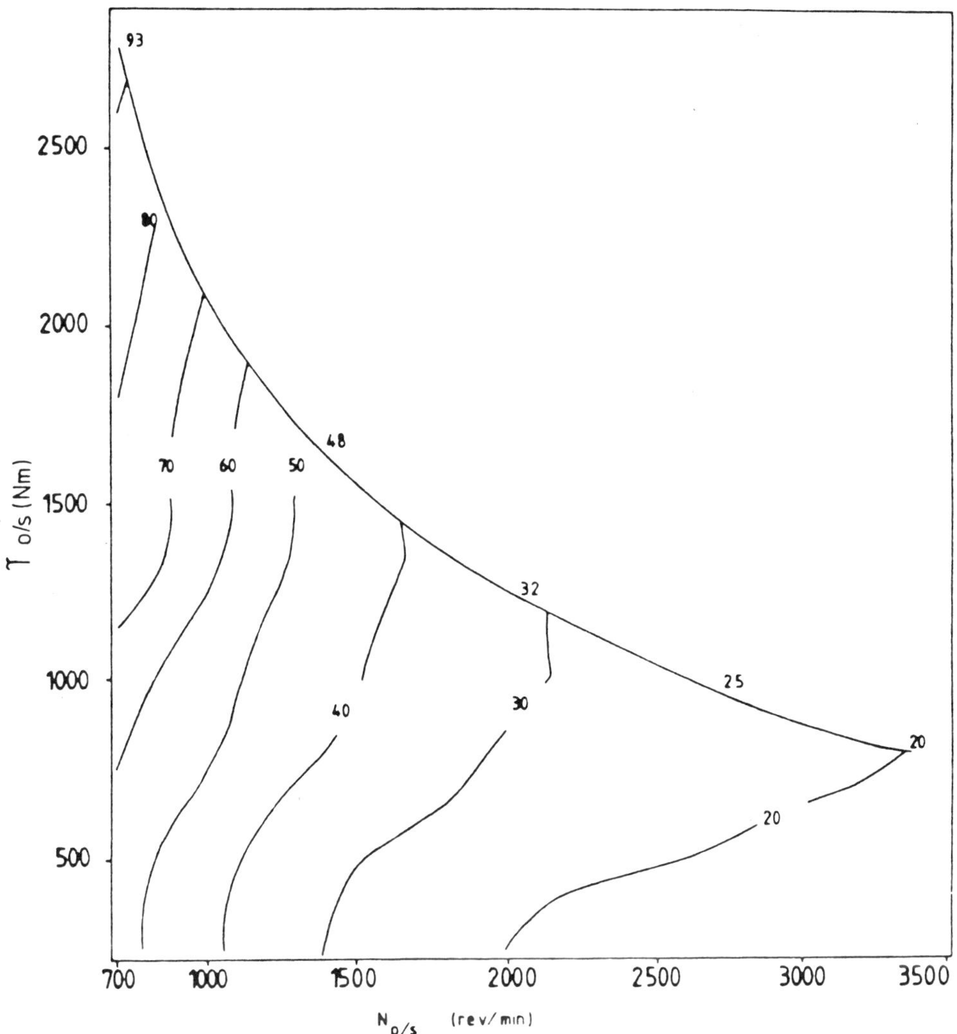


Fig. 7(c)  Turbine gear ratio contours


Table 6 summarises DCE operating characteristics on the limiting torque curve.  Except at the lowest speed, turbine power exceeds compressor power giving true compounding conditions, but gear losses are clearly substantial.


Control system - like all CVT systems, the DCE can be optimized in the sense that for any one output shaft condition, there is only one engine condition which gives best system efficiency.  These conditions can be established both by simulation and on the experimental rig.

Fig. 8 shows the control system in diagrammatic form.  Fuel rack position serves as a surrogate signal for output shaft torque which, together with output shaft speed, acts as feedback to the controls for turbine nozzle angle, (controlling boost), turbine CVT ratio (controlling $\eta_{turb}$) and injection timing (controlling $p_{max}$).  Once established, the appropriate schedules for these controlled variables are stored as lock up tables in the $\mu P$ and then output to the relevant actuators.  Such a system has already been implemented on the DCE rig.

## Table 6.  DCE Performance Characteristics

| | 682 | 1364 | 2045 | 2727 | 3409 |
|---|---|---|---|---|---|
| Output shaft speed | 682 | 1364 | 2045 | 2727 | 3409 |
| **Engine.** | | | | | |
| Speed (rev/min) | 1673 | 1930 | 2187 | 2388 | 2600 |
| Power (kW) | 229.0 | 251.8 | 265.7 | 267.1 | 265.8 |
| Brake thermal efficiency | 36.65 | 35.67 | 35.2 | 34.9 | 34.5 |
| A/F ratio | 27.52 | 26.46 | 26.08 | 26.3 | 26.7 |
| Max. cyl. pressure (bar) | 150.3 | 148.4 | 148.9 | 149.9 | 148.7 |
| Boost press. ratio | 3.79 | 3.59 | 3.35 | 3.12 | 2.92 |
| Exhaust temp. (K) | 888.8 | 931.6 | 946.8 | 941.8 | 932.8 |
| %age heat loss to coolant | 19.14 | 18.34 | 17.77 | 17.51 | 17.21 |
| **Compressor** | | | | | |
| speed (rev/min) | 11502 | 16501 | 9500 | 8004 | 6606 |
| press. ratio | 3.952 | 3.785 | 3.567 | 3.353 | 3.171 |
| efficiency | 71.4 | 71.7 | 72.5 | 73.1 | 74.6 |
| power (kW) | 173.2 | 150.6 | 126.8 | 98.51 | 73.97 |
| **Turbine** | | | | | |
| speed (rev/min) | 46500 | 48000 | 48000 | 50000 | 50000 |
| press. ratio | 3.952 | 3.785 | 3.567 | 3.353 | 3.171 |
| efficiency | 77.6 | 77.4 | 77.1 | 76.6 | 76.0 |
| power (kW) | 156.4 | 150.4 | 138.9 | 119.04 | 101.70 |
| **Output shaft** | | | | | |
| torque (Nm) | 2718 | 1627.8 | 1205.7 | 940.6 | 769.0 |
| power (kW) | 194.0 | 232.6 | 258.4 | 268.8 | 274.4 |
| Efficiency | 32.0 | 34.3 | 35.5 | 36.4 | 36.1 |
| **Gear Loss** (kW) | 18.2 | 19.0 | 19.2 | 18.83 | 19.12 |
| (%) | 9.4 | 8.2 | 7.4 | 7.0 | 6.9 |

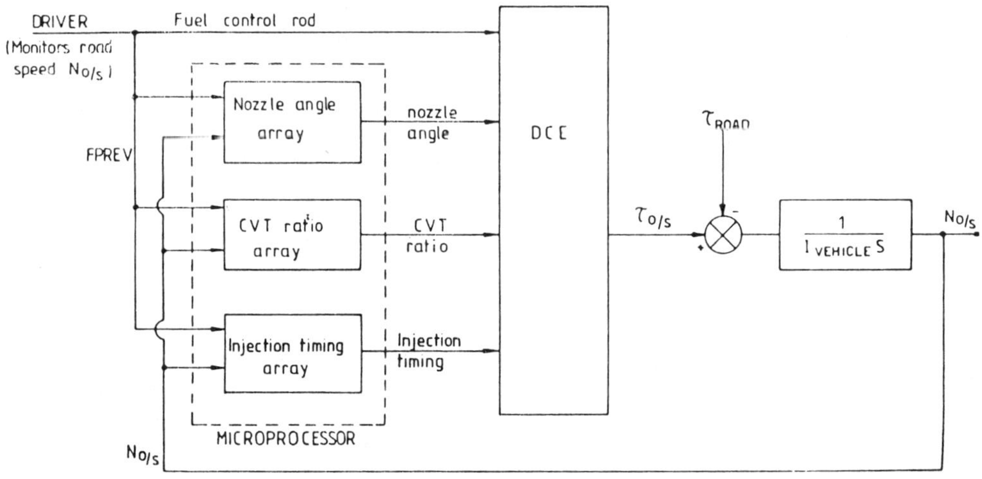


Fig. 8.  Steady-state control scheme

__Transient characteristics__ - space precludes a detailed discussion of
these, but outstanding transient response has already been demonstrated to
be a feature of the DCE.  This derives from the fact that through the
torque balance condition inherent in fully floating epicyclics, coupled
with the high inertia of the driven load, any increase in engine torque
following a fuel input step immediately leads to rapid acceleration of the
low inertia compressor, and hence rapid build up of boost.  Control
studies have shown that steps from 20% load to full load can be achieved
in approx. 1.5 sec. without exceeding the lower limit of fuel-air ratio.

CONCLUSIONS

Compounding has been shown to make possible very significant advances
in Diesel engine performance and operational flexibility beyond the limits
of the conventional turbocharged engine, driving the vehicle through a
multi ratio gearbox.

While the purely thermodynamic gains, expressed in terms of 1st and
2nd Law Analysis, are somewhat limited, the extension of the compounding
principle to cover not only thermodynamic but also transmission aspects,
holds out the promise of radically improved vehicle characteristics.

REFERENCES


1. Diesel Engine Reference Book (Chapter 3)
   Ed. Lilly, __Butterworth__ 1981.

2. Bruges, Available Energy & Second Law Analysis,
   __Butterworth__ 1959.

3. Kamo and Bryzik, Adiabatic Turbocompound Diesel Engines :
   Performance Prediction, SAE 780068, Detroit 1978.

4. Primus, Hoag, Flynn and Brands, An Appraisal of Advanced Engine
   Concepts using Second Law Analysis Techniques.
   SAE 841287, Detroit 1984.

5. Kamo and Bryzik, TARADCOM Adiabatic Turbocompound Engine Program,
   SAE 810070, Detroit 1981.

6. Toyamo, Yoshimitsu, Nishiyama, Shimauchi and Nakagaki,
   Heat Insulated Turbocompound Engine, SAE 831345,
   Milwaukee 1983.

7. Hatzopoulos, Thermal Insulation & Turbocompounding of Diesel
   Engines, M.Sc. Thesis, University of Bath 1983.

8. Wallace and Kimber, Optimization of the Differential Compound Engine
   using Microprocessor Control, SAE Paper 810 336, Detroit 1981.

9. Wallace, Tarabad and Howard, The Differential Compound Engine - a
   New Integrated Engine Transmission System Concept for Heavy Vehicles,
   1983, Proc.I.Mech.E., Vol 197A.

10. Wallace, Prince, Howard and Tarabad, Design and Performance
    Characteristics of the Laboratory Differential Compound Engine at
    Bath University, I.Mech.E. Conference on Integrated Engine Transmission
    Systems, Paper No. C196/86, 1986.

RECENT ADVANCES IN VARIABLE VALVE TIMING


T.H. Ma

Ford Motor Company
England


ABSTRACT

The application of variable valve timing (VVT) to spark ignition
engines is reviewed. In addition to higher performance, there
are opportunities to improve the ECE-15 vehicle fuel economy by
up to 15%. Added friction from the VVT system itself could
however significantly depreciate the potential benefit. A
selected number of VVT concepts developed by Ford are
described.


PERSPECTIVE

Modern high speed spark ignition automotive engines must satisfy a broad
range of performance requirements, and valve timing plays a significant
role in satisfying certain of these. Among the items most affected by
valve events are power output, low speed torque, light load driveability
and idle quality. There are areas of conflict among these requirements
because of the wide speed and load ranges that must be covered and a fixed
valve timing system usually requires a compromise to meet the extremes of
these conditions.

Variable Valve Timing (VVT) has the potential of improving both
performance and driveability by optimising for each specific mode of
engine speed/load operation. There are also opportunities to reduce fuel
consumption and exhaust emissions. Many ingenious VVT devices have been
proposed. These are however complex and cost effectiveness is always a
central issue. Volume production and reliability experience are also
lacking. Nevertheless, some of the simpler VVT devices such as those
improving the power output at the high and low ends of the RPM range, are
likely to appear in volume production during the 1980's and increasing
development effort on VVT is expected to gain momentum through the
1990's.

The pace and direction for the development of VVT are determined by a
number of factors which make up the cost effectiveness equation.
Different VVT concepts produce different VVT effects which may meet only
part of the total engine requirement. The device usually incurs
additional friction losses and the control system requires power to
operate. All these must be taken into consideration when evaluating the

VVT effectiveness. The cost and complexity of VVT must take into account the total system which includes the VVT mechanism itself and the control system which usually requires input devices to determine the engine operating condition, control functions in response to these inputs and actuators to select the VVT position. There are also packaging implications and possible side effects on related systems like the manifolds, the fuel system and emission controls. Noise and reliability are also important issues for customer acceptance. All these are the subjects of research and development at Ford Motor Company working in close cooperation with the Manufacturing areas to make sure that the cost and the manufacturing feasibility are fully addressed from the onset.

Perhaps the more significant reason for the renewed interest on VVT is its cost effectiveness in relation with other areas of development on the spark ignition engine. Because of pressures from the environmental air quality and the fuel crisis, the past two decades saw a leap forward in technology: first in emission control, followed by microporcessor management systems and advanced combustion chamber designs, and more recently, lean burn technology and knock control. Other fruitful areas of development are friction reduction, alternative materials and weight reduction. These are the obvious cost-effective areas of improving engine performance and fuel economy which will not be exhausted for a long time to come. However, as more and more progress is made, efforts in these established areas will be producing diminishing returns and new areas of development like Variable Compression Ratio, Variable Broad Band Tuning and Variable Valve Timing will become increasingly attractive. Indeed, efforts in these areas are gaining momentum and a number of recent publications (1-3) are clear indications that this new phase of engine development is already well on its way.

The objective of VVT is also undergoing a change in emphasis from the historical approach of VVT for performance to a new approach of VVT for fuel economy. Optimising for performance is basically maximising the volumetric efficiency over a wide engine speed range. Since the valve timing of an engine is very much part of the dynamics of gas exchange which include the intake and exhaust manifold systems and the port and valve designs, there are equal opportunities to improve volumetric efficiency either by VVT or by variable manifold tuning or by improving the valve flow time-area and opening and closing rates through multi-valve cylinder head designs using more than one intake and exhaust port. These alternative methods are effective, relatively simple to implement and reliable since they are based on established technology. Such improvements are however not additive, so any improvement by these means will diminish the potential benefit of VVT making it less cost effective if performance over the speed range is the primary objective.

The place for VVT in the next generation of advanced automotive engines therefore lies in its ability to optimise for fuel economy and driveability over the load range through improved idle quality and more efficient part-load operation with reduced throttling. This puts more emphasis on the direct control of the residual gas fraction through valve overlap, combined with much larger timing changes in the intake valve closing timing to achieve part-load power modulation. Within this large VVT range would still be retained the capability of optimising the timing for performance, thus realising all the potential benefits of VVT but with fuel economy as the primary objective.

In the following sections, the tradeoff between performance and fuel economy, and the VVT strategies to achieve them are discussed. Data is based on work carried out by Ford of Europe and Ford of America. In general, there are opportunities to improve fuel economy by 15% with VVT in the vehicle for the ECE-15 drive cycle compared with a fixed timing vehicle at the same level of performance. Higher economy gains are possible when VVT is combined in operation with VCR and lean burn. Added friction from the VVT system must be kept to the mininum in order to realise these improvements.


VARIABLE VALVE TIMING FUNDAMENTALS

The controlling parameters for VVT are the intake and exhaust valve event durations, the overlap and the position of these events relative to the piston motion. The objective is to optimise the engine operation in the volumetric efficiency, mechanical efficiency and residual gas fraction which would translate into improvements in the engine full-load performance, part-load fuel economy and light-load driveability/idle quality respectively.

## Volumetric Efficiency

This determines the maximum performance of the engine at WOT (Wide-Open-Throttle). The provision for both the intake and exhaust durations significantly longer than their respective piston events is crucial to achieving high volumetric efficiency at high engine speeds. In terms of the phased relationship between valves and pistons, this enables continued cylinder filling following the intake stroke prior to intake valve closing (IVC) (Fig.1). Likewise high speed operation is additionally facilitated by early exhaust valve opening (EVO) well before BDC between the expansion and exhaust strokes. This aids the scavenging process by exploiting blowdown to accelerate the column of out-flowing exhaust and thus reduce the pumping work associated with scavenging at high engine speeds.

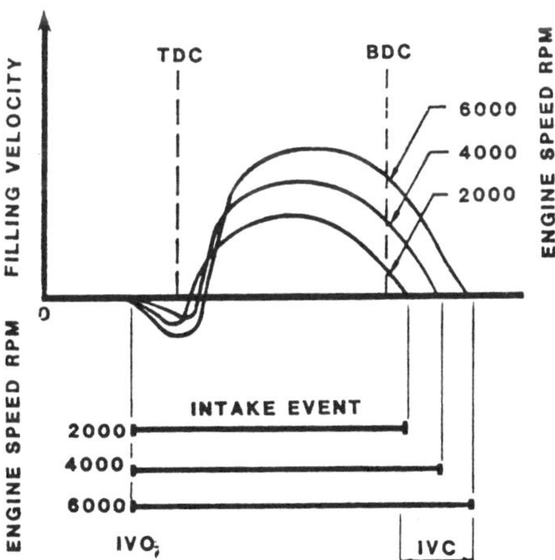


Fig. 1. Intake valve event relative to cylinder filling.

Though the above valve timing features significantly increase performance at high engine speeds, they incur torque penalties at low engine speeds. Late IVC allows some fresh charge to return to the intake system at the beginning of the compression stroke and thereby reduces low-speed volumetric efficiency. Early EVO lowers the expansion ratio and thus lowers cycle efficiency.

Exhaust valve closing (EVC) also has an effect on volumetric efficiency in that at WOT and high engine speeds it regulates how much exhaust is allowed to escape. Late EVC favours power at the expense of low speed torque. Delaying the intake valve opening (IVO) tends to isolate the intake manifold from the cylinder until the cylinder and the intake manifold pressures are more nearly equal. Engine performance is relatively insensitive to IVO and this can be delayed significantly without incurring penalties (5).

A typical performance curve from VVT is shown in Fig.2. In general, the range of VVT required to maximise volumetric efficiency over the engine speed range should allow a variation in the valve opening duration (excluding ramps) of between 225 degrees to 275 degrees and a change in the overlap between 25 degrees to 45 degrees. The timing points most beneficially to be altered in the order of effectiveness according to their respective sensitivity on volumetric efficiency are IVC (most effective), EVC, EVO, IVO. Further reduction in the overlap is required for part-load operation to control the residual gas fraction at idle and light loads and this will be discussed in the later section.

Finally, it should be kept in mind that the engine valve timing accounts for only part of the total gas exhange dynamics and VVT alone cannot improve volumetric efficiency beyond a certain maximum limit. There are other controls in the gas exhange process, like variable dynamic tuning

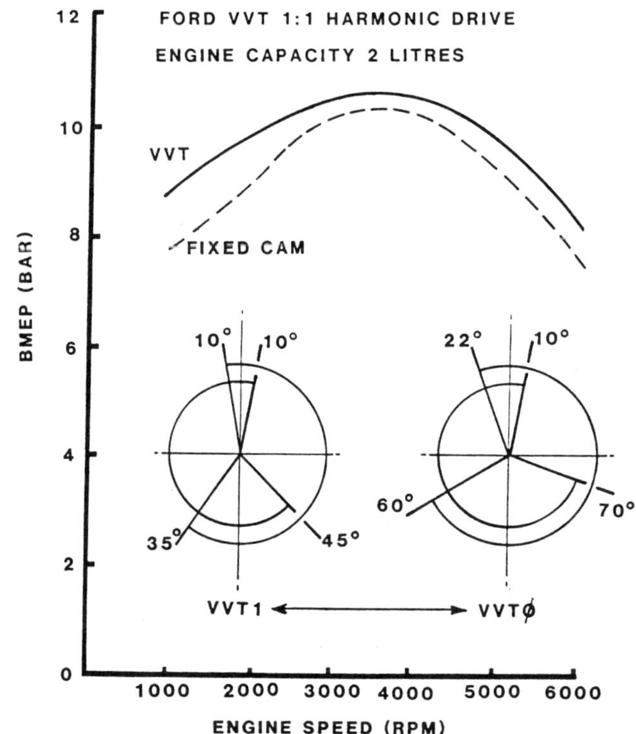


Fig. 2. Effect of variable valve timing on wide-open-throttle performance.

of the intake and exhaust systems and multi-valve designs, which can
increase the volumetric efficiency significantly above that. Compared
with these alternatives, VVT is not the best option if improved
performance is the only objective. There are of course additional
advantages from VVT, namely, improved idle quality and reduced throttling
at light-load, which the above mentioned alternatives cannot provide and
it is in these areas that VVT should be directed for improvements with
performance gained as a bonus.


Mechanical Efficiency

This influences the brake thermal efficiency and hence the fuel economy of
the engine. VVT alters the balance between the brake, friction and
pumping works (Fig.3 – BMEP, FMEP and PMEP respectively) which may either
improve or impair the brake thermal efficiency.

At WOT, there is a small improvement in brake thermal efficiency because
the increased BMEP from VVT is accompanied by a less than proportional
increase in mechanical losses. This however does not translate into fuel
economy at part-loads as the mechanical efficiency would revert back to
its normal level when the volumetric efficiency is deliberately reduced to
regulate part-load output by inlet throttling or other means.

FMEP will increase with the introduction of any VVT device simply because
of the added complexity and the extra power required to operate the
system. This mechanical loss might vary with engine speed, but stays the
same with engine load since the valve train would have to continue to
operate regardless of the engine output. This puts increasing penalty on
the mechanical efficiency as load is reduced, and in particular during
idle. BMEP for WOT performance is also reduced by the same amount as the

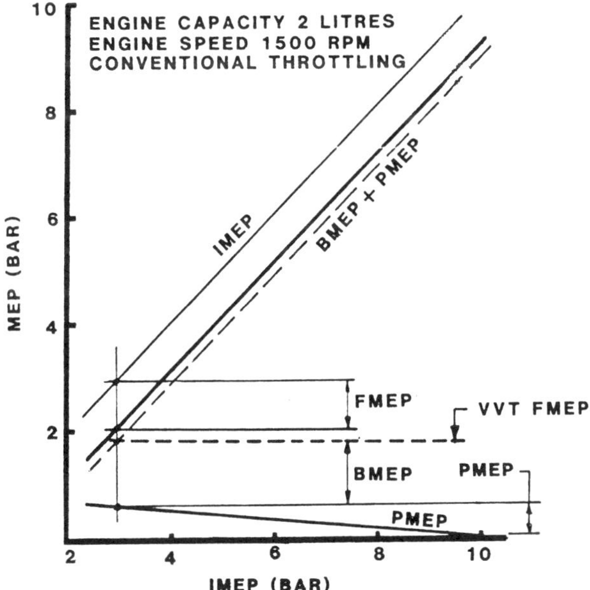


Fig. 3. Variation of mechanical mean effective
pressures with load.

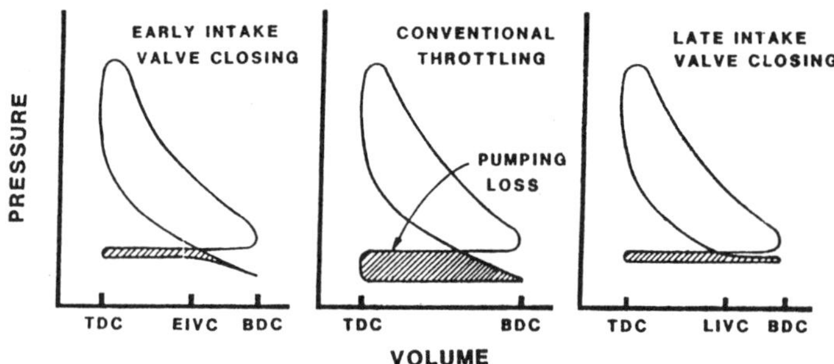


Fig. 4. Part load modulation by variable
intake valve closing.

FMEP increase. This added friction could substantially reduce the
effectiveness of VVT, particularly in the fuel economy at light loads, and
it is of utmost importance that the mechanical losses from VVT are kept to
the minimum through careful design.

PMEP during part-load operation can be reduced with VVT as a means of
improving fuel economy. Instead of regulating the amount of part-load
air-fuel mixture drawn into the cylinders by controlling the intake
pressure through inlet throttling which incurs the high pumping loss, the
same regulation can be achieved with much reduced pumping loss by
controlling the intake valve closing timing to effect partial air
admission by either early cutoff (EIVC) or late trapping (LIVC) of the
intake charge (7) (Figs.4). In practice, either method is suitable to
modulate power down to 30% load and some throttling is still necessary to
achieve idle and very light loads. The range of IVC timing change for
load modulation is considerably larger than that required for optimising
volumetric efficiency and the VVT design is correspondingly different.
Typically a range of 120 crank degrees in either direction is required,
which would of course provide ample flexibility for optimising volumetric
efficiency at WOT as well.

The nett improvement in brake thermal efficiency by EIVC or LIVC is
however considerably less than the equivalent reduction in pumping loss
(8). There is a deterioration in combustion for a number of reasons which
has impaired the indicated thermal efficiency. EIVC cuts off the air
admission early in the intake stroke from which point onwards the
trapped gas expands and recompresses nearly adiabatically until the same
timing position in the compression stroke is reached at which point the
gas pressure and temperature would return to the same initial levels when
it was first trapped. The compression stroke for the combustion cycle
therefore only begins from this point which means the effective
compression ratio is significantly less than the geometric swept
volume/clearance volume ratio of the engine. LIVC delays the trapping
of the air-fuel mixture by allowing a portion of the cylinder content to
be pumped back during the first part of the compression stroke.
Compression for the combustion cycle only begins from the point of LIVC
which again results in an effective compression ratio considerably less
than the geometric volume ratio of the engine. Thus in either case of
EIVC or LIVC, the compression temperature is lower than that of the

240

equivalent part-load condition by inlet throttling and the indicated thermal efficiency is lower. Other factors lower the cycle temperature even further partly because of the reduced heating by the smaller residual content and partly because of the reduction in mixture turbulence which results in slower burn. In spite of these factors, there is still worthwhile improvement in fuel economy by EIVC or LIVC in the order of 7% at 30% load. NOx emissions in the exhaust gas is reduced by up to 20% because of the lower in-cycle temperatures but there is no change in HC and CO because the exhaust temperature is relatively unaffected.

The above thermodynamic cycle is akin to the Atkinson cycle where the expansion ratio is greater than the compression ratio but unfortunately with EIVC and LIVC the compression ratio is decreasing at constant expansion ratio. It is therefore logical to combine EIVC or LIVC with variable compression ratio (VCR) in such a way that the clearance volume is reduced with decreasing load to maintain a constant high effective compression ratio, in which case the geometric expansion ratio will be even higher giving a high indicated thermal efficiency (10,11). An improvement in fuel economy in the order of 15% should be possible with a combined LIVC/VCR system.

The reduced manifold vacuum that comes with EIVC or LIVC considerably reduces the amount of residual gas fraction. This would make the engine less sensitive to valve overlap effects.


Residual Gas Fraction

This determines the idle quality and smoothness of operation at low speed light load conditions. It also has significant effects on the exhaust emissions.

The residual gas fraction is extremely sensitive to the valve overlap duration and timing at light loads and low speeds. A minimum overlap is clearly desirable but this has to reconcile with the conflicting requirement of larger overlaps for volumetric efficiency and performance. VVT therefore offers the opportunity of optimising for both requirements between the extreme ends of the load range. In fact the system which aims explicitly to control overlap is one of the most cost-effective methods of VVT for good driveability and fuel economy and already has limited production applications (17).

Engine combustion at idle is inherently slow owing to excessive exhaust dilution, low pressure and quiescent fluid motion because of the low engine speed. Intake throttling reduces the intake manifold pressure to about half that of the exhaust manifold with the result that the retained exhaust must be expanded by this pressure ratio prior to any net induction of the fresh charge. The presence of any valve overlap provides a leakage path by which additional quantities of exhaust can back-flow from the exhaust manifold through the combustion chamber and into the intake manifold where it can mix with the fresh charge and be re-induced into the cylinder. This latter source of residual can be regulated by controlling the amount of valve overlap. As additional diluent can least be tolerated at idle, a minimum overlap is often beneficial. Improvements in the idle condition will result in improved stability, tolerance for weaker mixtures, tolerance for lower speed and tolerance for optimum spark advance. From this greater margin of combustion stability afforded by reduced valve overlap, improved idle quality and more fuel efficient idle control strategies may be employed (Fig.5).

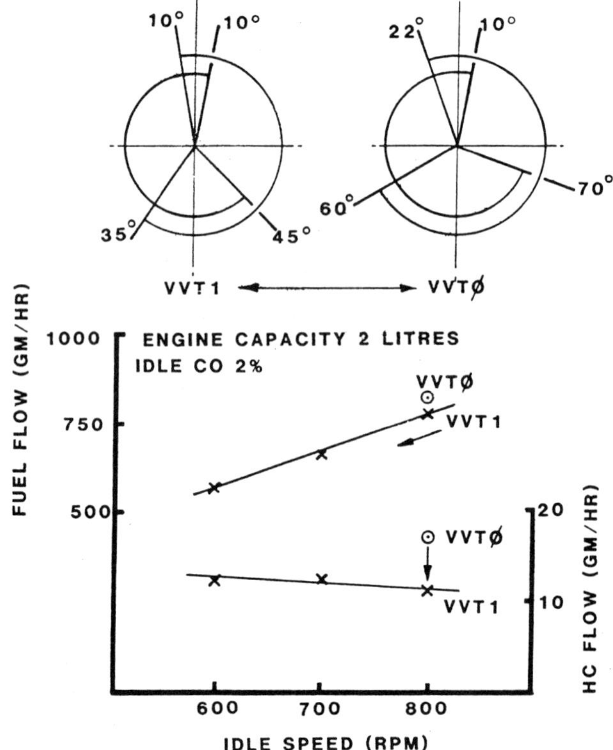


Fig. 5. Effect of valve overlap on idle speed ford VVT 1:1 harmonic drive.

Exhaust emissions are sensitive to overlap changes, but the effect is different with the method by which the overlap is altered (12-14). Reducing overlap by late IVO increases NOx because of the reduced internal EGR. HC and CO are not changed significantly. Reducing overlap by early EVC increases NOx up to the optimum overlap for least residual. Further advancing of EVC causes the residual to increase again resulting in a decrease of NOx and HC. The HC reduction is due to the trapping and recycling of the end of stroke exhaust fraction which contains a higher proportion of unburnt hydrocarbon.

The options available for modifying valve overlap to reduce the residual gas fraction are advancing EVC, delaying IVO or both simultaneously. Apart from emissions considerations, advancing the EVC tends to be detrimental to WOT performance which requires a slightly later EVC even at low speeds. Delaying the IVO while fixing EVC at the earliest possible timing would be more advantageous as WOT performance is relatively insensitive to IVO. The latter option is preferrable in some VVT systems where the overlap and duration changes are interlinked (for example, Harmonic Drive). In systems where overlap is explicitly controllable (for example, double camshaft engine), either timings may be changed individually or in a ganged manner. The overlap required to minimise residual gas fraction at idle and light loads is in the region of 10 degrees to 25 degrees. This adds to the overlap range previously selected for volumetric efficiency making a total range of 10 degrees to 45 degrees for the ideal VVT system.

Indicated thermal efficiency is expected to improve with reduced residual gas fraction. However, comparison at identical speed, load and air-fuel ratio between engines with and without VVT shows little change in the indicated thermal efficiency. The implication is that opportunities for

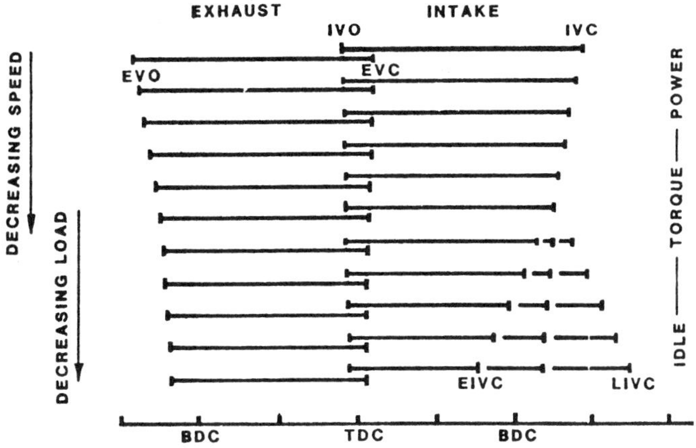



Fig. 6. Summary of VVT requirements.

VVT to improve fuel economy have to be found not directly from ISFC but indirecly from improving the mechanical efficiency with reduced throttling and changing the operation strategy of the vehicle by taking advantage of the improved idle quality and higher performance and run with reduced idle speed and lower axle ratio.

## Summary of VVT Requirements

Fig.6 shows the ideal range of valve timings to provide all the above VVT functions. Each timing should be independently adjustable but in practice this is almost impossible. Most VVT systems change one or two of the timing points explicitly and leave the rest to follow implicitly.

There are a number of VVT strategies which are particularly effective:

- Variable event durations of the exhaust, intake and overlap, all three together in combination in close agreement with the ideal requirement.

- Variable valve overlap from small overlap through a large angular range to achieve best stability at idle and light loads and best performance at WOT over the engine speed range.

- Variable IVC over a large angular range to modulate power by partial air admission with minimum throttling at part-loads, also optimise WOT performance over the engine speed range.

## VARIABLE VALVE TIMING STRATEGIES

Different VVT concepts produce different VVT effects. One system may match the engine requirement more closely than another. Usually a system is designed with a particular valve event change in mind and compromises have to be accepted with regard to the remaining parameters which simply happen as a consequence of the characteristics of the system. The availability of more VVT functions and the provision of large variable ranges are of course desirable but they must be balanced against the complexity and cost and very importantly the FMEP overhead.

Three VVT concepts have been considered by Ford to be effective for
improving fuel economy. The mechanical details of these will be described
in a later section. The strategies and their measured benefits are
discussed here first.

Harmonic Drive (Fig.7)

This provides an interlinked variable control of all the valve events
(exhaust, intake and overlap) in one action caused by superimposing an
angular oscillation on top of the camshaft rotation which makes the valve
events stretch and shrink in a dynamic manner. All the valve event
points, namely, IVC, EVO, IVO, EVC, are altered in their respective
desired directions such that the ideal VVT requirements are very closely
matched.

Variable Overlap (Fig.8)

For double camshaft engines, direct control of the valve overlap period is
provided by separately phase-shifting the intake and exhaust camshafts in
opposite directions (3). In spite of the fixed valve durations, this
system is capable of several useful VVT functions because of the large
phase-shift range available: variable overlap (for idle quality), variable
IVC (for performance) and moderate LIVC (for reduced throttling at
light-loads).

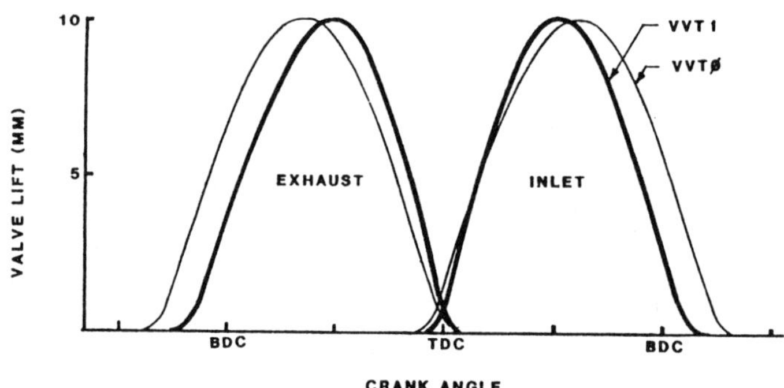


Fig. 7. Ford harmonic drive variable valve timing.

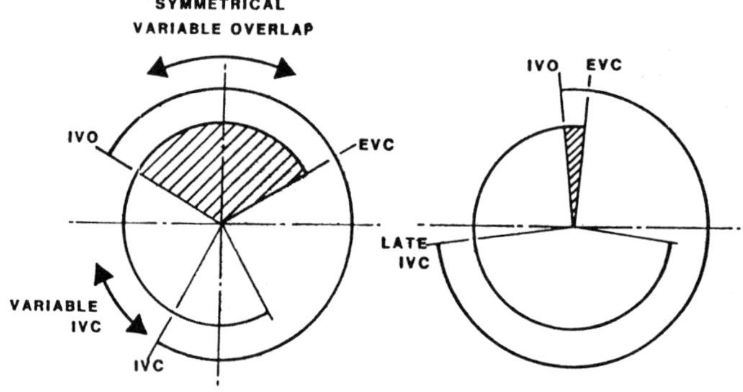


Fig. 8. Ford variable overlap for DOHC engine.

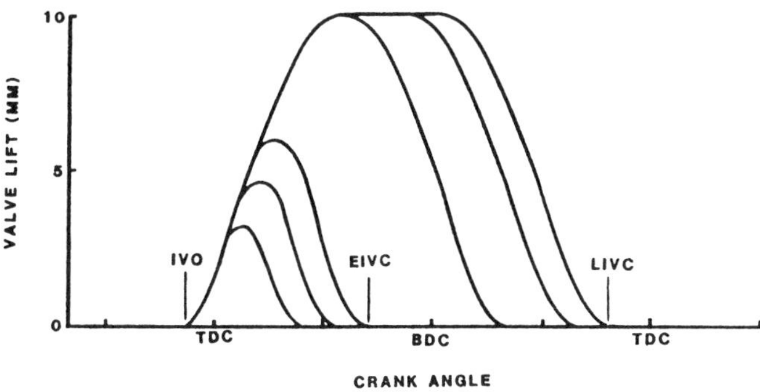


Fig. 9. Early and late intake valve closing.

Variable Air Admission (Fig.9)

This is provided by either EIVC or LIVC caused by large changes in the
intake opening period but with fixed IVO. The VVT function is therefore
solely variable IVC which meets both the volumetric efficiency requirement
and the part-load modulation requirement. Residual gas fraction during
idle is low in this system because of the reduced throttling and is
relatively insensitive to valve overlap.


## Performance and Economy Projection

Dynamometer test results from these VVT systems are used in the Ford
Vehicle Performance and Economy Projection Program to calculate the
on-the-road vehicle benefits from VVT. The 2L Ford Sierra vehicle with
fixed valve timing is used as baseline. The analysis shows the tradeoff
between performance and fuel economy, the effect of added friction from
the VVT system and the opportunities of some VVT functions, whilst taking
full account of the realistic vehicle usage using the ECE-15 fuel economy
cycle as the drive standard. This provides a useful framework defining
the relative benefit/penalty of each VVT function and enables various VVT
systems to be evaluated and compared according to the extent by which each
of these functions has been exploited. A summary of the results is shown
in Table 1 at the end of this section.

## FMEP Overhead

This includes the added friction from the VVT working parts and the power
consumption of the control system. The effect of VVT FMEP on the vehicle
performance is relatively small since it reduces the full load torque
curve only marginally. However it can rapidly depreciate the vehicle fuel
economy to unacceptable levels because of its increasingly adverse
influence on the mechanical efficiency as the engine is operated at light
loads. The penalty on vehicle performance is 0.8% for 0.1 bar FMEP
increase measured by the 50-100 KPH top gear acceleration but the
corresponding penalty on the ECE-15 fuel economy is 2.6% for 0.1 bar FMEP
increase. It is clear that great care must be taken in the design of any
VVT system to reduce the FMEP overhead to the minimum.

## Tradeoff between Performance and Fuel Economy

The improved WOT torque curve from VVT can be used either as a performance
advantage or as a fuel economy advantage. Direct application of VVT to

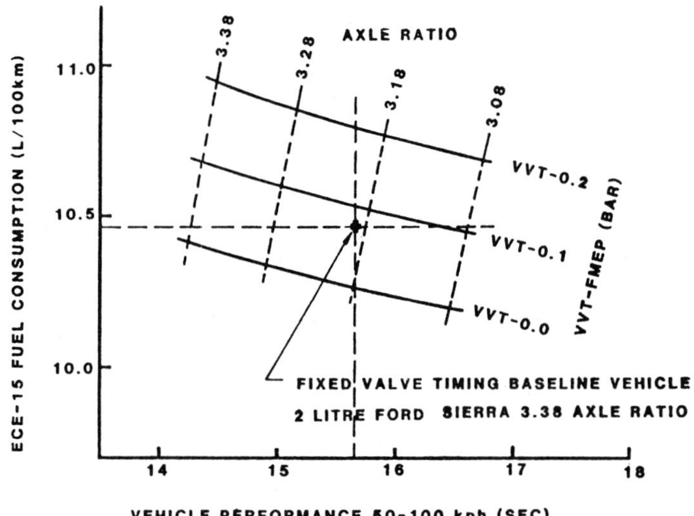


Fig. 10. Relationship between vehicle performance,
fuel economy and VVT friction.

the baseline vehicle without any modificiation to the drive train would
result in 9% higher vehicle performance and 0.5% better fuel economy if
the VVT friction is not included. Alternatively, this performance
advantage can be converted into a fuel economy advantage by lowering the
axle ratio to make the vehicle performance the same as that of the fixed
valve timing baseline vehicle in which case there is a gross improvement
in the vehicle fuel economy of 1.9%. However, after deducting the VVT
friction penalty (typically 0.05 bar FMEP), there is a nett fuel economy
penalty of (0.8%) in the performance mode. Trading the performance for
fuel economy would turn the nett fuel economy back to an advantage of just
0.6%. Figure 10 shows the relationship between vehicle performance, fuel
economy and VVT-FMEP. It is clear that any significant improvement in
fuel economy would have to come from additional strategies other than just
improving the engine WOT performance.

## Reduction in Idle Speed

The improved idle quality from reduced overlap should allow the engine to
operate at lower idle speeds. 200 rpm reduction from the baseline idle
speed translates to 6.1% improvement in the ECE-15 vehicle fuel economy.
This large improvement reflects the heavy weighting of the idle period
within the ECE-15 drive cycle which makes it a particularly effective
strategy to vary valve overlap explicitly to control idle quality.

## Power Modulation at Part-Load

Power modulation by either EIVC or LIVC saves about 0.3 bar PMEP at 1500
rpm engine speed and 2.62 bar BMEP engine load which corresponds to
approximately 30% power modulation. BSFC is reduced by 7% at this load
but the advantage diminishes progressively to zero as the engine
approaches full load. Some throttling is still required for idle and very
light load operation. The calculated improvement from EIVC and LIVC in
the ECE-15 vehicle fuel economy is 6.5%.

## Combined Strategies

The above VVT opportunities can be added together to a first approximation
since they address different aspects of the engine operation, namely,

friction, volumetric efficiency, residual fraction and pumping loss.
Table 1 therefore can be used as a guide to quantify the various functions
available in a VVT system and to compare one system against another in
terms of their relative cost-effectiveness. A VVT system which aims only
at improving performance will be doing so at the expense of fuel economy
because of the inevitable added friction. Systems that can provide in
addition to performance specific features to improve idle quality and/or
reduce pumping loss will have sufficient margin in the fuel economy
improvement to allow a range of vehicle strategies to either retain some
performance improvement or trade it for maximum fuel economy. Reducing
the FMEP overhead must be the most important design objective in any VVT
system.

In the context of the above, Variable Overlap is very effective if double
camshafts can be justified. The Harmonic Drive can improve performance
and idle quality but carries a slightly higher friction penalty (0.07 bar
FMEP) because of the continously oscillating motion. EIVC or LIVC adds
another large economy advantage which can be further improved with VCR but
the FMEP penalty may be high depending on the design. The benefit from
lean mixture calibration has not been included in Table 1 because data is
not available.

TABLE 1

VEHICLE PERFORMANCE AND FUEL ECONOMY FROM VVT

| VVT<br>STRATEGY | AXLE<br>RATIO | PERFORMANCE | | FUEL ECONOMY | |
|---|---|---|---|---|---|
| | | 0-100<br>KPH<br>(SEC) | 50-100<br>KPH<br>(SEC) | ECE-15<br>FUEL CON<br>L/100KM | BETTER/<br>(WORSE)<br>% |
| BASE<br>VEHICLE | 3.38 | 11.17 | 15.65 | 10.465 | – |
| VVT/BASE<br>PERFORMANCE | 3.38 | 10.52 | 14.25 | 10.416 | 0.5 |
| VVT-0.1<br>BAR FMEP | 3.38 | 10.60 | 14.37 | 10.680 | (2.1) |
| VVT-0.2<br>BAR FMEP | 3.38 | 10.67 | 14.49 | 10.935 | (4.5) |
| VVT/BASE<br>FUEL ECONOMY | 3.18 | 10.66 | 15.65 | 10.269 | 1.9 |
| VVT-0.1<br>BAR FMEP | 3.20 | 10.74 | 15.65 | 10.540 | (0.7) |
| VVT-0.2<br>BAR FMEP | 3.22 | 10.79 | 15.65 | 10.796 | (3.2) |
| OVERLAP/BASE<br>IDLE-200 RPM | 3.38 | 11.17 | 15.65 | 9.829 | 6.1 |
| EIVC/BASE<br>PART-LOAD | 3.38 | 11.17 | 15.65 | 9.784 | 6.5 |
| LIVC/BASE<br>PART-LOAD | 3.38 | 11.17 | 15.65 | 9.784 | 6.5 |

## VARIABLE VALVE TIMING DESIGNS

The design of VVT systems has been the subject of numerous publications (15-24). These may fall into one of four categories: Phase shifters, event stretchers/shrinkers, lost motion and variable geometry. The former two are considered to be more practical and between them are the majority of VVT designs applied to a large variety of valve train configurations. Usually, a high performance camshaft is selected as the base cam and VVT is applied to it to correct for low speed torque and idle quality. This approach is consistent with the design practice of valve trains where high lifts with long periods are desirable to keep the accelerations and stresses under control.

### Ford Harmonic Drive

This is suitable for single camshaft engines where the valve events are dynamically shrunk or stretched by causing the camshaft to rotate at varying angular velocities. By phasing the angular oscillation in relation with the valve event (Fig.11), it is possible to cause the two ends of the event to move relative to each other at the appropriate times in opposite directions thereby producing effectively variable event durations. Incidentally, this oscillation applied to the intake and exhaust cams together on the same camshaft would cause an interlinked VVT effect whereby all the three events (exhaust, intake and overlap) are simultaneous shrunk or stretched. This produces timing change combinations of IVC, EVO, IVO and EVC in close agreement with the ideal VVT requirements.

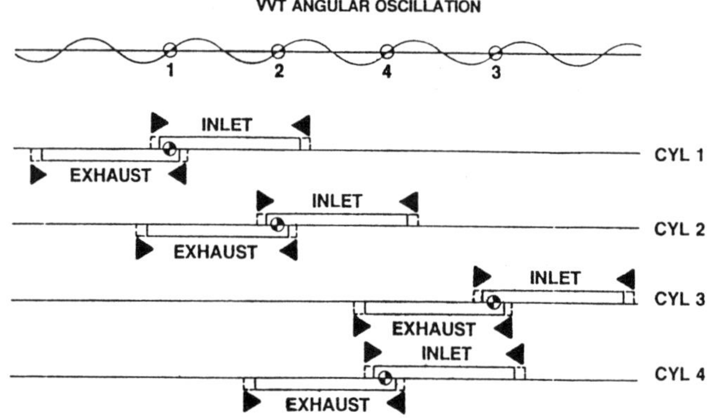


Fig. 11. Effect of VVT angular oscillation on valve events.

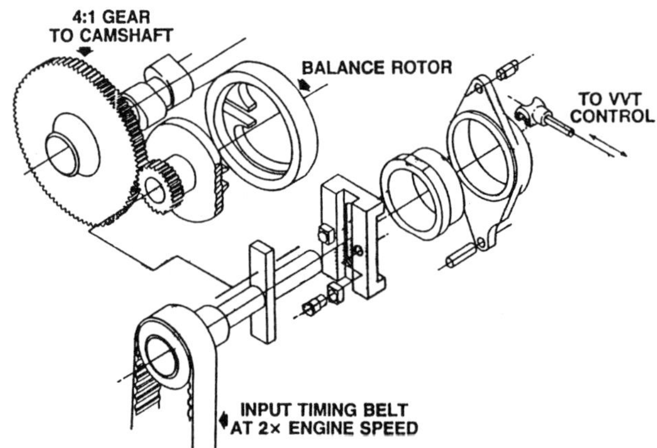


Fig. 12. Ford VVT 4:1 harmonic drive

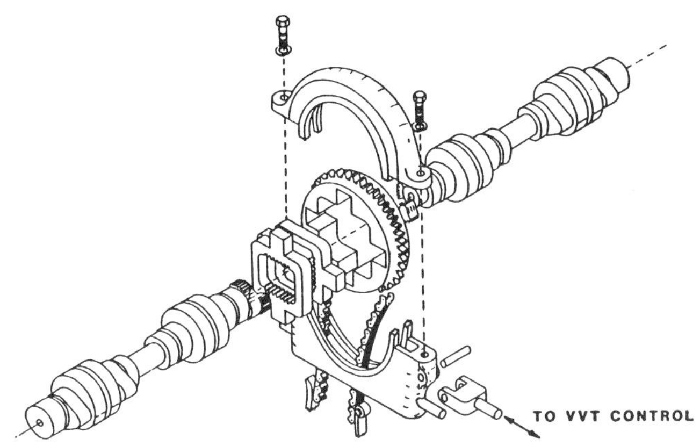



Fig. 13. Ford VVT 1:1 harmonic drive

There are two design options: 4:1 Harmonic Drive which produces four angular oscillations per camshaft revolution and 1:1 Harmonic Drive which produces one angular oscillation per camshaft revolution. The 4:1 Harmonic Drive (Fig.12) operates at four times camshaft speed to give the correct number of oscillations which synchronises with the cylinder frequency of a 4-cylinder engine. This therefore requires only one harmonic coupling driving via a 4:1 reduction gear a single integral camshaft carrying all the intake and exhaust valves for all cylinders. The 1:1 Harmonic Drive (Fig.13) operates at camshaft speed and requires the oscillation phasing to be timed individually for each cylinder hence a separate harmonic coupling driving a short camshaft for each cylinder. In a 4-cylinder engine, the design is integrated into a compact centrally driven unit with just 2 sliding components.

The Ford Harmonic Drive is unique compared with other similar systems (18-20) in that the variable angular oscillation amplitude is created by an eccentric coupling with a tranverse slider mechanism which produces identical but opposite angular changes to two oppositely-phased VVT members, thereby performs the VVT function and at the same time maintains perfect inertia balance. This design is extremely compact, robust and potentially of low cost. The friction penalty however can be high because of the continuously oscillating motion. The 4:1 Harmonic Drive friction is unacceptable at 0.4 bar FMEP because of the high speed and the high loads, but the 1:1 Harmonic Drive friction has been significantly reduced to 0.07 bar FMEP through careful design for low friction and reduced number of moving parts.

### Ford Variable Overlap

This is suitable for double camshaft engines where the intake and exhaust cams can be individually phase shifted. The Ford Variable Overlap System changes the intake and exhaust phase positions symmetrically in opposite directions to change overlap and is unique by virtue of its very wide control range (54 degree from each camshaft) compared with other similar systems (32 degrees from the intake camshaft only) (16,17). This offers in addition to the direct overlap change further control functions which take advantage of the fact that IVC automatically also has 54 degrees variable range sufficient for optimising performance over the engine speed range in spite of compromising IVO which has secondary effects anyway. At very small overlap which is used for idle and light loads, IVC is automatically very late and this produces some LIVC effect giving better BSFC through reduced throttling.

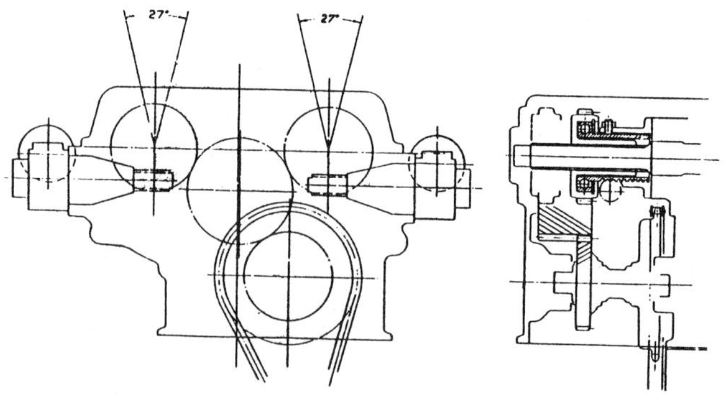

Fig. 14. Ford variable overlap for DOHC engine

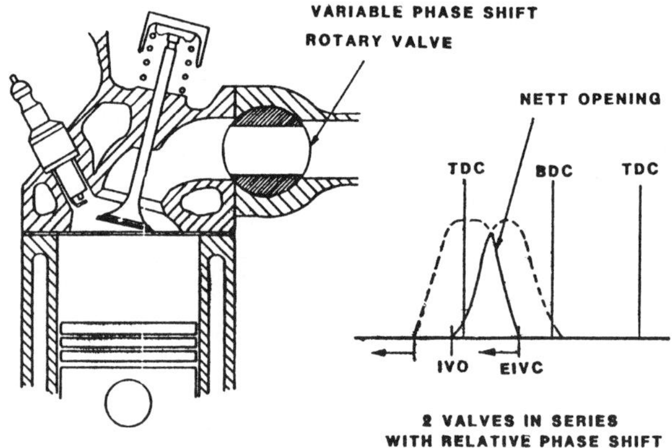


Fig. 15. Early intake valve closing

The mechanical system (Fig.14) consists of a helical centre gear which is
driven via intermediate shafts from the crankshaft, and this drives two
helical camshaft gears which can be moved axially along straight splines
on their respective camshafts. The axial movement causes the gears to
rotate relative to the helical centre gear and relative to each other.
The friction overhead is approximately 0.05 bar FMEP.

<u>Ford Variable Air Admission</u>

Early Intake Valve Closing by variable air cutoff is created by two valves
in series along the intake port. This can be either a rotary valve in
series with the intake valve (Fig.15) or a reed valve in series with the
intake valve. One of the valves is phase-shifted relative to the other
and the intake event is variable and stays effective only when both valves
are open. The nett effect is that IVO is relatively stationary and IVC is
progressive advanced with phase shift to give EIVC. A large phase shift
range of 120 degrees is provided by an epicyclic gear train which
unfortunately carries a high friction penalty (0.15 bar FMEP).

Late Intake Valve Closing requires a large extension of the intake opening
period well beyond BDC to allow some of the air to be expelled back out of
the cylinder for variable air trapping. One method (9) uses split-cams
where the opening and the closing cams can be phased apart from each other
by means of an electrically driven gearbox which is mounted on and rotates
with the camshaft. Another method which is suitable for engines with two

250

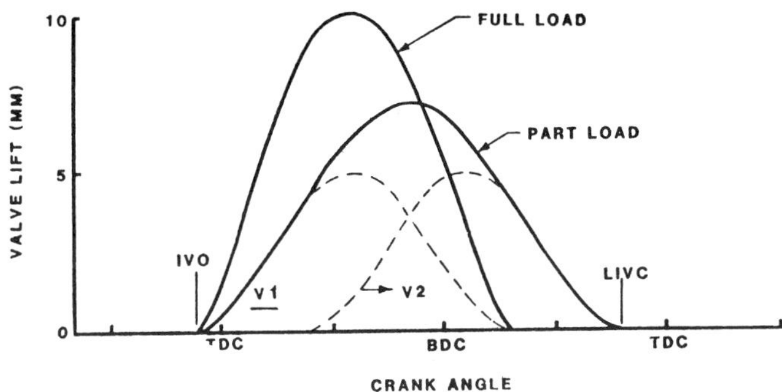


Fig. 16. Late intake valve closing by 2 intake
valves with relative phase shift.

intake valves per cylinder is to phase shift one of the valves relative to
the other as they operate in parallel. The intake event is variable and
stays effective as long as one of the intake valves remains open. This
produces a combined intake opening period which can be extended over a
large range and a progressive reduction in the intake opening area with
LIVC (Fig.16).

This latter method and the Harmonic Drive and the Variable Overlap have
been the subject of numerous patent applications made by Ford and are
currently under active development.


CONCLUSIONS

1. Variable Valve Timing is one of the advanced engine concepts which can
significantly improve fuel economy and driveability in vehicles of the
1990's.

2. Improved idle quality and reduced throttling at part-loads are the
important VVT strategies for fuel economy.

3. Added friction from the VVT system must be kept to the minimum in order
to realise the potential benefits.

4. For engines with single camshaft, the 1:1 Harmonic Drive offers an
interlinked control of all the VVT parameters which follow very closely
the ideal VVT requirements.

5. For engines with double camshafts, Variable Valve Overlap with a large
control range is very effective by virtue of its multi-functions: variable
overlap, variable IVC and some LIVC effect.

6. Early or Late Intake Valve Closing combined with Variable Compression
Ratio is the future VVT system likely to be developed.



REFERENCES

1. P.Walzer, P.Adamis, H.Heinrich and V.Schumacher  "Variable Valve Timing
and Variable Compression Ratio in SI Engine"  MTZ Motortechnische
Zeitschrift 47 (1986) 1
2.  T.Bowman, A.Chopra and K.Kunisch  "Performance and Powertrain
Development for Integrated Powertrain Control"  FISITA paper 865028

3.  D.Stojek and  A.Stwiorok  "Valve Timing with Variable Overlap Control" FISITA paper 845026

4. C.R.Stone  and E.K.M.Kwan  "Variable  Valve  Timing  for  IC  Engines" Automotive Engineer Aug/Sept 1985

5. T.W.Asmus  "Valve Events and Engine Operation"  SAE 820749

6.  D.L.Stivender  "Intake  Valve Throttling  - A  Sonic Throttling Intake Valve Engine"  SAE 680399

7.  J.H.Tuttle  "Controlling  Engine Load  by Late  and Early Intake Valve Closing"  SAE 800794 and SAE 820408

8.  S.Hara,  Y.Nakajima and  S.Nagumo  "Effects  of  Intake  Valve  Closing Timing on SI Engine Combustion"  SAE 850074

9.  A.C.Elrod and  M.T.Nelson  "A Variable Valve Timed Engine to Eliminate Pumping Losses"  SAE 860537

10.  D.Luria, Y.Taitel  and A.Stotter   "The Otto-Atkinson  Engine - A New Concept in Automotive Economy"  SAE 820352

11. R.J.Saunders   "Atkinson Cycle  Spark Ignition  Engines"  Int.  J.  of Vehicle Design, IAVD Congress 1984 Vol 4

12.  G.B.K.Meacham  "Variable Cam Timing as an Emission Control Tool"  SAE 700673

13. R.M.Siewart   "How  Individual  Valve  Timing  Events  Affect  Exhaust Emissions"  SAE 710609

14.  M.A.Freeman and R.C.Nicholson  "Valve Timing for Control of Oxides of Nitrogen"  SAE 720121

15. G.A.Torazza   "A Variable  Lift and  Event Control  Device  for  Valve Operation"  Paper 2/10 FISITA Congress 1972

16. C.A.Schiele, S.F.DeNagel and J.E.Bennethum  "Design and Development of a Variable Valve Timing Camshaft"  SAE 740102

17. Alfa Romeo  "Variable Inlet Valve Timing Aids Fuel Economy"  Chartered Mechanical Engineers April 1984

18.  D.A.Parker and  M.A.Kendrick  "A Camshaft with Variable Lift-Rotation Characteristics"  Paper B-1-11 FISITA Congress 1974

19.  D.Scott   "Eccentric  Cam  Drive  Varies  Valve  Timing"  Automotive Engineering Oct 1980

20.  G.E.Roe  "Variable  Valve  Timing  Unit  Suitable  for  IC  Engines" I.Mech.E Proc 1972

21. J.Pichard  and M.Gratadour   "Ballistic Valve  Timing  System"  B-1-6 FISITA Congress 1974

22.  F.Nagao,  K.Nishiwaki  and  S.Kajiya   "Valve  Timing  Control  by  a Hydraulic Tappet with a Leak Hole"  Bulletin JSME Nol 11 No 48 1968

23. R.J.Herrin  and D.J.Pozniak  "A Lost-Motion  Variable  Valve  Timing System for Automotive Engines"  SAE 840335

24. A.Titolo  "Variable Valve  Control From  Fiat"  MTZ Motortechnische Zeitschrift 47 (1986) 5

THE OUTLOOK FOR CONVENTIONAL AUTOMOTIVE ENGINES


Bernard I. Robertson

Director - Powertrain Engineering
Chrysler Motors


ABSTRACT

This paper deals with the outlook for conventional automotive
engines in the face of continuing competition from a variety of
alternative powerplants.  The primary areas considered are
technological attributes, entry barriers, and cost.  A brief
review of the technological improvements of the recent past is
presented and the overall results in powerplant attributes are
reviewed, together with forecasts for the immediate future.
Promising developments which may support the projected future
improvements are briefly summarized.  A major hurdle presented to
any alternative powerplant is that of entry barriers.  The
specific entry barriers presented by capital investment, the
current infrastructure and public acceptance are discussed.
A breakdown of typical costs of today's conventional engine is
presented together with a review of the likely areas for future
improvement.  An overall reduction in the area of 30% is forecast
over the next five years.  It is concluded that, in the final
analysis, the most critical test for any powerplant, conventional
or alternative, is that of customer value.  The marketplace will
ultimately select the successful candidate.  In order to be this
successful candidate, any replacement powerplant must match the
conventional engine in most of its characteristics and beat it
in the rest.  Furthermore, the target to be met by any replacement
candidate is not that represented by a 1986 powerplant, but by an
engine which is continually improving in all technological
respects and corresponding cost at a very significant annual rate.

INTRODUCTION

It's my task in this paper to state the case for the conventional
reciprocating internal combustion engine and define just how formidable
a target I believe it will be in the future.

Back in the very early days of the automobile, the fledgling
reciprocaing IC gasoline engine was itself an "alternative", challenging
established diesel, steam and electric propulsion. It won the resulting
shakeout, based upon the same key issues we face today, namely specific
output, efficiency, manufacturability and cost, durability and adaptability.  This last has been perhaps the most important attribute of today's
conventional powerplant:  the ability to adapt to widely changing conditions
in terms of economics, environment, and customer demands.

There is always a danger, when one speaks on behalf of the status quo,
of being branded a neanderthal with little vision of the future.  I do not
propose to defend my position, but I would just like to point out that
Chrysler has, over the years, been very active in the development of alternative powerplant candidates and an array of those candidates on which
we have expended a considerable amount of effort is shown in Fig. 1.  We
pursued these various candidates to differing degrees.  In the case of the
diesel engine, we developed a complete family of diesels and started tooling
our Windsor, Ontario, plant for them before the diesel market fell apart.
When we introduced our new four cylinder gasoline engines in 1981, we tooled
the engine plant to build a stratified charge derivative, but elected at the
last minute not to produce it, due to the marginal benefits it represented
at considerable additional cost.

Two of the other more celebrated examples of our alternate powerplant
development were, of course, the Chrysler gas turbine, on which we probably
expended the most effort over the longest period, and also electric
vehicles, including a cooperative effort in which we engaged with G.E.  My
point here is simply, that Chrysler has a long history of taking seriously
the promise of alternative candidates, and we certainly plan to continue
that posture.  Please do not misunderstand my spokesmanship for the conventional IC engine to suggest otherwise.

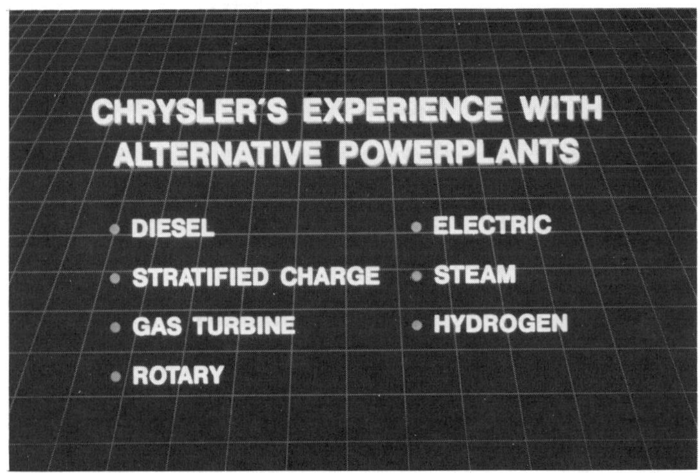


Fig. 1

Incidentally, in preparing for this talk, I had a little difficulty deciding just where to draw the line between alternative engines and varying degrees of refinement of the existing engine, and you may well feel when I'm done that you would have drawn the line elsewhere.  I can only say that it is something of a judgement call.

One other point I should make is that I am based in the U.S., and most of the data readily available to me is therefore U.S. oriented.  I should stress, however, that Canada is an extremely important manufacturing center and market to us, second only to the United States, and virtually everything I shall say from here on applies equally well to the U.S. and Canadian markets, even if the data happens to be U.S. for convenience.

The Challenges
-----

I would like to illustrate the challenges to any alternative powerplant presented by the conventional reciprocating internal combustion engine. I'll break my remarks down into three broad categories, namely technological attributes, structural entry barriers, and cost.

In the technological area, the conventional engine has made dramatic improvements over time, and I'd like to briefly illustrate some of those. I'll cover briefly the topics of specific output and response, efficiency over the load range, and adaptability.

First of all, looking at output and response, there have been dramatic gains in the last few years in the area of breathing capacity, with such changes as valve and port design, multiple valves (Maserati has just recently raised the ante to 6 valves per cylinder), pressure charging - predominantly turbocharging, but some supercharging, and fuel injection. The result has been that despite energy crises and the continuing drive for fuel efficiency, the average performance of the entire U.S. car fleet, if you can visualize such a parameter, has been steadily improving over time, as illustrated in Fig. 2.  Meanwhile, efficiency has been dramatically improving at the same time.  Some of the more notable methods to achieve this have been refinements in combustion characteristics, introduction of swirl to promote fast burn, friction reduction, and overall control sophistication.  One good example of all this, unfortunately Japanese, is

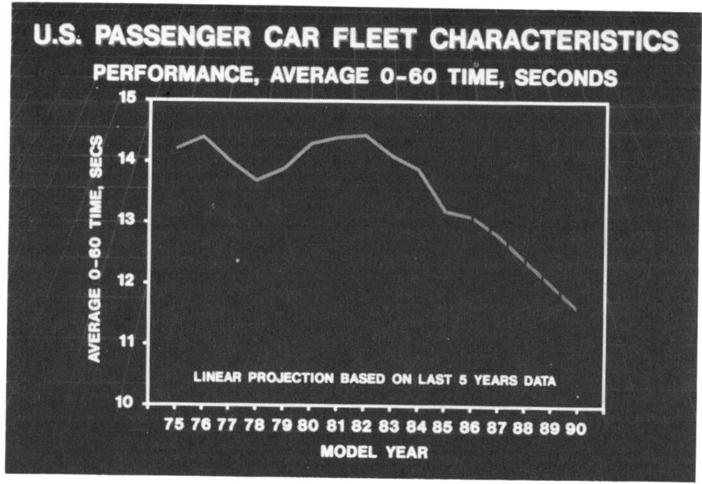


Fig. 2

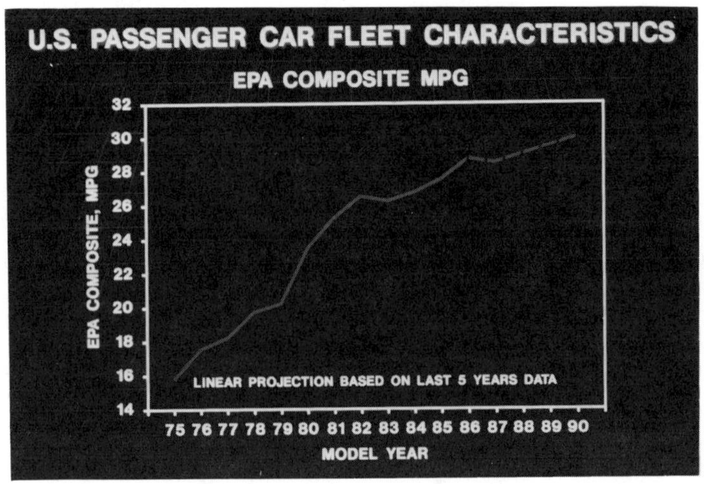



Fig. 3

the Honda Legend V6 configuration, which is sort of the state of the art in sophistication, if not cost control, with single overhead cam, four valves per cylinder, hydraulic lash adjusters, horizontal push rods, and individual intake tracts for all twelve inlet valves. The result of all this industry refinement is shown in Fig. 3. The U.S. passenger car fleet has improved dramatically in fuel economy over the same period that the performance has also improved. There has, of course, been a fair amount of vehicle down-sizing and weight reduction during this period, which has undoubtedly helped the fleet miles per gallon.

In order to separate out powerplant efficiency from vehicle weight effects, we use a measure called TFC/IW, or total fuel consumed divided by inertia weight, which is dimensionally equivalent to gallons per ton mile, to normalize for vehicle weight and thereby impute efficiency. It is a very convenient way of characterizing the entire vehicle population without running individual tests on every model within it. The fleet miles per gallon plot is reproduced in Fig. 4 in its TFC/IW form; as you can see,

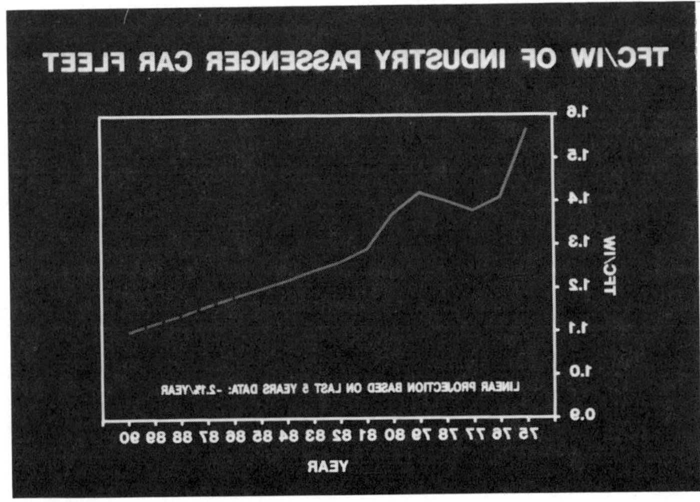



Fig. 4

there has been a steady, inexorable improvement in fuel efficiency, quite apart from any benefit resulting from weight reduction.

In the rather complicated chart in Fig. 5, I've shown the story on one specific model, the Chrysler K car, since its introduction in 1981 through its replacement with our new A car in 1989. There has been in this case a dramatic improvement in efficiency over the intervening years, and the key factors responsible are also included. I think it's worth noting that the 18% overall fuel economy improvement has been achieved despite a growth in vehicle weight. This increase, of three weight classes, is due to the vehicle growing in size and opulence over the decade. It has also improved noticeably in driveability, performance and NVH (noise, vibration and harshness) at the same time that this 3.3% annual improvement in efficiency was achieved. Fig. 6 shows the entire U.S. midsize passenger car fleet in miles per gallon and the corresponding track for the K car.

### 2.2 LITER ENGINE - K-BODY/A-BODY FUEL ECONOMY TREND

| MODEL YEAR | TEST WEIGHT | EPA COMP. MPG | TECHNOLOGY NOTES |
|---|---|---|---|
| 1981 | 2,750 | 26.9 | INTRODUCTORY YEAR |
| 1982 | 2,750 | 27.1 | LOW DISPLACEMENT POWER STEERING PUMP |
| 1983 | 2,750 | 29.4 | INCREASED C.R., LOW LOAD VALVE SPRINGS, IDLE FUEL CONSUMPTION, SLIPPERY OIL |
| 1984 | 2,750 | 28.4 | 260K TORQUE CONVERTER, TIRES, POLY-VEE BELT |
| 1985 | 2,875 | 28.4 | ECU CONTROLLED FAN, AERO, POWDERED METAL INSERTS |
| 1986 | 2,875 | 30.4 | A515 CYLINDER HEAD, LPTBI, 220K TORQUE CONVERTER |
| 1987 | 2,875 | 30.4 | LOW TENSION RINGS, 5W30 ENGINE OIL, TIRES |
| 1988 | 2,875 | 32.3 | LOCK-UP TORQUE CONVERTER, ROLLER FINGER FOLLOWERS, LOW OVERLAP CAM |
| 1989 | 3,000 | 31.3 | AERO, A-BODY REPLACES K-BODY, 3.02 OTGR |
| 1990 | 3,125 | 31.7 | 2.5 LITER ENGINE, A604 4 SPEED AUTO. TRANS. |
| | | 4.8 MPG | |
| | | 17.8% | |

ON THE BASIS OF TFC/IW THE 1981 TO 1990 FUEL ECONOMY GAIN AVERAGES 3.3% PER YEAR COMPOUNDED

Fig. 5

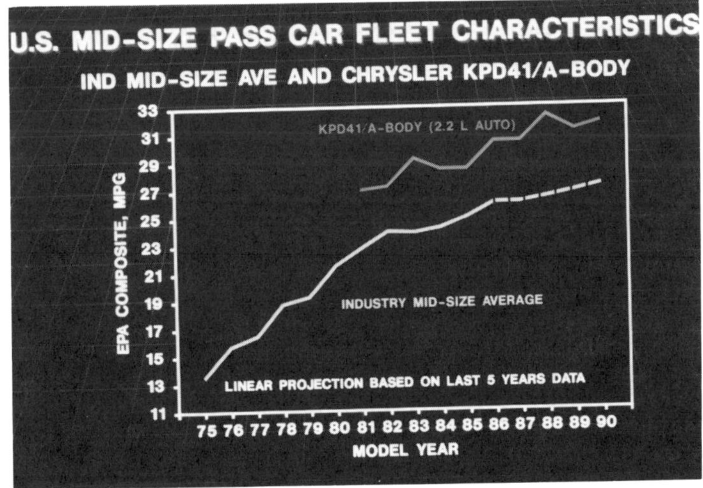


Fig. 6

The other technological criterion I listed was adaptability.  This, of
course, means different things to different people, but in our case, it has
meant "how can we satisfy the widest array of product needs with the same
basic investment?"  A good case in point is the 2.2L east-west four cylinder
engine that we introduced in 1981.  It has formed the basis for a 1.8L
naturally aspirated derivative, a 2.5L derivative with balance shafts, a
2.2L north-south derivative for use in our new Dakota light duty truck, a
turbocharged version for our sports cars, a higher boost intercooled turbocharged version, a 1.5L three cylinder counter balanced version, a 16 valve
naturally aspirated and turbo sub-family, and even a pre-chamber stratified
charge version.  As you know, not all of these derivatives have yet made it
into production, but nevertheless, I think this illustrates the ubiquitous-
ness of the basic configuration, and the appeal to a manufacturer with
somewhat limited capital resources but a wide array of product demand.

I've talked up to this point about improvements to the engine up to
today, and in the illustrations you will have noticed a projection over
the next several years of continuing improvement.  I've shown in Fig. 7 a
brief list of some of the technologies which we fully expect will be introduced widely in the next several years and which underwrite that projected
incremental future improvement.  You might notice one or two of the entries
have already appeared in isolated cases so this does not require a great
deal of foresight on my part.  You've heard about all of these concepts
already from other participants, so I will simply observe here that we are
actively pursuing them all, and I'll skip the details.  I would like to
dwell momentarily on the last item, however.  Chrysler has made a very
strong market statement with turbocharging and we plan to continue that
dominance.  In fact, this year we expect to sell over 250,000 turbo-equipped
cars in the U.S. and Canada.

We will be introducing intercooling very soon, and we anticipate
continued improvements in lag reduction through such techniques as variable
geometry and reduced inertia.  We are looking at ceramics as an approach to
inertia reduction, but we are still concerned about their durability.  At
the same time, we expect to make substantial reductions in system cost and
to further improve the reliability through dramatic changes in the lubrication system of the turbo.  I should add that our turbo reliability is
already excellent, as indicated by our 5 year/50,000 mile warranty, but one
of our goals is to eliminate the engine oil dependency while achieving zero
defects.

TECHNOLOGY TRENDS FOR THE 1990'S
LEAN, CLOSED LOOP FUEL CONTROL
ELECTRONIC EGR
VARIABLE VALVE TIMING
VARIABLE MANIFOLD TUNING
2 SPEED ACCESSORY DRIVE
LOAD SHEDDING
ENERGY MANAGEMENT
- COOLANT
- EXHAUST
ADVANCED TURBOCHARGING

Fig. 7

We anticipate alternative materials playing a very significant role in the future, for both  friction reduction and mass reduction reasons.  We will be expanding our use of roller tappets to cover essentially all engines by 1988, and we are working hard on replacing the steel roller and needle bearings with a ceramic roller for further friction reduction and reliability improvement.

In the area of mass reduction, we have a major effort underway looking at such diverse products as metal matrix composite connecting rods, polymer pistons with a variety of different piston crowns, ceramic piston pins, ultra lightweight aluminum pistons versus composite pistons and, of course, we, like many others, are looking hard at graphite fiber composites such as the AMOCO Torlon$^R$ material for such applications as 2.2L connecting rods. There is no doubt that this whole area is a very fertile one for future engine improvements.

We at Chrysler have been outspoken advocates for improvements in the quality of existing automotive fuels.  We have actively discouraged the use of low level alcohol blends because of their deleterious impact on driveability and component life.  However, we remain convinced that ultimately, neat methanol represents a major opportunity in alternate fuels and we therefore have a vigorous development activity underway on vehicles dedicated to neat or near-neat methanol.  We are also collaborating with the U.S. Government in fleet evaluation of K cars on M85, and we have lobbied hard for pro-alcohol legislation.  One of our new dedicated methanol K cars is currently on display at the Alternate Fuels Exhibit at EXPO 86.

Based on progress to date, and the activity I know to be underway throughout the industry at the moment, I have great confidence in projecting that by 1995, a conventional reciprocating gasoline engine will average a TFC number around 1.0.  There are already a few examples at that level in the marketplace today.  Meanwhile, I believe the average specific power will be up to 55-60 brake horsepower per litre for the entire fleet.  At the same time, versus today's engine, the future conventional IC engine will offer better NVH, better emissions, better durability and have a wider fuel capability.  Nor will the process end there, because there are many somewhat more remote possibilities currently under study.  These include automatic stop-start, cylinder deactivation, two stroke cycle engines, which offer reduced size and weight while presenting challenges in the emissions and fuel economy area; integral block and head with rotary valves, hydrogen fuel, wobble plate mechanisms, and so on.  Obviously, depending upon the price of fuel and demands of the marketplace, these technologies could well be developed to commercial applicability in order to continue the trend in improvement of what is still basically the conventional IC engine in the second half of the 1990's.

Entry Barriers

I'd like to turn now to what I called the structural entry barriers, namely capital investment, infrastructure and public acceptance.  Incidentally this is where I really start taking some risks with my neanderthal image, but please bear with me.

Looking at capital investment, modern engine tooling for high-volume manufacture is both sophisticated and expensive.  We have a rule of thumb in the industry that for totally new engine tooling, the cost in 1986 dollars is roughly $1,000 per unit of annual production capacity.  That is, for a typical increment of say 400,000 engines capacity, the bill is roughly $400M.  The approximate capacity for vehicles available to the U.S. market, car and truck, today is roughly 20 million, and consequently it is backed by 20 million units of engine capacity.  With our current rule of thumb,

that translates to roughly $20 billion worth of equivalent investment in
replacement cost.  Obviously, while some of that tooling is far from new,
the fact remains that it represents an enormous capital investment.  In
order to justify replacement with something totally different, the
challenger must be very demonstrably superior.  Even then, replacement will
be very gradual, due both to the expense, and to the need to continue
production uninterrupted during the changeover period.

Another factor worth considering is the excess capacity in the marketplace.  Fig. 8 illustrates the combination of domestics, imports and
transplants (import nameplates built in North America) in the U.S. market.
There existed in 1985 roughly 20 million units of capacity to service a
demand for 17.3 million cars and trucks, or unused capacity of 2.6 million.
By 1991, by which time we anticipate a significant growth in the number of
imports and transplants, the U.S. market will be attacked with about 23.1
million units of supply, while at the same time we anticipate demand being
down slightly at 16.4 million, yielding 6.7 million units of excess
capacity.  So not only is there roughly $20 Billion worth of investment in
place for the status quo, over 6 million units of that capacity are essentially unused.  Clearly, the opportunity for an alternative powerplant is
not going to be created by any kind of demand for additional powerplant
capacity.  It will have to go head to head against the existing product and
ultimately justify direct replacement.

The next barrier I'd like to touch on briefly, is that of the infrastructure.  There are over 170 million vehicles in the U.S. fleet today
and they are supported by an enormous infrastructure, comprising manufacturers of test equipment, tool suppliers, replacement parts manufacturers,
and re-manufacturers.  These people all supply a vast service industry which
in turn is supported by a vocational education network.  Some 71% of
surveyed car owners described themselves as "do-it-yourselfers" in a recent
study.  Nevertheless, 83% of them prefer to have their car serviced professionally, by this enormous network of repair facilities.  There are, in the
USA, around 140,000 service stations and another 115,000 automotive repair
shops.  These facilities employ about a million people and perform about 2
<u>billion</u> repair jobs annually.  These jobs yielded a reported $60 billion
revenue in the last year for which we have data.  Incidentally, the most
popular repair service is replacing spark plugs - there are about 400

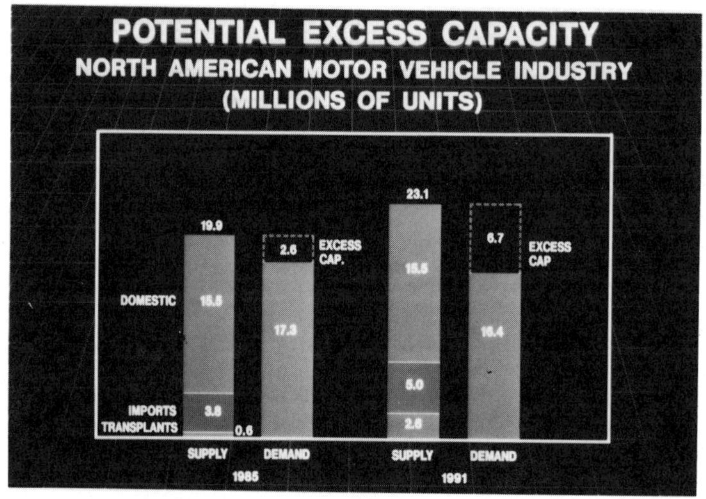


Fig. 8

million of these jobs every year.  Some 50 million replacement batteries are
sold every year, but only half of them are professionally installed, so the
do-it-yourself segment cannot be ignored either.

My point here is, the established pattern of service is deeply
ingrained in our culture and there is hardly anyone who doesn't know
something useful about maintaining or repairing the existing technology:
technology which many researchers are trying to replace!  While none of this
should discourage or inhibit development of alternate powerplants, it does
not auger well for a dramatic change to some new and unconventional
product.  When and if such a change takes place, it will necessarily be a
gradual one and must be accompanied by a determined effort to address the
needs of the infrastructure.

Public acceptance is the other category of great importance in terms of
structural entry barriers.  When Chrysler ran a field trial with 50 gas
turbine vehicles back in the 1960's, they were extremely well received and
we still get letters from participants asking us when they can purchase a
production vehicle.  There's little doubt in that case that, assuming it was
executed properly, that product would have had little difficulty with public
acceptance.  On the other hand, when Chrysler attempted to radically change
the standard technique of automatic transmission gear selection in the early
1960's, no one was interested and we were forced to revert to the industry
standard, the selector quadrant, which has remained with us to this day.

Another example of this point is the diesel engine.  During a period of
extreme concern over fuel availability and price, the 20% fuel efficiency
advantage of the diesel spurred a substantial growth in the market. However,
as soon as that fuel anomaly abated somewhat, the other characteristics of
diesel engines, including tardy cold starting, fuel and exhaust smell and
overall noise level, came to the surface and severely impacted the public
acceptance of the product, to the point where the passenger car diesel
market has almost evaporated.

One could cite other examples, like the ill-fated Cadillac V8-6-4
system, which we always felt deserved a better reception than it got.  But I
think the point is clear - public acceptance is a critical factor and unfortunately, is hard to predict without a rather substantial field trial before
mass production.

<u>Cost</u>

The other general category I had planned to touch on is that of cost.
This, of course, is a very sensitive area and I cannot cite specific
dollar values on production costs of engines, but Fig. 9 will serve to
illustrate the breakdown of today's engine costs into its major components:
labor, material and fixed burden.  Incidentally, most people are surprised
to learn that actual material costs in a typical engine are only half of the
total.  Chrysler has a widely publicized and very real goal of reducing
our costs by 30% by 1990.  Looking at those three categories of expense
within the engine, we expect to achieve our goals in the labor category
with new innovative flexible labor agreements and flexible automation.
In the material area, obviously redesign and innovation where appropriate
along the lines I touched on earlier are major opportunities, and of late
we have much more vigorously searched world-wide for the highest quality
suppliers at the most attractive economics. In the fixed cost area, we,
just like everyone else, are aggressively pursuing every category of fixed
costs.  We are well on the way towards our objectives and I can say with
some confidence that we fully expect to achieve the 30% reduction on
schedule.

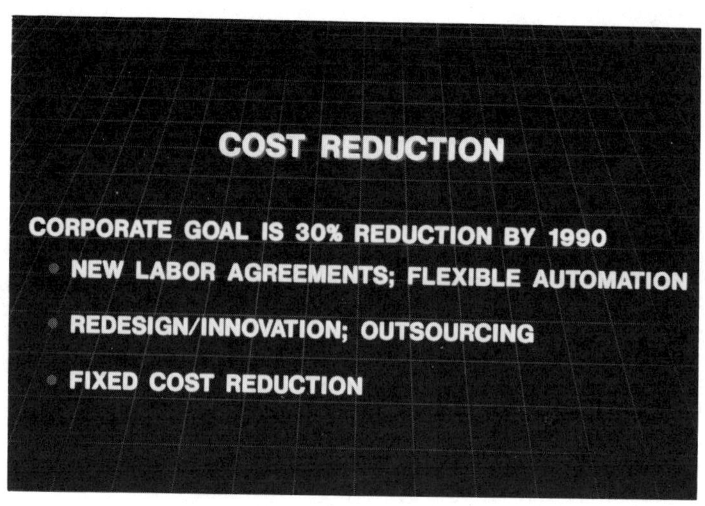


Fig. 9

<u>Customer Value</u>

The final thought which I'd like to leave with you may be the most important one of all.  That is, as an automobile manufacturer, the most compelling reality we must observe is that the customer is the boss.  In the final analysis, whatever we do must provide value to the customer: that is the acid test.  We as engineers may love to dabble with radical alternatives to the current product, but if the end result is not more value for the customer, it simply will not succeed.  One of the several measures we at Chrysler use to interrogate our future proposals is something we call customer value index, or CVI.  It's really aimed at efficiency, which of course lends itself more readily than some other parameters to a quantitative analysis.  In essence, we fully account all piece cost and investment expense and calculate miles to payback at some presumed cost of fuel and cost of money.  There is no hard and fast rule on what is an acceptable mileage to payback, but it is a useful measure for comparing different future proposals.  I've illustrated just three such proposals in Fig. 10 to show you how it works.  It's not perfect,

| ITEM | CVI | VALUE COMMENTS |
|---|---|---|
| ROLLER FOLLOWERS - 1988 4 CYLINDER ENGINES | 3,800 MILES | EXCELLENT VALUE BASED ON FUEL ECONOMY ALONE. ALSO FEATURES ADDED VALUE THROUGH IMPROVED PERFORMANCE, IMPROVED NVH, LESS HEAT REJECTION, AND IMPROVED TECHNICAL IMAGE |
| 4-SPEED AUTOMATIC TRANSAXLE WITH ADAPTIVE ELECTRONIC CONTROL | 85,000 MILES | NOT JUSTIFIED BY FUEL ECONOMY VALUE ALONE. REQUIRED TO IMPROVE OUR COMPETITIVE MARKETING POSITION, THIS TRANSAXLE OFFERS IMPROVED PERFORMANCE, IMPROVED NVH, AND UNEXCELLED SHIFT QUALITY |
| 2-SPEED ACCESSORY DRIVE | 139,000 MILES | POOR FUEL ECONOMY VALUE. THIS FEATURE WOULD ONLY BE CONSIDERED AS PART OF A GAS GUZZLER AVOIDANCE STRATEGY |

Fig. 10

as you can see.  The four speed automatic transaxle, for example, has a
depressing CVI value but its other attributes, in terms of NVH at highway
cruise and simply competitive posture make it an urgent priority for us,
despite the CVI.  It's worth noting by the way, that even in the heyday of
diesel engines, we never had a diesel engine program with a CVI below
100,000 miles.  That was a major indicator to us that as soon as the fuel
situation stabilized, the product would probably collapse.

<u>Summary</u>

    I believe that the conventional reciprocating IC engine, as it stands
today, represents a formidable target.  Furthermore, it has been steadily
improving through its life and I'm perfectly certain this will continue.
Specifically, we anticipate improvements in specific output, efficiency,
NVH, emissions, durability and cost.  Any replacement candidate in our
judgement must match the conventional engine in most of its characteristics and beat it in the rest.  As I've tried to illustrate, over and above
the technological criteria, the entry barriers to any replacement candidate are high and will ultimately control the rate of change to any form
of new technology.  Finally, adaptation of the infrastructure will
certainly be required and must be planned for.  Having said all of that, I
would just like to say it was certainly not my intention to discourage
anyone.  There are a number of very promising looking alternatives out
there, and we're trying to get our arms around as many of them as we can.
We will participate vigorously in whatever change takes place.

    The core message that I wanted to convey is "don't look at the 1986
conventional engine as the target":  you've really got to look at where it
will be in <u>1996</u> and ask yourself "can I beat it?"